Global Climate and Ecosystem Change

NATO ASI Series

Advanced Science Institutes Series

A series presenting the results of activities sponsored by the NATO Science Committee, which aims at the dissemination of advanced scientific and technological knowledge, with a view to strengthening links between scientific communities.

The series is published by an international board of publishers in conjunction with the NATO Scientific Affairs Division

A	**Life Sciences**	Plenum Publishing Corporation
B	**Physics**	New York and London
C	**Mathematical and Physical Sciences**	Kluwer Academic Publishers Dordrecht, Boston, and London
D	**Behavioral and Social Sciences**	
E	**Applied Sciences**	
F	**Computer and Systems Sciences**	Springer-Verlag
G	**Ecological Sciences**	Berlin, Heidelberg, New York, London,
H	**Cell Biology**	Paris, and Tokyo

Recent Volumes in this Series

Volume 234—Constructive Quantum Field Theory II
edited by G. Velo and A. S. Wightman

Volume 235—Disorder and Fracture
edited by J. C. Charmet, S. Roux, and E. Guyon

Volume 236—Microscopic Simulations of Complex Flows
edited by Michel Mareschal

Volume 237—New Trends in Nonlinear Dynamics and Pattern-Forming Phenomena: The Geometry of Nonequilibrium
edited by Pierre Coullet and Patrick Huerre

Volume 238—Physics, Geometry, and Topology
edited by H. C. Lee

Volume 239—Kinetics of Ordering and Growth at Surfaces
edited by Max G. Lagally

Volume 240—Global Climate and Ecosystem Change
edited by Gordon J. MacDonald and Luigi Sertorio

Volume 241—Applied Laser Spectroscopy
edited by Wolfgang Demtröder and Massimo Inguscio

Series B: Physics

Global Climate and Ecosystem Change

Edited by

Gordon J. MacDonald

University of California, San Diego
La Jolla, California

and

Luigi Sertorio

University of Turin
Turin, Italy

Plenum Press
New York and London
Published in cooperation with NATO Scientific Affairs Division

Proceedings of a NATO Advanced Research Workshop
on Model Ecosystems and Their Changes,
held September 4-8, 1989,
in Maratea, Italy

Library of Congress Cataloging-in-Publication Data

NATO Advanced Research Workshop on Model Ecosystems and Their Changes (1989 : Maratea, Italy)
Global climate and ecosystem change / edited by Gordon J. MacDonald and Luigi Sertorio.
p. cm. -- (NATO ASI series. Series B, Physics ; vol. 240)
"Published in cooperation with NATO Scientific Affairs Division."
Includes bibliographical references and index.
ISBN 0-306-43715-5
1. Climatic changes--Congresses. 2. Ecology--Congresses.
I. MacDonald, Gordon J. (Gordon James), date. II. Sertorio, Luigi, 1933- . III. North Atlantic Treaty Organization. Scientific Affairs Division. IV. Title. V. Series: NATO ASI series. Series B, Physics ; v. 240.
QC981.8.C5N36 1989
574.5'222--dc20 90-47793
CIP

A Division of Plenum Publishing Corporation
233 Spring Street, New York, N.Y. 10013

Printed in the United States of America

SPECIAL PROGRAM ON CHAOS, ORDER, AND PATTERNS

This book contains the proceedings of a NATO Advanced Research Workshop held within the program of activities of the NATO Special Program on Chaos, Order, and Patterns.

PREFACE

Humankind's ever-expanding activities have caused environmental changes that reach beyond localities and regions to become global in scope. Disturbances to the atmosphere, oceans, and land produce changes in the living parts of the planet, while, at the same time, alterations in the biosphere modify the atmosphere, oceans, and land. Understanding this complex web of interactions poses unprecedented intellectual challenges.

The atmospheric concentrations of natural trace gases—carbon dioxide (CO_2), methane (CH_4), nitrous oxide (N_2O), and lower-atmosphere ozone (O_3)—have increased since the beginning of the industrial revolution. Industrial gases such as the chlorofluorocarbons (CFCs), which are not part of the natural global ecosystem, are increasing at much greater rates than are the naturally occurring trace gases. All these gases absorb and emit infrared radiation and thus have the potential for altering global climate.

The major terrestrial biomes are also changing. Although world attention has focused on deforestation, particularly in tropical areas, the development of agriculture, the diversion of water resources, and urbanization have all modified terrestrial ecosystems in both obvious and subtle ways. The terrestrial biosphere, by taking up atmospheric carbon dioxide, acts as a primary determinant of the overall carbon balance of the global ecosystem. Although the ways in which the biosphere absorbs carbon are, as yet, poorly understood, the destruction (and regrowth) of forests certainly alter this process. Significant changes of the marine system on a global scale are less well documented, but it has been clearly demonstrated that man-made pollutants are invading even the deep sea, with inevitable implications for climate.

Climate change and ecosystem response were the topics of a NATO Advanced Research Workshop conducted at Maratea, Italy, September 4-10, 1989, that attempted to bring together the two most active fields of research: climate dynamics and ecosystem change. Within the two main themes of the workshop, modeling and predictability received special attention. In both climate and ecosystem change, tension exists between those who advocate models that attempt to include all potentially relevant processes and those who believe that simplified representations can capture the essential elements of the dominant processes. Models in both fields can rapidly become so complex that their behavior is as difficult to understand as the real systems they are intended to depict, and cascading uncertainties undermine efforts to make predictions. While it is clear that atmospheric concentrations of important greenhouse gases are changing, the links between compositional variations, climate, and ecosystem behavior remain a mystery. The difficulty in prediction arises partly from imperfect knowledge about key variables, but also from the fundamental nonlinear nature of climatic and ecological systems.

The opening chapter by G. MacDonald provides an overview of what is known and not known about climate systems. Insight into future climate change can be obtained by examining past climate regimes and the conditions that fostered them. However, an underlying problem in understanding past climate and possible future change is disentangling the intrinsic natural variability of the system (due to its nonlinear character) from the effects of imposed, external forcings. Ice age climate exhibits synchrony with orbital fluctuations as well as with variations in atmospheric composition. What part of ice age climate change was due to celestial mechanics, and what part was due to internal changes in the global ecosystem? The same question applies to current and future climate. Available data strongly indicate that global average temperatures have increased by about 0.5°C over the past century, but it is not possible to conclude decisively that this increase is due to changes in atmospheric composition.

D. Hartmann reviews the traditional approach to numerical modeling of climate in the second chapter. This approach assumes that the climate of the earth can be modeled using both explicit physics equations for processes that occur on the scale of the model grid resolution (greater than 100 km in the horizontal dimension) and simplified parameterization for processes that take place over shorter distances. These models reproduce features such as seasonal changes in surface temperature, but the question of "tuning" the free parameters to achieve agreement remains controversial.

Terrestrial ecosystems play a major role in fixing the carbon cycle and thus determine one of the major feedbacks to climate change. In Chapter 3, M. Crawley argues that changes in vegetation not only affect the carbon balance but also directly influence such climate parameters as albedo, surface roughness, and evapotranspiration.

By contrast, the oceans provide a major sink for carbon introduced into the atmosphere by the burning of fossil fuels and by deforestation. The driving force moving carbon from the atmosphere into the oceans is the difference in partial pressures of CO_2. Biological activity reduces the partial pressure of CO_2 in the surface layer by removing carbon detritus through gravitational action and vertical mixing. The oceans also provide a sink for heat and thus slow the warming of the atmosphere. J. McGlade examines certain responses of the ocean biota to climate change in Chapter 4.

In Chapter 5, H. Abarbanel returns to some of the issues that were raised in Chapter 1 concerning the nonlinear behavior of climate. The principal issue addressed in this chapter is how to analyze a time series generated by nonlinear processes. The discussion examines current views of chaotic systems whose deterministic structures are revealed by analysis of the underlying strange attractor. The emphasis on analysis of large complex systems is continued in two Appendices by A. Trevisan and B. Andresen that summarize the seminars they presented at Maratea.

Acknowledgments

The workshop, under the direction of Luigi Sertorio, was supported by grants from NATO and by the Istituto Nazionale Fisica Nucleare (INFN). Preparation of the proceedings was supported by The MITRE Corporation. The manuscripts of the papers came from diverse sources and required heavy editing, for which Margaret Jennings, Malinda Kerr, and Ione Swenson deserve deep appreciation. Thanks are also due to Dorothy Oliver for preparing various drafts of the manuscript.

Luigi Sertorio
Torino, Italy

Gordon MacDonald
McLean, Virginia USA

CONTENTS

CLIMATE DYNAMICS

ECOSYSTEM CHANGE

NONLINEAR SYSTEM ANALYSIS

GLOBAL CLIMATE CHANGE

Gordon J. MacDonald

Vice President and Chief Scientist
The MITRE Corporation
7525 Colshire Drive
McLean, VA 22102, USA

INTRODUCTION

Changes in the composition of earth's atmosphere due to human activities now dominate changes that occur naturally. Theory which links compositional change to climate change, together with observations of the atmospheric compositions and climates of earth's neighboring planets, Venus and Mars, compels us to consider seriously the possibility that future changes in climate due to Man's activities will be damaging. This article reviews the scientific evidence relating to past and present changes in earth's atmospheric composition and climate, and the evidence for future climate shifts.

Climate change has become a scientifically fashionable subject over the past decade, partly because of the great public and political interest. Yet despite increased research activity, controversy abounds: Because the atmosphere is a chaotic, dynamical system, how can trends in climate be positively detected against its great natural variability? Has a warming trend actually been detected? To what extent can nature's recuperative powers overcome imposed alterations in atmospheric composition? And to what extent can computer models reliably predict future conditions? These questions illustrate areas in which scientific disagreement exists today, and where controversy is likely to persist into the foreseeable future.

In addressing these questions, I will first review the evidence pertaining to current and past changes in concentrations of gases that have a major influence on the thermal budget of the atmosphere. Following this, in the third section, I examine the sources and sinks for these gases. Two factors establish the anthropogenic origin of increased atmospheric carbon dioxide (CO_2): the coincidence of the onset of escalating CO_2 levels and the beginning of the industrial revolution, and the rough balance between the amount of carbon dioxide emitted as a result of fossil fuel combustion and the increased carbon dioxide concentration in the atmosphere and ocean. Methane is increasing at a greater rate than carbon dioxide, but its sources and sinks are poorly understood, as are the sources for nitrous oxide. Despite numerous observations on ozone made in support of pollution control programs, the long-term variations in ozone concentration are poorly determined. On a per molecule basis, the strongest absorbers of outgoing infrared radiation are chlorofluorocarbons (CFCs). These compounds have no known natural sources, but atmospheric concentrations of CFCs continue to increase rapidly, even though some western nations have imposed controls on their use.

Global Climate and Ecosystems Change, Edited by G. MacDonald and L. Sertorio, Plenum Press, New York, 1990

A guide to possible future climate change is provided by the analysis and modeling of links between past climate and atmospheric composition. The post-ice age provides the best data on paleoclimates. Ice age climate changes are considered in the fourth section; models for past climate change are covered in the fifth section. Observations of past climates demonstrate the importance of the biosphere in controlling the composition of the atmosphere and thus global climate.

The final section deals with two issues: the statistical determination of a trend in a climate parameter (annual-mean global temperature), and the attribution of that trend to forcing from an externality, such as changing atmospheric composition. The statistics of atmospheric change are such that the currently observed trend of a rise in global temperature of about 0.6°C per hundred years is statistically robust. While statistical requirements for a warming trend are satisfied, the issue of attribution of this trend to greenhouse gases remains; with its long-term persistence, the atmosphere's natural variability may enhance or diminish externally imposed trends.

CHANGING ATMOSPHERIC COMPOSITION

Introduction

As the precision of atmospheric chemical analysis has improved over the past century, the concept of a chemically static atmosphere has been replaced by the view that the chemistry of the atmosphere changes on a variety of time and spatial scales, and that such changes are due to both natural and man-made causes. With the recognition that the atmosphere is chemically variable has come the realization that changes in climate are inevitable. The gases whose heat absorption properties influence earth's climate and their variation over time are the subject of this section.

Certain constituents of the atmosphere provide a warming blanket by absorbing thermal infrared radiation emitted by the earth's surface and atmosphere that would otherwise escape into space. The two dominant constituents of the atmosphere, reactive oxygen and inert nitrogen, do not absorb in the infrared region of the electromagnetic spectrum and play no direct part in the greenhouse phenomenon. Next in abundance are argon, carbon dioxide, and water, which are present at concentrations of the order 0.01%; the latter two absorb thermal energy. In discussing alteration in atmospheric composition, global average concentrations of the greenhouse gases are used when appropriate. For water, high variability in concentration in the lower atmosphere (troposphere) makes the concept of a global average for water vapor content inapplicable. Carbon dioxide, however, mixes within the atmosphere on a time scale of one year; for an annual time period, then, the concept of a global average is useful. At the parts per million (ppmv) level, methane (CH_4), neon, and krypton follow in order of abundance, but only methane is an infrared absorber. At still smaller concentrations, hundreds of atmospheric species have been identified, and most are infrared absorbers. Among these, nitrous oxide (N_2O), the CFCs, and ozone (O_3) have high enough abundances and particular absorption properties to make them significant contributors to the atmospheric greenhouse. CH_4, N_2O, and the CFCs all have long atmospheric lifetimes compared with the global mixing time, but for ozone, high variability in the troposphere makes consideration of a global average ozone concentration inappropriate.

The atmosphere's chemical composition is an important determinant of its thermal budget. When atmospheric composition is altered, shifts occur in the thermal budget. These shifts are reflected in changes in the global, equilibrium average temperature of the atmosphere, where the averages are taken spatially over the globe and temporally over some period, usually a year. In terms of climate change, measures other than temperature, such as precipitation, soil moisture, or river flow, may be of higher economic significance, but at present are more difficult to analyze. Similarly, the frequency and intensity of extreme events, such as droughts, floods, and severe storms, are economically and societally more significant than the global averages of atmospheric constituents, but these statistical quantities have not yet received the attention that averages have in discussions of climate.

The radiative heating or cooling that results from a shift in atmospheric composition is computed by calculating upward and downward fluxes through unit horizontal areas of the atmosphere. Within a vertical column, account is taken of the distributions of temperature and the concentrations of radiatively active gases, particularly water vapor. The infrared-absorbing properties of the various species are well known, owing to laboratory studies. Also, because the overlap of the blackbody radiation curves for the sun and earth is not great, solar radiation may be separated from terrestrial radiation on the basis of wavelength, which simplifies the computation. However, analyzing the extent of climate change that follows an alteration in atmospheric chemical composition is complicated by feedback processes. For example, warming at the surface leads to increased water vapor content, and because water vapor is an infrared absorber, this portion of the feedback leads to further warming. But increased water vapor can also lead to alterations in the coverage, thickness, and height distribution of clouds. Clouds influence the atmospheric thermal balance by trapping upwelling infrared radiation, as well as by reflecting incoming sunlight back into space, which changes the earth's albedo. For thin, low-lying stratus clouds, the cooling albedo effect swamps the heating due to the trapping of infrared radiation. With thick, mid-level cumulus clouds, outgoing radiation is trapped, and the atmosphere warms. Surface warming can alter snow and ice cover, changing both the earth's albedo and atmospheric water vapor content. Alterations in water vapor, cloudiness, and snow cover are important but are not the only feedbacks at work. Because of their complexity, feedback processes can be modeled only approximately, leaving large uncertainties as to the total climatic impact of changing atmospheric composition.

Variation in Atmospheric Concentration of Carbon Dioxide (CO_2)

The modern era of climate-related atmospheric chemistry began in 1958, when C. D. Keeling commenced his highly precise measurements of carbon dioxide at Mauna Loa, Hawaii, as part of the activities of the International Geophysical Year. In 1958, the annual average concentration of CO_2 was 315 ppmv; by 1988, the annual average had risen to 350 ppmv. The increase in CO_2 concentration from 315 to 350 ppmv over the 30-year period corresponds to the addition of 74.3 Gt of carbon to the atmosphere.[1] The current exponential rate of increase for carbon dioxide is 0.4% per year (see Fig. 1). A further feature of the variation of CO_2 concentration is its seasonal oscillation, which is due to biological activity—mostly in the Northern Hemisphere. Atmospheric CO_2 decreases during the growing season (May through October) and increases during the colder months through respiration. At Mauna Loa the peak-to-peak amplitude is about 6 ppmv, or 12.7 Gt of carbon.

The exponential rise of CO_2 in the atmosphere has been detected at stations around the world. Keeling et al. (1989) have described a set of observations that span the Central Pacific (see Fig. 2). The trend in concentration for the stations in the Northern Hemisphere is shown in Fig. 3, and the Southern Hemisphere stations in Fig. 4. The figures clearly show that the rise in CO_2 concentration is worldwide, with the South Pole stations lagging behind those in the Northern Hemisphere (where most fossil fuel emissions occur) by about a year.

Figures 3 and 4 also show that the seasonal oscillation is greatest at high latitudes in the Northern Hemisphere, while in the Southern Hemisphere the

[1] 1 ppmv of CO_2 in a dry atmosphere of total mass 5.119×10^6 Gt corresponds to 2.1228 Gt of carbon.

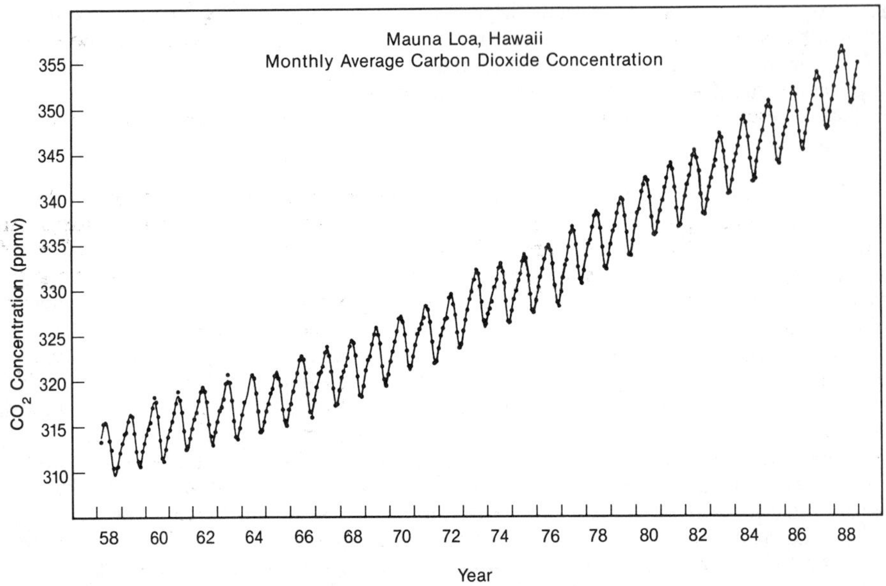

Fig. 1. Variation of atmospheric concentrations of carbon dioxide at Mauna Loa Observatory, Hawaii (source: Keeling et al., 1989).

seasonal oscillation has a very small amplitude beyond 10°S of the equator. The latitudinal variation of the seasonal oscillation is further illustrated in Fig. 5, which is based on a larger number (15) of stations over the Northern Hemisphere. The observed amplitude lies approximately on a curve that represents the difference between summer and winter levels of solar radiation. The proximity of the seasonal CO_2 oscillation with the solar insolation curve signifies the dominance of photosynthesis and land-plant respiration in fixing the amplitude of the annual cycle of CO_2 variation. The large amplitude of the oscillation in the northern, land-covered hemisphere and the low amplitude in the southern, ocean-covered hemisphere demonstrates the overriding importance of the terrestrial biosphere in determining the seasonal fluxes of CO_2 into the atmosphere.

The record of the atmospheric concentration of carbon dioxide at high latitudes can be extended back in time by analyzing the composition of air trapped during the formation of glacial ice in Greenland and Antarctica. Because the conversion of snow into ice takes at least a decade, the dating of past air samples cannot be very precise; nevertheless, trapped air bubbles give the composition of the atmosphere, averaged over a few decades, for the past few hundred years. The historical variation of CO_2 derived from air bubbles trapped in glaciers is shown in Fig. 6. The preindustrial concentration of atmospheric CO_2 lay in the range of 270 to 280 ppmv and was at about that level from the end of the last glacial period. If the preindustrial value is taken as 277 ppmv for the year 1740, then 80.7 Gt of carbon were added to the atmosphere in the 218 years between 1740 and 1958. This amount contrasts with the net addition of 74.3 Gt of carbon during the last 30 years.

Using an ice core from Vostok, Antarctica, a joint French-Soviet team extended the record of atmospheric carbon dioxide concentration farther back into glacial times, reaching past the interglacial period (~ 140,000 years ago) prior to the present one. Their research shows that during times of maximum ice coverage in the Northern Hemisphere, atmospheric CO_2 concentrations reached a minimum

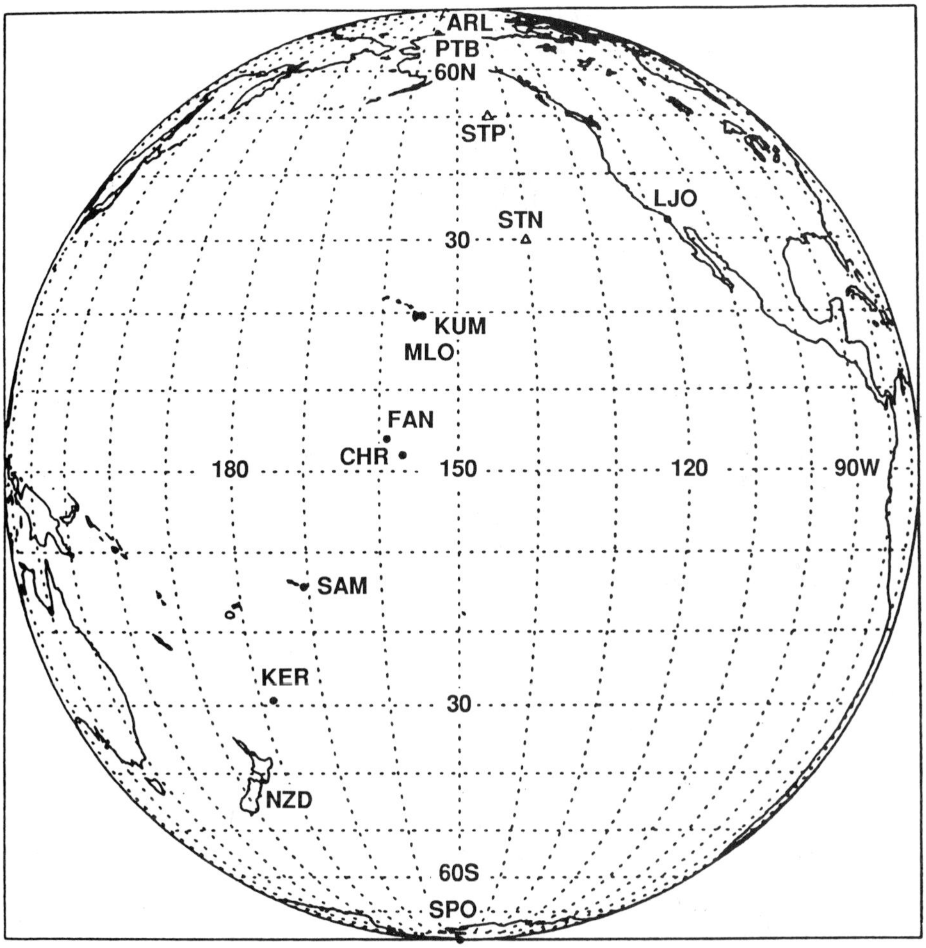

Fig. 2. Locations of air sampling stations on land and over water. Code names: ARL, Ice Floe Station Arlis (position approximate); PTB, Point Barrow, Alaska; STP, Ocean Station 'P'; LJO, La Jolla, California; STN, Weather ship 'N'; KUM, Cape Kumukahi, Hawaii (moved from Hilo, Hawaii, after 1963); MLO, Mauna Loa Observatory, Hawaii; FAN, Fanning Island; CHR, Christmas Island; SAM, Cape Matatula, Samoa; KER, Raoul Island, Kermadec Islands; NZD, Baring Head, New Zealand; SPO, South Pole (source: Keeling et al., 1990).

value of about 180 ppmv (Barnola et al., 1987), as is shown in Fig. 7. During the prior interglacial warm period, the atmospheric CO_2 concentration reached 290 ppmv, a level comparable to the preindustrial concentration.

At the onset of the last two interglacial periods, the atmospheric carbon dioxide concentration underwent changes of about 100 ppmv on a time scale of 10,000 years as a result of, as yet, poorly understood natural processes. Since the industrial revolution began, sufficient carbon dioxide has been added to the atmosphere to increase the concentration by about 70 ppmv on a time scale of 250 years. During the last 30 years alone, the CO_2 concentration has increased by about 35 ppmv. The recent increases in concentration are smaller on a percentage basis (25% over 250 years) than during the glacial periods (60% over 10,000 years). However, these comparisons are significant; changes at smaller concentrations have a greater impact on the radiative balance than do changes at higher concentrations, because certain CO_2 absorption bands (15 μm) become saturated at higher CO_2 concentration levels.

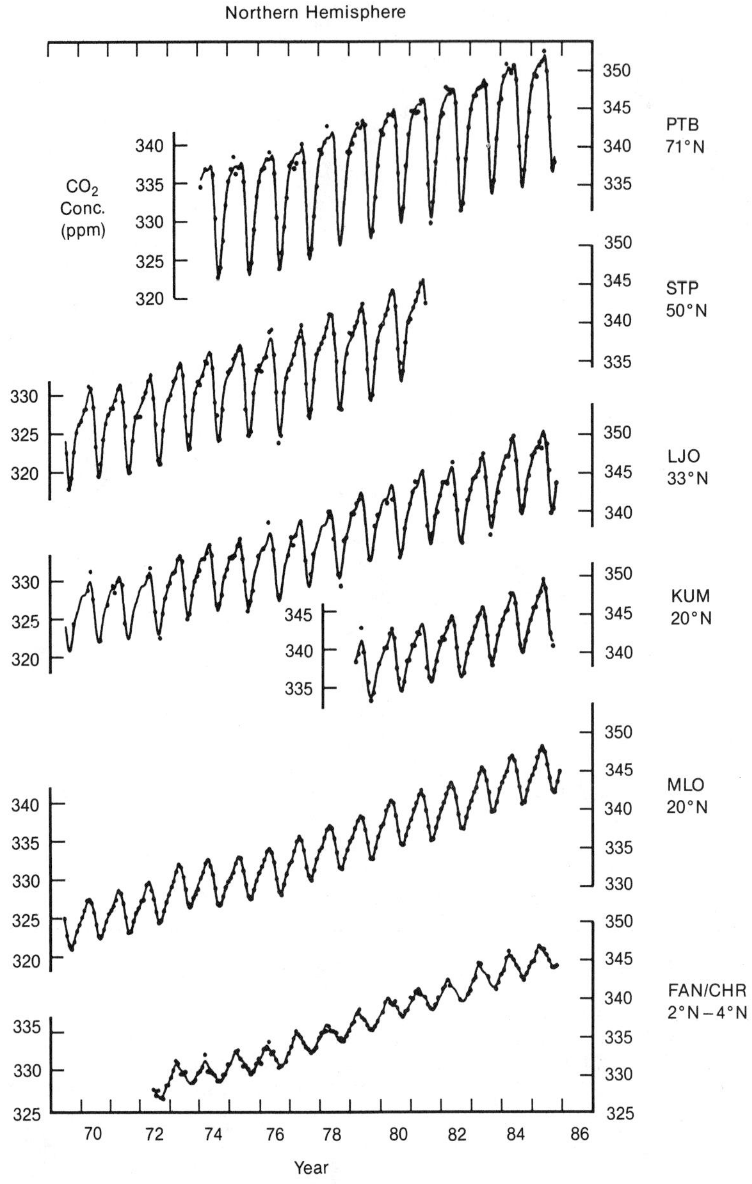

Fig. 3. Long-term variation in CO_2 concentration for the Northern Hemisphere stations shown in Fig. 2 (source: Keeling et al., 1990).

Variations in the Atmospheric Concentration of Methane (CH_4)

Methane is a strong infrared absorber; it is also the most abundant reactive trace gas in the troposphere. Methane plays a major role in tropospheric chemistry, determining in part the concentrations of two important oxidants: ozone and the extremely reactive hydroxyl radical. Because of its reactivity, the residence time for methane is short (~ 10 years) as contrasted to the carbon dioxide residence time of several hundred years. Methane is also the source of about one-half the hydrogen and water in the stratosphere and thus is a major determinant of the stratosphere's chemical and thermal regimes.

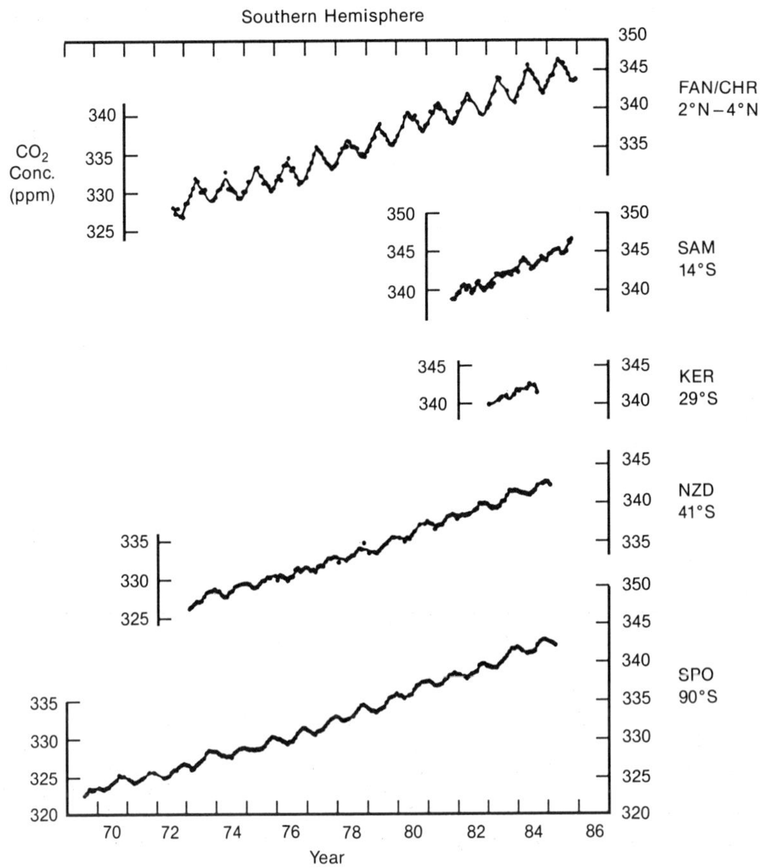

Fig. 4. Long-term trends in atmospheric carbon dioxide concentration for Southern Hemisphere stations shown in Fig. 2 (source: Keeling et al., 1990).

Various studies have established that atmospheric methane is currently increasing at a rate of 1% per year, or almost three times the rate at which carbon dioxide is increasing (Rasmussen and Khalil, 1981; Fraser et al., 1984; Blake and Rowland, 1988). In 1988, the global methane concentration was 1.7 ppmv. A reliable record for methane measured directly from the atmosphere is available only for the 1978 to 1988 period; earlier measurements are not of the same quality as those taken during the past decade. Rinsland et al. (1985) reanalyzed solar absorption spectra taken on Jungfraujoch in 1951 and deduced a methane concentration for that year of 1.14 ppmv. This measurement is consistent with an exponential increase of methane at the rate of 1.08% per year for the period 1951 to 1984. However, there is no extended time series for atmospheric methane concentration comparable to that obtained by Keeling for carbon dioxide.

Evidence from the analysis of gases trapped in ice from both Greenland and Antarctica shows that the atmospheric methane concentration has doubled in the last 200 years but remained at a level between 0.6 and 0.8 ppmv over the previous 3,000 years (see Fig. 8). The Vostok ice core extends the record for methane back 160,000 years (see Fig. 7) to the time of maximum ice coverage. Methane concentrations during glacial peaks were low—about 0.35 ppmv, or almost one-half the value observed during interglacial times (Stauffer et al., 1988). The Vostok ice core data show that methane increased from 0.32 ppmv to 0.62 ppmv between the end of the last glaciation and the subsequent interglacial period, again paralleling shifts in the CO_2 concentration.

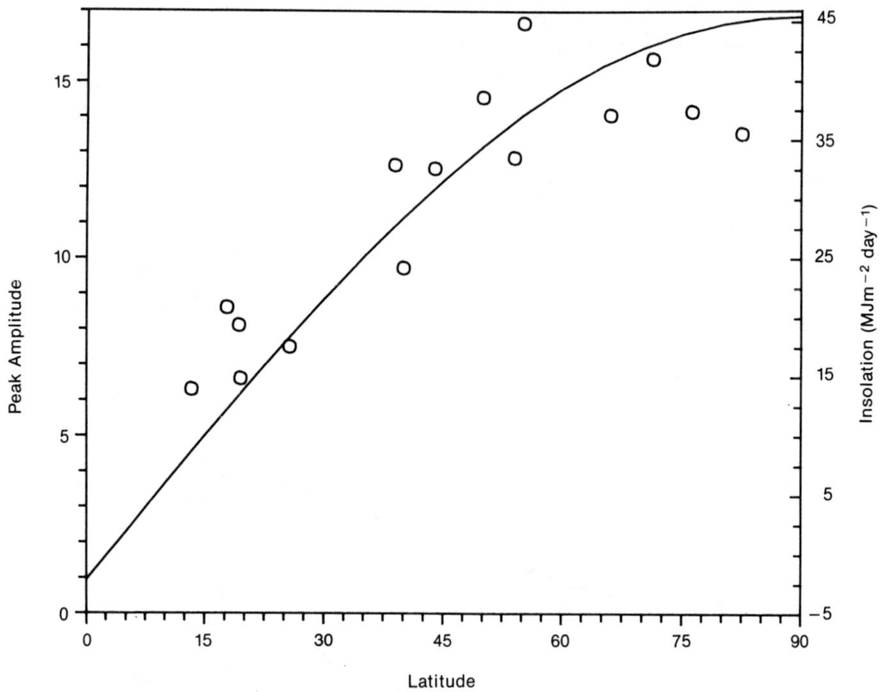

Fig. 5. Latitudinal variation of the seasonal cycle of atmospheric carbon dioxide. Mean amplitudes compiled from data reported by Fraser et al. (1983), Chung (1988), Bacastow et al. (1985), Halter and Peterson (1981), Pearman and Hyson (1981), and Komhyr et al. (1985). The curve represents the difference between clear-sky solar radiation on June 21 and December 21 in order to provide an indication of the latitudinal variation caused by the growth and decay of organic matter.

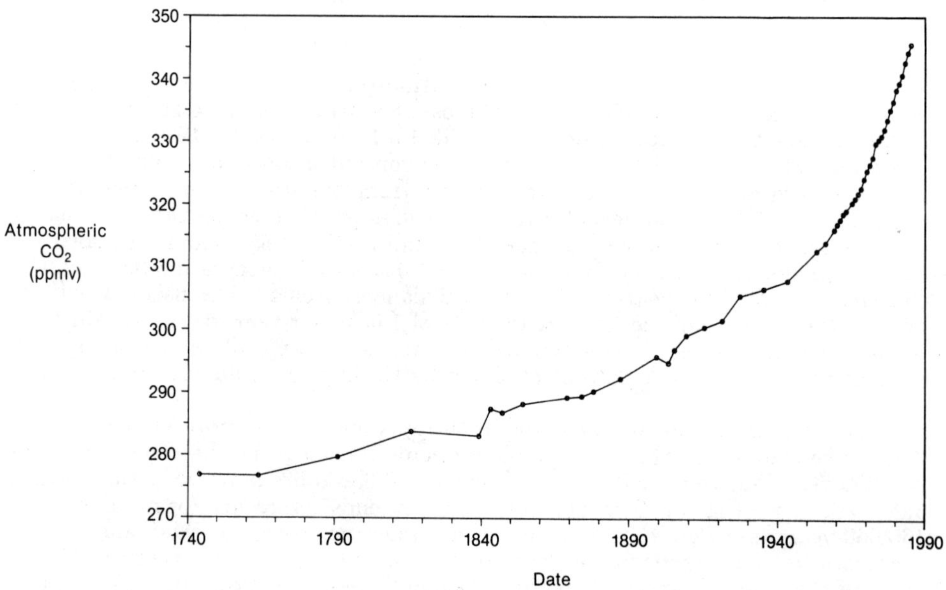

Fig. 6. Historical variation of CO_2 concentration in the atmosphere. From measurements made on Antarctic ice cores (after Neftel et al., 1985; Friedli et al., 1986). Points for 1958 to 1988 are from Keeling (1990).

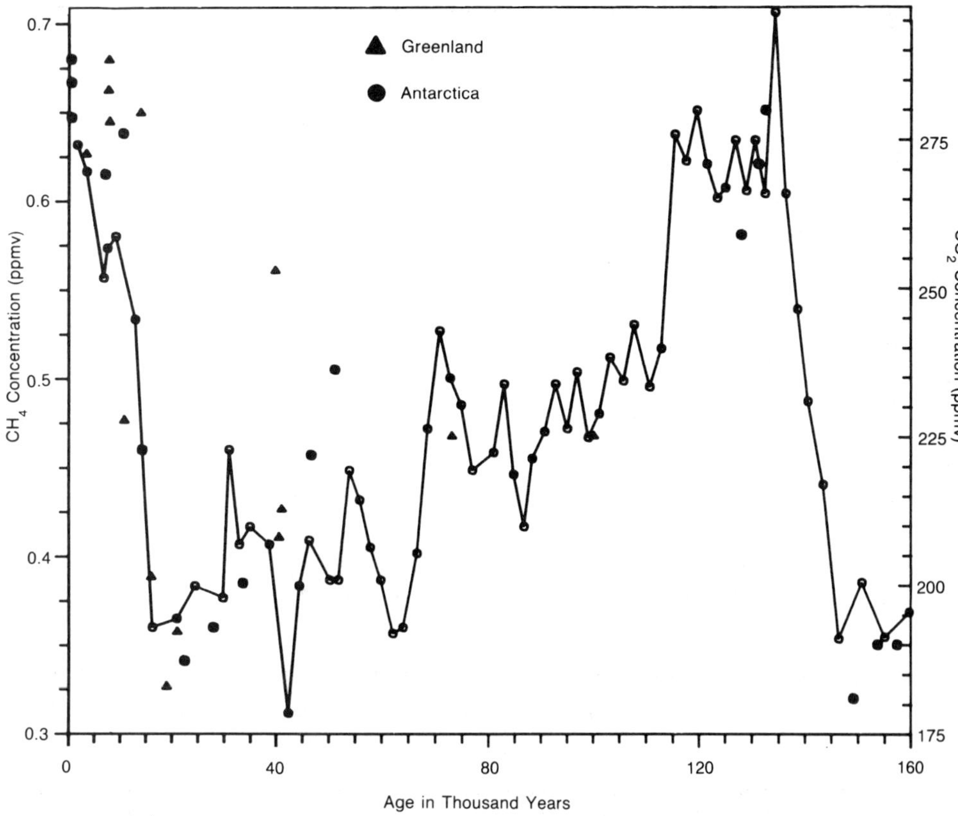

Fig. 7. Variation of methane and carbon dioxide during the past 160,000 years. The methane data are computed from Stauffer et al. (1985), Stauffer et al. (1988), Raynaud et al. (1988), and Craig and Chou (1982). The solid line representing the CO_2 variation in air trapped in the Vostok ice core is from Barnola et al. (1987).

Concentrations of Atmospheric Halocarbons (CFCs)

Unlike methane and carbon dioxide, there are no known natural sources for halocarbons such as $CFCl_3$, CF_2Cl_2, CH_3CCl_3, etc. Industrial production of halocarbons began late in the industrial revolution (1930s), but today they are the most rapidly increasing species in the atmosphere. Inert in the troposphere, halocarbons undergo photodecomposition when they reach the stratosphere, producing compounds and free radicals that destroy stratospheric ozone.

The most abundant chlorine- and fluorine-containing halocarbon is CF_2Cl_2 (CFC-12), which has a calculated lifetime in the atmosphere of about 140 years and a current concentration of 0.4 parts per billion (ppbv), a quantity roughly 400 times lower than the concentration of methane. Despite this small concentration, CF_2Cl_2, like other halocarbons, is a potent infrared absorber. One CF_2Cl_2 molecule has the approximate equivalent greenhouse effect of 10,000 CO_2 molecules. Worldwide, CF_2Cl_2 is used primarily as a refrigerant and as a propellant in aerosol cans. During the decade of 1975 to 1985, CF_2Cl_2 levels increased at an annual average global rate of about 10% per year, but restrictions on production have recently reduced the annual rate of increase to about 5% (SORG, 1988).

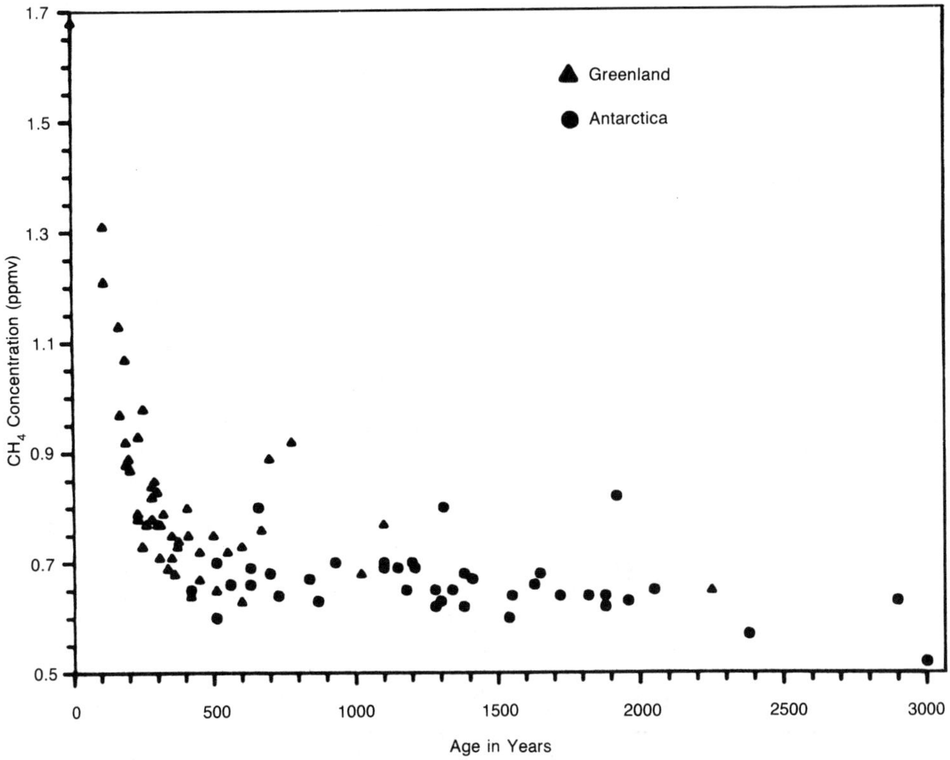

Fig. 8. Historical concentration of atmospheric methane. Data obtained from measurements of air taken from dated ice cores (Rasmussen and Khalil, 1984). The value for the present concentration is taken from Blake and Rowland (1988).

The next most abundant halocarbon, $CFCl_3$ (CFC-11), is used both in the blowing of packaging foams and in aerosol cans. The atmospheric concentration of $CFCl_3$ has paralleled that of CF_2Cl_2 and is currently at a level of about 0.11 ppbv; its atmospheric residence time, however, is less than that of CF_2Cl_2—about 100 years.

Methyl chloroform (CH_3CCl_3) is primarily used as a solvent for degreasing and is rapidly growing at a rate of about 6% per year, despite a short atmospheric lifetime of about six years (Prinn et al., 1987). Many other halocarbons are strong infrared absorbers but have not yet reached high enough concentrations to affect the atmospheric radiation balance significantly.

Concentration of Atmospheric Nitrous Oxide (N_2O)

Nitrous oxide is an infrared absorber that has both natural and man-made sources such as fertilizer and fossil fuel combustion. The current concentration of N_2O is about 0.31 ppmv and is increasing at an annual rate of 0.8 ppbv (Khalil and Rasmussen, 1983, 1988). Measurements on ice cores show a preindustrial level of 285 ppbv for N_2O and no large changes in concentration during the prior 3,000 years. A puzzling feature contained in the ice core data is the late onset of change in the N_2O concentration, which began rising around 1940, rather than earlier in the industrial revolution (Friedli and Siegenthaler, 1988).

Concentration of Tropospheric Ozone (O_3)

Ozone is a short-lived constituent in the lower atmosphere that is produced when photolysis of nitrogen dioxide occurs in the presence of hydrocarbons and carbon monoxide. A principal component of smog, ozone has become a regional pollutant in many parts of the world.

Atmospheric ozone concentrations are measured on a routine basis in support of national air pollution control programs; despite this, data on long-term trends is remarkably sparse. The longest, nearly continuous record of ozone levels has been taken by the German Democratic Republic Meteorological Service at Cape Arkona on the Baltic Coast (Feister and Warmbt, 1987), where the average rate of increase of ozone was 2.6% per year from 1956 to 1984. Working on the assumption that the Arkona record is typical of the variation of regional, lower-level ozone in Europe, Volz and Kley (1988) compared the more recent Arkona measurements with ozone observations taken between 1896 and 1910 at the Montsouris Observatory in Paris. Their comparison shows that in the 1980s, the ozone concentration at Arkona was double the level recorded in Paris during the late 1800s. Further, no long-term trend appears in the Paris data over the sampled interval, though the year-to-year variations are comparable to those observed at Arkona. The Paris and Arkona data are consistent with the supposition that tropospheric ozone in Europe has undergone a rapid increase over the last 40 years.

Increases in concentrations of lower-level ozone have also been measured at remote sites; increases of 1.4% and 0.8% per year were recorded at Mauna Loa, Hawaii, and Point Barrow, Alaska, respectively. A negative (−0.9% per year) trend has been recorded at Samoa, while no trend is detectable at the South Pole (Oltmans and Komhyr, 1986). The above data suggest an increasing concentration of tropospheric ozone in the Northern Hemisphere but are insufficient to provide a reliable estimate of global change.

Summary of Changes in Atmospheric Chemistry

Table 1 provides a summary of recent changes in the concentration of atmospheric constituents that are capable of altering the thermal budget of the atmosphere.

Table 1. Major shifts in composition of the atmosphere.

Gas	Residence Time (years)	Concentrations (ppmv) Pre-1850	Concentrations (ppmv) 1988	Observed Trends 1975-1985 (%)
CO_2	500	280	350	4.6
CH_4	10	0.7	1.8	11.0
Tropospheric O_3 (< 12 km)	(hours)	?	10-100*	?
N_2O	100	0.28	0.31	3.5
$CFCl_3$	140	0	0.23*	103
CF_2Cl_2	100	0	0.40*	101
CH_3 CCl_3	6	0	0.13*	155

*ppbv

Although discussions of climate change generally focus on variation in global average temperature, data on a number of other variables are available for the past century (see Table 2). For many of these variables, the records are sketchy, and data collected with instruments are limited; e.g., upper-air temperature observations are available only after World War II. For periods before about 1880, discussion of climate change must be based on proxy data, such as historical documents, tree ring widths, time of harvest, price of wine, ice breakup in harbors, etc. (Lamb, 1977).

Determining the history of surface temperature from past observations is a difficult undertaking, given the inhomogeneity in sampling (both spatial and temporal), as well as in instrumentation and observers. Because only about 30% of the globe is land area, observations taken on land alone may not provide a true picture of global temperature change. Land-station records contain fluctuations that result from non-climatic factors, such as station moves, changes in instrumentation, changes in protocol, and so on. A major program to correct station records, where possible, and to omit those records that are unreliable was undertaken by Jones et al. (1986a). Jones et al. (1986b) have attempted to extend land records by using tens of millions of sea-surface temperature readings obtained by ships at sea.

One reconstruction of temperature over the past 108 years is shown in Fig. 9 (Hansen and Lebedeff, 1987; 1988). This record shows the features common to various compilations of temperature records: all agree on a warming of about 0.5°C over the past century; also, the Southern Hemisphere record shows a

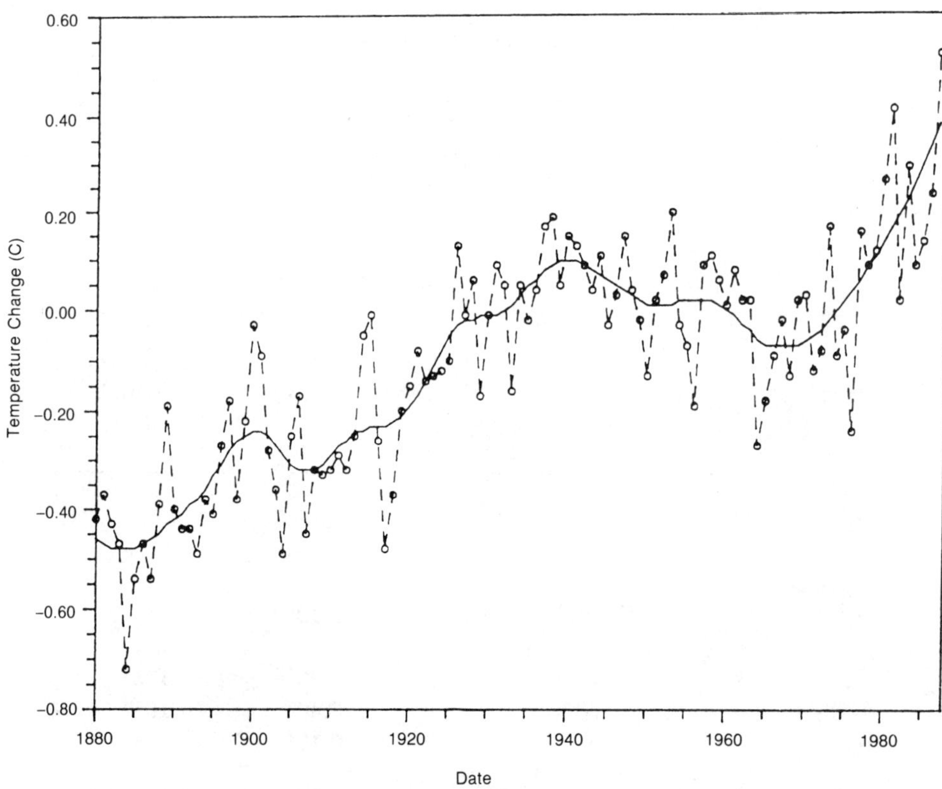

Fig. 9. Variation of global average temperature. The solid curve is a filtered version corresponding to a five-year running average (after Hansen and Lebedeff, 1987; 1988).

Table 2. Variables that may reflect climate change.

Atmospheric Surface Temperature	Lake Levels
Subsurface Ground Temperature	Sea Level Pressures
Sea-Surface Temperature	Winds
Upper-Level Atmospheric Temperature	Snow Cover
Moisture Content of Atmosphere	Sea Ice Extent
Soil Moisture	Subsurface Ocean Temperature
Precipitation	Sea Level
River Flows	

gradual rise in temperature in the 20th century, while the Northern Hemisphere record is characterized by two warming periods (1910 to 1940 and 1975 to 1988), with a cooling in between.

An obvious feature of the temperature curve in Fig. 9 is that the six warmest years on record—1980, 1981, 1983, 1986, 1987, and 1988—occurred during the decade of the 1980s, and preliminary data suggest that 1989 will join the list (Jones et al., 1988). A continuing uncertainty in the temperature records is the bias introduced by the urban heat-island effect. Using a data set for the United States, Karl and Jones (1989) found an urban bias of between 0.1° and 0.4°C over the 1901 to 1984 period, which is as large as or larger than the overall 0.16°C temperature increase in the same period. Analysis of temperature records for other countries that have undergone urbanization will be required before a confident statement can be made as to the overall urban heat-island bias in global average temperatures.

A completely independent method of obtaining changes in temperature has been used by Lachenbruch and Marshall (1986), who examined the variation of temperature with depth in the permafrost region on the North Slope of Alaska. Because the ground is permanently frozen to a depth of several hundred meters, subsurface temperatures remain undisturbed. In the absence of water movement, heat is carried out of the earth solely by thermal conduction. The temperature at depth can be extrapolated to the surface if the thermal conductivity of the sediments is known. Lachenbruch and Marshall found that the observed subsurface temperature was consistently 2° to 4°C warmer than that obtained by upward extrapolation of the thermal gradient. Because the depth to which the temperature perturbation reaches gives an approximate time scale over which the surface warming has taken place, Lachenbruch and Marshall concluded that a warming of 2° to 4°C has occurred during the past century over the region sampled. This observation is consistent with a global average warming of 0.5°C, owing to the amplification of postulated greenhouse warming at high latitudes. Unfortunately, data are available only for a limited region in the Arctic.

Marine-surface sea and air temperatures, like land data, reveal long-term trends. According to marine measurements, recent decades have been the warmest of the past 120 years, while the period from about 1890 to 1920 was distinctly colder than average. Most marine measurements are taken by ships of opportunity. Such ships follow preferred routes and thus leave large areas of the ocean inadequately sampled. For sea-surface temperatures, problems of data interpretation arise from the many changes made in measurement techniques over the years. Several different kinds of buckets have been used for bringing up sea water, and different (and often unknown) instructions were followed regarding the amount of time to expose the thermometer to the water before recording the reading. An even greater discontinuity took place when the shift to steam engines resulted in a transition to the measurement of water from the intake pipes rather than buckets.

Measurement of upper-air temperatures using radiosondes also shows long-term trends (Angell, 1988). Radiosonde observations have the advantage of being free from urban heat-island effects. However, there have been type and design changes of radiosondes, and a reliable station-by-station record of these changes over time is not available. Since 1958, a 63-station network has been used to monitor layer-mean temperature by measuring heights of fixed pressures. Although sparse, the network appears to give reasonable estimates of temperature changes. The network record shows that the period from 1964 to 1976 was cooler than average, and the past decade was warmer than average, both by a few tenths of a degree Celsius. Further, observations of stratospheric temperature show a recent sharp cooling that is to be expected from the buildup of CO_2, which increases radiative cooling into space.

Another variable with which to measure climate change is sea level. Tide-gauge observations reflect an average global sea level rise of 1.0 to 1.5 mm per year since 1880 (Gornitz et al., 1982; Peltier and Tushingham, 1989). However, individual tide-gauge records can display considerable variation over short distances, and the network of measuring stations is widely scattered, with many fewer stations operating in the Southern Hemisphere. When interpreting sea level data in terms of climate, a major issue is the disentanglement of the effects of isostatic rebound and tectonic activity. With such geologic activity in mind, and taking the effect of melting Alpine glaciers into account (Wigley and Raper, 1987), the observed rise is consistent with a warming of 0.5°C.

Summary of Changes in Atmospheric Composition and Climate

Major changes in the composition of the earth's atmosphere have taken place since the beginning of the industrial revolution. Atmospheric concentrations of two principal greenhouse gases, carbon dioxide and methane, are increasing at exponential rates of 0.4% and 1.0% per year. Accompanying these shifts in abundance of greenhouse gases have been trends in several climate parameters. One of these parameters, the global average or surface temperature, has increased about 0.5°C over the past century. An examination of the possibility that climate trends are linked to altered atmospheric chemistry requires an analysis both of natural trends in climate and of the magnitude of expected shifts in climate arising from chemical changes.

SOURCES AND SINKS FOR GREENHOUSE GASES

Introduction

Through various activities such as burning fuels, flaring natural gas, making cement and plastics, using industrial chemicals, cultivating crops, clearing and fertilizing land for agriculture, and adding nutrients to lakes and oceans through fertilizer and sewage runoff, man brings about changes in the composition of the atmosphere. Nature attempts to adjust to man's superimposed changes, but the shifts observed in atmospheric composition show that nature cannot respond rapidly enough to adapt to changes induced by human activity on time scales of decades. Changes in atmospheric composition are also brought about through natural fluxes among the biosphere, atmosphere, and oceans. Both natural and anthropogenic changes affect the concentrations of greenhouse gases and thus have the potential to alter climate. If society decides that remedial action must be taken to mitigate the greenhouse effect, it will be necessary to identify and determine the magnitudes of the various anthropogenic and natural sources and sinks of the greenhouse gases.

Carbon Dioxide Released by Human Activities

The coincidence of the dawning of the industrial revolution with the onset of major changes in the composition of the atmosphere (see Fig. 6) signifies that man's activities play an important role in changing the composition of earth's atmosphere. Three kinds of human activity release carbon dioxide to the atmosphere in a major way: burning fossil fuels, converting tropical and other forest areas to alternative land use, and manufacturing cement. The production of one metric ton of cement releases 0.136 metric tons of carbon as CO_2; this figure does not include the fuel burned in calcining the limestone. Secondly, when carbon-saturated forest areas are cleared and converted in use to lands that have smaller inventories of carbon in trees, surface litter, and soil, there is a net release of carbon to the atmosphere that can be large but is often unknown. In addition, for every cubic meter of timber burned, about 0.26 metric tons of carbon are released as CO_2. Releases from the third major CO_2 source, fossil fuels, are very large, with annual emissions exceeding 5 billion tons.

The fossil fuel sources of carbon dioxide can be estimated from the records of global industrial and economic activities collected by the United Nations (Keeling, 1973; Rotty, 1987). The accuracy of data on fossil fuel emission depends on the reliability of information in three areas: quantities of fuel used, the carbon content of each fuel, and the fraction of the fuel that is actually burned. The reliability of the relevant data improved markedly after the oil crisis of 1973, which greatly increased international attention in the area of energy management.

The amount of carbon dioxide produced by the use of fossil fuels depends not only on the quantity of energy used but also on the mix of fuels. The amount of carbon generated in delivering a fixed amount of thermal energy depends on the hydrogen-to-carbon ratio of the fuel. Methane, with a high hydrogen-to-carbon ratio, releases less carbon dioxide than does coal in delivering the same amount of thermal energy, as is indicated in Table 3. When "synthetic" fuels are burned, the energy used in converting the feedstock into fuel must be included in the total amount of carbon dioxide these fuels contribute to the atmosphere. Synthetic fuels thus produce large quantities of carbon dioxide as compared to natural fossil fuels (see Table 4) (MacDonald, 1982).

Figure 10 illustrates the historical variation in the amount of carbon added to the atmosphere through the burning of fossil fuels. Fuel combustion over the last 128 years can be separated into five major periods, each of which has a distinct rate of exponential growth (see Table 5). Between 1860 (when the United Nations data set begins) and 1913, fossil fuel use increased at an average exponential rate of 4.3% per year. Coal was the dominant fossil fuel during that period, though large and uncertain amounts of wood were burned. With the advent of World War I, the global growth rate of energy use slowed down markedly to 1.5% per year. This slower rate continued through the world turbulence of the 1920s, the depression of the 1930s, and World War II. During the 23-year period beginning in 1950, rapid growth in energy use resumed. Oil overtook coal as the dominant fuel during this period, with the growth rate for both oil and gas use far exceeding that of coal (see Fig. 11). The oil price shock of 1973 began another low growth period for energy use. After 1973, the growth rate in energy use sank to 1.2% per year, largely as a result of oil price increases. This rate was lower than the growth rate that prevailed during the World Wars and the intervening depression. The use of coal began to grow at a greater rate than the use of oil during the post-1973 period, reversing a 50-year trend. The collapse of oil prices in the early 1980s triggered another jump in the growth rate of energy use, beginning in 1984. Since then, world energy use has been increasing at an average rate of about 2.4% per year, with coal use growing most rapidly.

Changes in the global fuel mix have been accompanied by major shifts in the regional distribution of carbon dioxide sources. In 1950, North America was responsible for 45% of carbon dioxide emissions, while the developing world

Table 3. Carbon dioxide emission from the direct combustion of various fuels.

Fuel	CO_2 Emission Rate (kg C/10^9J)	Ratio Relative to Methane
Methane	13.5	1
Ethane	15.5	1.15
Propane	16.3	1.21
Butane	16.8	1.24
Gasoline	18.9	1.40
Diesel Oil	19.7	1.46
No. 6 Fuel Oil	20.0	1.48
Bituminous Coal	23.8	1.73
Subbituminous Coal	25.3	1.87
Wood	25.3	1.87

Table 4. Carbon dioxide emissions from the production and burning of various synthetic fuels.

Fuel	CO_2 Emission Rate (kg C/10^9J)	Ratio Relative to Methane
Shale Oil		
In situ 28 gal/ton Shale	48	3.5
High-Temperature 25 gal/ton Shale	66	4.9
High-Temperature 10 gal/ton Shale	104	7.7
Liquids from Coal		
Sasol Technology; Eastern Coal	42	3.1
Exxon Donor Solvent; Eastern Coal	39	2.9
Gasoline from Methanol from Coal	51	3.8
High-Btu Gas from Coal		
Lurgi	41	3.0
Hygas	40	3.0
Methanol		
From Natural Gas	21	1.5
From Coal	36-44	2.7-3.2

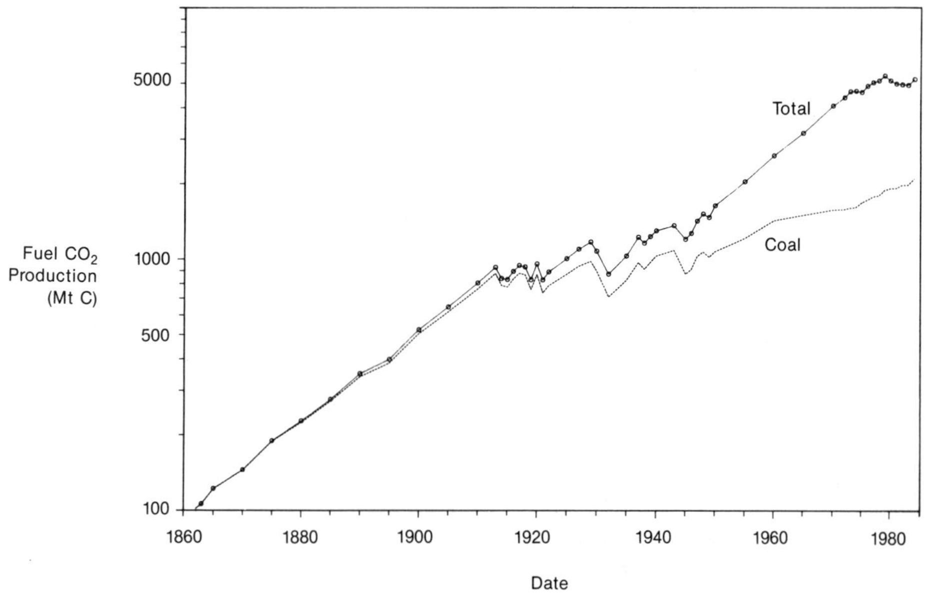

Fig. 10. Historical variation in carbon dioxide emission from the burning of fossil fuels. Data are from Keeling (1973), Rotty (1987), and Marland and Rotty (1984).

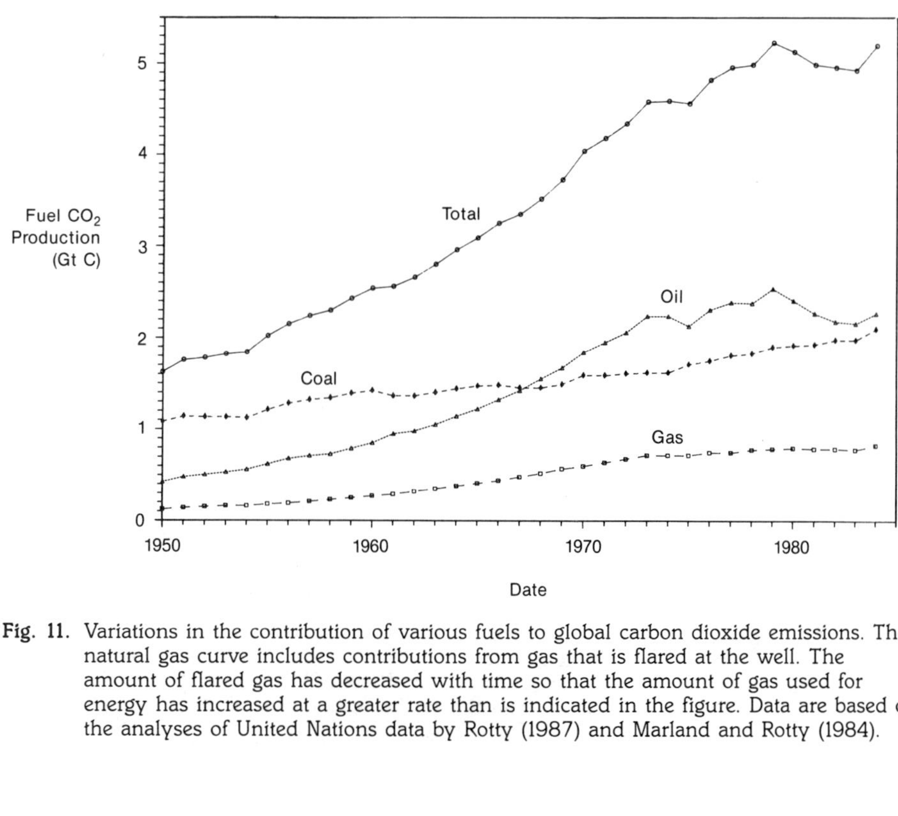

Fig. 11. Variations in the contribution of various fuels to global carbon dioxide emissions. The natural gas curve includes contributions from gas that is flared at the well. The amount of flared gas has decreased with time so that the amount of gas used for energy has increased at a greater rate than is indicated in the figure. Data are based on the analyses of United Nations data by Rotty (1987) and Marland and Rotty (1984).

Table 5. Historical rates of exponential growth of fossil fuel use (in percent per year).

Fuel/Period	1860-1913	1913-1950	1950-1973	1973-1984	1984-1988*
All Fuel	4.3	1.5	4.5	1.2	2.4
Coal	4.2	0.55	1.8	2.4	3.1
Oil	-	-	7.3	0.1	1.5
Natural Gas	-	-	8.0	2.3	2.0

*Preliminary estimates based on 1984-1986 data and preliminary estimates for 1986-1988 (Energy Information Administration).

contributed only 6%. By 1984, North America's share had dropped to 25%, with the developing world emitting 15%, as is illustrated in Fig. 12. In the U.S., 1986 CO_2 emissions were 1.8 times higher than those of 1950, though the U.S. share of total global emissions declined from 42% to 22% over the same interval. Despite these growth differentials, more than half the global CO_2 emission can still be attributed collectively to three countries—the United States, the Soviet Union, and China.

The shift of CO_2 sources is further illustrated in Fig. 13, which gives the percentage change from 1986 to 1987 for the world, various regions, and the United States and Canada. On a global basis, the increase in CO_2 emissions from 1986 to 1987 was 1.6%, while emission from the United States increased by 1.9%.

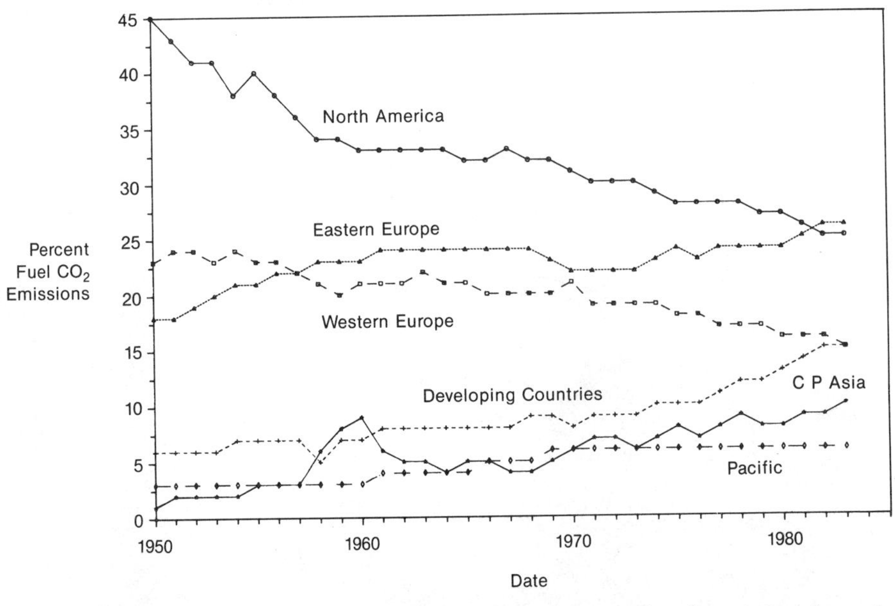

Fig. 12. Changes in the fractional contribution to global carbon dioxide emissions by major regions of the world. Eastern Europe refers to the centrally planned economies of Eastern Europe, including the entire Soviet Union. CP Asia includes the centrally planned economies of Asia, Peoples Republic of China, North Korea, Mongolia, North Vietnam, etc. Japan, Australia, New Zealand, and the Pacific Island nations comprise the Pacific region.

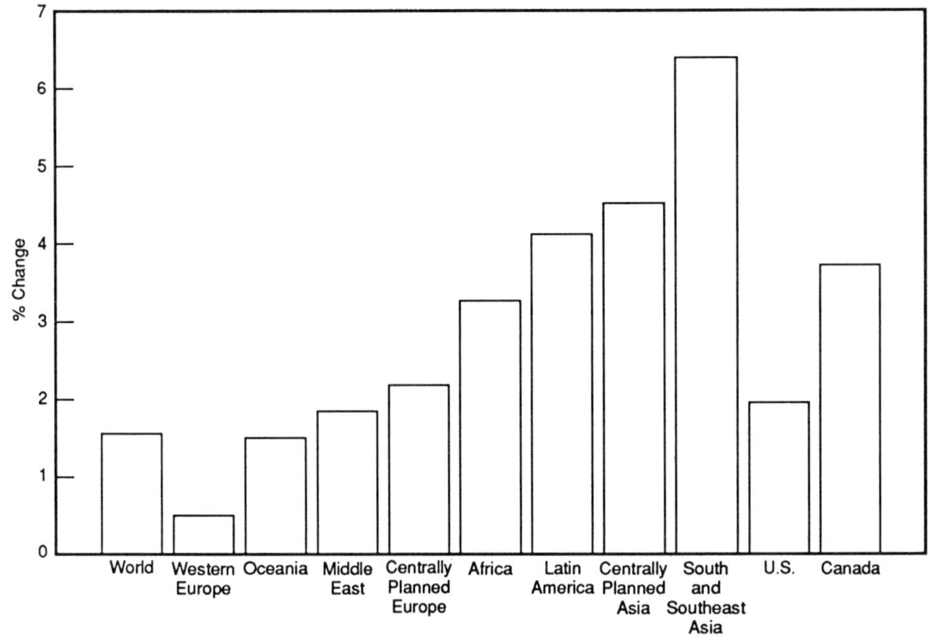

Fig. 13. Percentage change in carbon dioxide emission from 1986 to 1987. Emissions include fossil fuel combustion and cement manufacturing (after Marland and Boden, 1989).

Although CO_2 emissions in the developing world have been growing rapidly, the oil price shock of 1973 led to a leveling off of emission in much of the industrialized world, as illustrated by emissions in the United States (see Fig. 14) and in Japan (see Fig. 15). The collapse of oil prices during the 1980s led to a resumption of growth in energy use in the United States and, to a lesser extent, in Japan. In contrast, carbon dioxide emissions in France have dropped by about 30% since 1979 (see Fig. 16), owing to the strong nuclear energy policy that country has pursued.

In 1987, the United States emitted about 1.24 Gt of carbon as CO_2, roughly 10 Mt of which was contributed by cement manufacturing. The emission total does not include wood burning, either intended or natural, though wood burning in the United States probably amounts to no more than 1% to 2% of the CO_2 released by fossil fuel burning. The fossil fuels used to power the electrical utility and transportation sectors contribute about 66% of the total; the other third is divided between the industrial and residential/commercial sectors (see Table 6).

Data on the amount of carbon released to the atmosphere through agriculturally related activities are much less secure than fossil fuel-related data. Estimates for such activities are based largely on agricultural and forestry statistics (Houghton et al., 1983; Houghton and Woodwell, 1989). On the basis of alternative scenarios for agricultural expansion, Houghton et al. (1983) obtained a range of global releases for 1980 (the date of the last FAO survey) of 1.8 to 4.7 Gt of carbon. With additional data, Houghton et al. (1987) have computed a lower limit of release of 0.133 Gt of carbon from temperate and boreal regions, and 1.659 Gt from the tropics, for a total release of 1.791 Gt of carbon from agricultural lands in 1980.

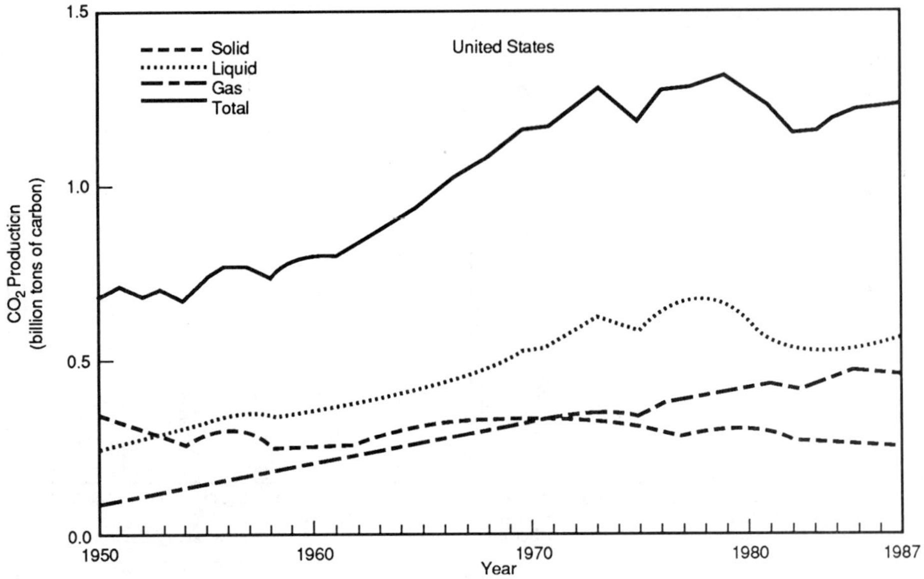

Fig. 14. Emissions of carbon dioxide to the atmosphere from the United States as a result of fossil fuel combustion and cement manufacture (after Marland and Boden, 1989).

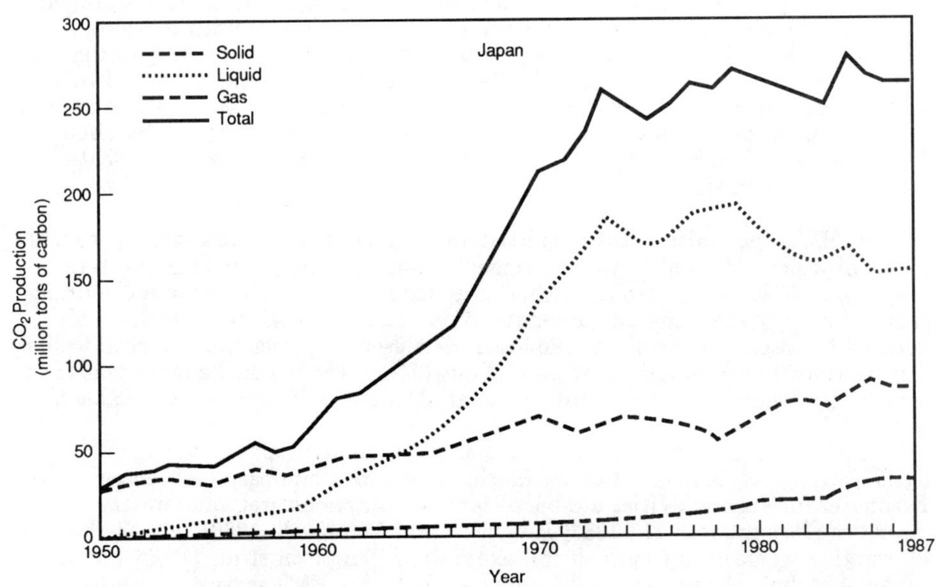

Fig. 15. Emissions of carbon dioxide to the atmosphere as a result of fossil fuel combustion and cement manufacture in Japan (after Marland and Boden, 1989).

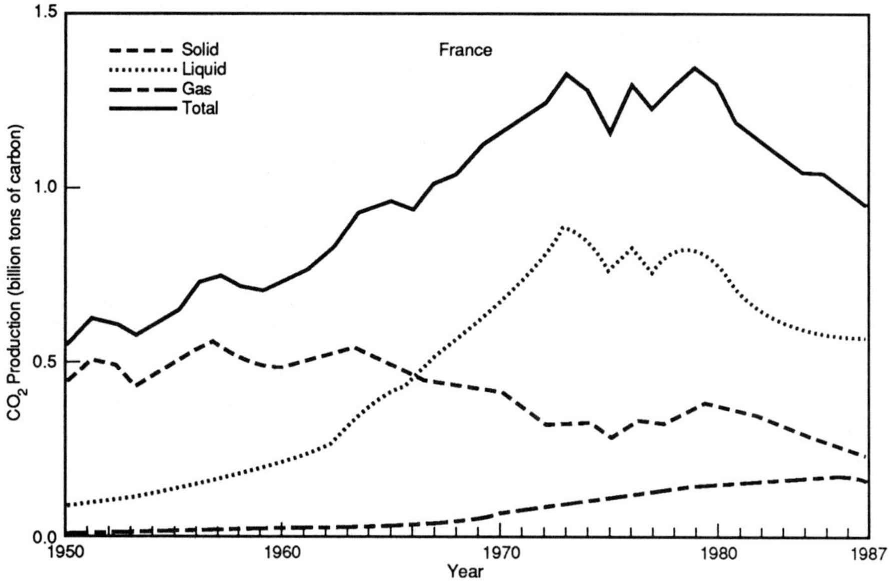

Fig. 16. Emissions of carbon dioxide from fossil fuel burning and cement manufacture in France (after Marland and Boden, 1989).

Sources of Methane

Methane is the most abundant organic gas in the atmosphere; its current concentration of 1.7 ppmv corresponds to 3.61 Gt of carbon. Calculations based on the photochemistry of methane yield an atmospheric residence time for this gas of 8.1 to 11.8 years (Cicerone and Oremland, 1988), which implies a quasi-steady state source of 0.4 to 0.59 Gt of carbon per year. At present, the amount of methane in the atmosphere is increasing at a rate of about 0.036 Gt per year.

Table 6. Estimated carbon dioxide emissions in the United States for 1987 by sector of economic activity and fuel type*.

Sector/Fuel	Coal	Natural Gas	Petroleum	Total	Percent
Residential/ Commercial	4.4	93.2	49.7	147.3	12.1
Industrial	68.4	93.3	108.8	270.5	22.3
Transportation	-	6.8	357.5	364.3	30.1
Electric Utilities	361.2	41.6	28.0	430.8	35.5
Totals	434	235	544	1213	100
Percent	35.8	19.4	44.8	-	100

*units are millions of metric tons

In order to determine the sources of this methane, isotopic ratios can be used. An analysis of the ratio of $^{14}CH_4$ to $^{12}CH_4$ provides a measure of the amount of added methane that is of relatively recent biological origin, because methane released from old deposits will not contain any ^{14}C, which is generated by cosmic rays, bomb tests, and nuclear reactors. Lowe et al. (1988) estimated that only 68% of the methane sources were biogenic. This estimate has been recently revised by Manning et al. (1989) to 74%. Wahlen et al. (1989) found that at the end of 1987, 21% ± 3% of atmospheric methane was derived from fossil carbon sources. Given the 20% to 30% range for fossil carbon in atmospheric methane, modern biogenic sources must annually release between 0.28 and 0.47 Gt of carbon as methane.

Conventional discussions on the origin of the increase in atmospheric methane focus on enteric fermentation in animals, rice paddies, termites, biomass burning, land fills, and other sources that tend to be concentrated in the mid latitudes and tropics (see Table 7). These primarily anthropogenic sources account for three-quarters of the postulated methane source, while the more widely distributed natural wetlands sources represent only one-fourth of methane emissions in standard methane budgets. However, the observed geographical distribution of atmospheric methane shows higher levels at higher northern latitudes, where wetland sources are abundant (see Fig. 17).

Table 7. Annual methane release rates for identified sources (after Cicerone and Oremland, 1988).

	Annual Release in Mt of C	Range of Values in Mt of C
Man-made Sources		
Enteric Fermentation (animals)	60	50-75
Rice Paddies	80	45-130
Biomass Burning	40	40-75
Landfills	30	20-50
Coal Mining	25	20-35
Gas Drilling, Venting, Transmission	35	20-40
Subtotal	270	195-405
Natural Sources		
Termites	30	7-75
Natural Wetlands (forested and unforested bogs, swamps, tundra, and alluvial formations)	85	75-200
Ocean	7	3-15
Freshwater	5	1-25
Methane Clathrates	4(?)	1-75
Subtotal	131	87-390
TOTAL	401	282-795

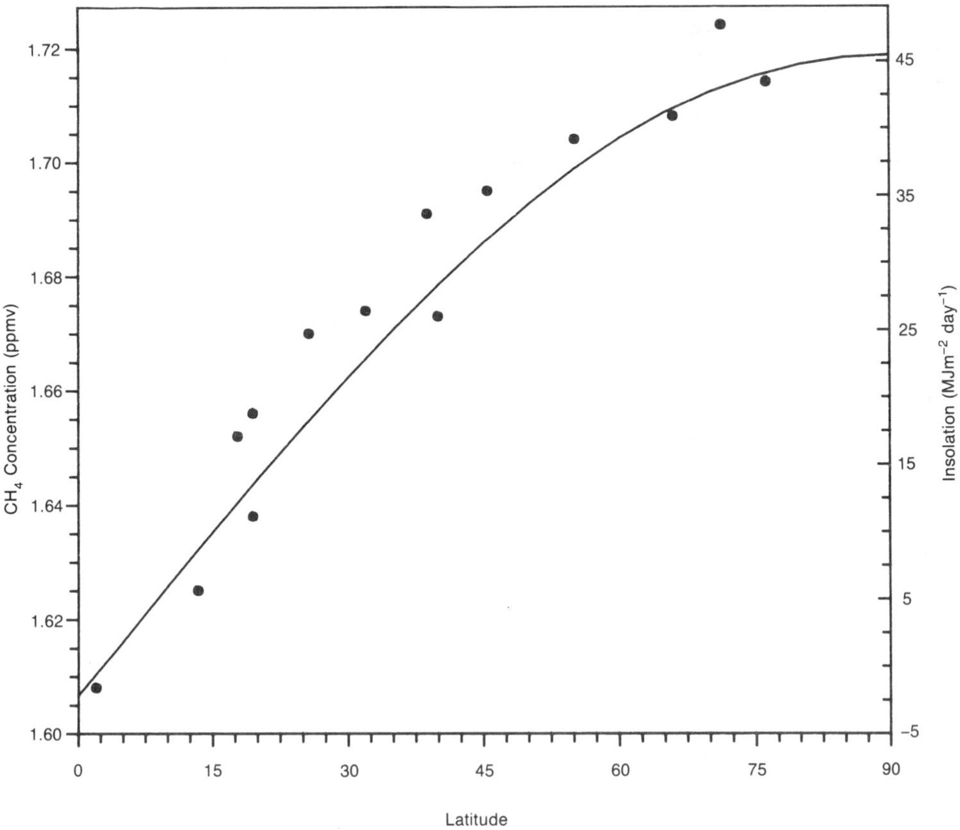

Fig. 17. Latitudinal variation of atmospheric methane concentration in the Northern Hemisphere for 1984 (after MacDonald, 1989).

Geographically, observed concentrations of methane mimic the latitude variation in the amplitude of the CO_2 seasonal cycle, roughly paralleling the difference between summer and winter insolation. In contrast, atmospheric methane concentrations in the Southern Hemisphere show no regular latitudinal variation, remaining approximately constant. For 1984, the annual mean concentration at Ascension Island (7.9°S) was 1.589 ppmv, while that at the South Pole was 1.575 ppmv. This strong asymmetry between the Northern and Southern Hemispheres indicates a terrestrial origin for the methane. Further, close adherence of the latitudinal distribution to the solar insolation curve clearly indicates that much of the methane is of biological origin, in accordance with the $^{14}C/^{12}C$ ratios observed in atmospheric methane.

The evidence for a major northern biologic source of methane in the high northern latitudes is very strong, as indicated in Fig. 17. Possible northern sources of CH_4 include releases from natural wetlands and bogs in the boreal forests of Siberia and Canada, releases from methane clathrates in coastal and Arctic settings, and emissions from extensive development of gas fields and coal mines, particularly in Siberia. The increased release of biogenic methane could result from changes in the abundance of wetlands, or from thermal stimulation of methane-generating bacteria. These sources are discussed below in some detail in order to illustrate the complexities of the interaction of the biosphere with the atmosphere.

Methane is produced in swamps and bogs where anaerobic conditions exist. In the farming areas of Eurasia, Africa, and North America, a recent and rapid reduction in wetlands has occurred (Zaloumnis, 1987). In contrast, swamps and shallow lakes remain abundant at northern latitudes. Nisbet (1989) has suggested that fluctuations in the beaver population has had a marked influence on methane generation in northern wetlands. In North America, beavers disappeared over much of the United States and were rare in parts of Canada by the mid-1850s. The population of beavers in Europe, including European Russia, was wiped out by 1900 (Tyurnin, 1984). In the early part of the 20th century, beavers were reintroduced in Canada and Europe in an attempt to build up populations. Since 1950, the Canadian beaver population has increased dramatically in response to conservation measures. The buildup is illustrated by the number of beaver pelts sold in Canada: in 1928, 57,000 pelts were sold; in the 1970s, 350,000 to 450,000 pelts were sold annually (Nisbet, 1989). In the Soviet Union, the beaver has returned, but at a lesser rate of increase.

Beavers influence methane production in two ways. They create swamps out of forest land, and they bring carbon in the form of trees into anaerobic swamp conditions, which are conducive to the formation of methane. The first step in conversion of forest area to swamp is the formation of a beaver colony. The density of colonies in favorable locales is 0.5 to 3 colonies per km^2. Each colony forms a pond. Beaver ponds persist from 5 to 20 years, with new ponds of open water having a depth of 1 to 2 m. As the ponds mature and become swamp, they are abandoned. The amount of open water continues to decrease over a time period of 20 years, until the swamp becomes a bog. Large ponds can persist for longer times depending on the history of the dam and the colony's activities. The bogs and swamps are eventually colonized by new trees, and the cycle repeats itself.

The importance of the beaver to methane generation is not only in creating swamps and carrying carbon into the swamp: every year, each beaver may introduce to its pond a metric ton of wood, which is selected for its high nutritional value (Francis et al., 1985). The carbon that is introduced into wetlands by the beavers is converted much more efficiently than carbon in wetlands without beavers. In studies of wetlands, Naiman et al. (1986) found that beaver ponds produce 33 times the methane as equivalent water surfaces where there are no beavers. The recovery of the beaver in Canada and, to a lesser extent, in the Soviet Union may have resulted in a major increase in the release of methane. Not only has the total area of wetlands increased as a result of the return of the beaver, but the efficiency of conversion of carbon to methane has increased with expanded beaver activity.

In addition to enhanced methane production from beaver ponds, it is possible that the rapid growth of atmospheric methane is a direct result of global warming, the evidence for which was described in the previous section. Methane production in wetlands is strongly linked to temperature, because of the high temperature-dependence of metabolism in methanogenic bacteria. The large seasonal fluctuation in methane concentration that peaks in October reflects the sensitive temperature dependence of methanogenesis; for example, in a Danish red alder swamp, methane production increases tenfold for a 0° to 15°C rise in temperature. If this value is representative, a global average warming of 0.5°C corresponds to a warming in the northern latitudes of about 1°C, which in turn corresponds to a 16% increase in methane production due to the warming alone.

Global Carbon Cycle

Atmospheric carbon dioxide and methane are participants in the overall global carbon cycle. We may crudely represent the balance sheet for the carbon cycle by:

$$\mathbf{A} = \mathbf{F} + \mathbf{M} + \mathbf{D} - \mathbf{B} - \mathbf{S},$$

where

A = annual increase of atmospheric CO_2 (3.69 Gt in 1987)

F = the annual release of carbon dioxide in fossil fuel burning and cement production (5.67 Gt in 1987)

M = annual rate of oxidation of methane (0.5 ± 0.1 Gt)

D = decrease in biomass and soil cover produced by deforestation and soil erosion (~ 1.79 Gt)

B = net increase of biomass and soil carbon due to increasing atmospheric CO_2 (?)

S = net annual flow of CO_2 into the oceans (~ 2.2 Gt)

The value for **A** is accurately known from observations at Mauna Loa and elsewhere. The addition of CO_2 to the atmosphere **F** is probably correct to within a few percent, while the annual rate of methane oxidation **M** is known to within 0.1 Gt. The net rate of uptake of carbon dioxide by the oceans depends on the details of the model in which CO_2 is transferred from the surface waters to the deep ocean. An estimate of the effective eddy diffusion can be made by using ^{14}C generated before and after bomb tests as a tracer (Siegenthaler, 1983). On this basis, **S** is estimated to be 2.2 Gt (Keeling et al., 1989). The values of **D** and **B** are highly uncertain. Using the listed values for **A**, **F**, **S**, and **M**, we obtain a positive net flux to the biosphere **B** − **D** :

$$\mathbf{B} - \mathbf{D} = 0.28 \text{ Gt},$$

which implies that the uptake of CO_2 by the biosphere due to increased carbon dioxide is large, 2.07 Gt, if values obtained by Houghton et al. (1987) for forest destruction are accepted.

Questions concerning the effects of the recent rise in atmospheric carbon dioxide on plant growth are controversial. Houghton et al. (1983), who found little ecological or physiological basis for expecting stimulation of growth, point to the lack of direct evidence supporting an increase in biomass or soil carbon as evidence against stimulation. They support this view by noting that the increase in global average temperature during the 20th century has probably promoted additional releases of CO_2 and opposed carbon dioxide-stimulated growth. Peterson and Melillo (1985) further argue that a lack of available nitrogen and phosphorus inhibits increased plant growth. On the other hand, Detwiler and Hall (1988), after summarizing numerous opinions expressed by others, conclude that "the possibility that CO_2 fertilization occurs in undisturbed ecosystems cannot be dismissed."

Additional information on the relative roles of the oceans and biosphere as sources and sinks for carbon can be obtained by analyzing variations in the ratio of the two stable isotopes of carbon, ^{12}C and ^{13}C. Approximately 1% of stable natural carbon has an atomic weight of 13. The ratio of ^{13}C to ^{12}C is expressed in the form

$$\delta^{13}C = \gamma/\gamma_S - 1,$$

where γ denotes the ratio for the sample, and γ_S is the ratio for a standard material ($\gamma_S = 0.0112372$). By custom, only the reduced ratios are multiplied by 1,000 and expressed in "per mil," or as o/oo. The product of $\delta^{13}C$ and carbon mass is very nearly an additive function when CO_2 samples are mixed. If $\delta_1^{13}C$ and $\delta_2^{13}C$ denote ratios of samples containing N_1 and N_2 moles of CO_2, respectively, then the mixture has very nearly the $\delta^{13}C$ value

$$\delta^{13}C = ({}^{13}\delta_1 N_1 + {}^{13}\delta_2 N_2)/(N_1 + N_2).$$

The isotopic ratio for the atmosphere is about −8%, as is the isotopic ratio for carbon in the ocean. However, the isotopic ratio for plant carbon (and fossil fuels) is about −25%. The addition of 1 ppmv of biospheric carbon to an atmosphere that has a CO_2 concentration of 350 ppmv alters $\delta^{13}C$ and brings about a change in concentration of

$$\frac{\Delta(\delta^{13}C)}{\Delta C} = [((350)(-8\%) + (-25\%))/351] - 1(-8\%) = 0.048\%/\text{ppmv}.$$

Figures 18 and 19 show variations in the stable isotopic composition of atmospheric carbon dioxide in the Pacific for the Northern and Southern Hemispheres, respectively. The high-amplitude oscillation in seasonal concentration of CO_2 found at high northern latitudes (see Fig. 20) is reflected in the large-amplitude fluctuation in isotopic ratio (Fig. 21). The oscillation represents the biospheric contribution and is not due to seasonal breathing by the oceans.

In addition to the seasonal oscillations, the time series for the $\delta^{13}C$ show a trend toward more negative values. Unfortunately, because fossil fuels and plant materials have similar isotopic ratios, these data cannot be used to separate out the contributions of the biosphere from industrial activities.

A representation of the overall carbon cycle is given in Fig. 22. Of particular interest to considerations of future carbon dioxide levels are those reservoirs of carbon that may be vulnerable to changes in surface conditions. The largest of these reservoirs is the ocean, which contains some 39,000 Gt of carbon, although much of this carbon is locked away in a stable inorganic form at ocean depths that are remote from the atmosphere. The carbon in this reservoir will respond to surface warming only on time scales of a thousand years, the calculated mixing time of the deep ocean. Coal, lignite, and shale oil contain large, though uncertain, amounts of carbon (~ 5,000 Gt), but this store of carbon is impervious to surface warming and remains in its geologic repositories, except when human activities intervene. Smaller quantities of carbon (~ 1,400 Gt) are stored in near-surface soils. Changes in land use promote the release of carbon from the soil reservoir, as has been discussed. Further, increases in surface temperature will speed up the metabolism of soil bacteria, which increases the release of both carbon dioxide and methane.

A fourth large store of carbon has recently been recognized in the form of methane clathrates (Kvenvolden, 1988; MacDonald, 1990). Methane clathrates are solids in which a rigid cage of water molecules surrounds methane. Clathrates are stable in cold regions and under the oceans down to depths of 1,500 to 2,000 m. They currently hold very large quantities of carbon (~ 11,400 Gt) and are potentially vulnerable to surface warming, a topic I will return to in the section on Modeling Ice Age Climate.

Halocarbons, Nitrous Oxide, and Ozone

As previously noted, there are no known natural sources for CFC-11 and CFC-12. The observed rise in the concentrations of these halocarbons is clearly due to widespread industrial use. For nitrous oxide, the origin of the increased concentration is not as clear. During the combustion of fossil fuels containing nitrogen (principally coal), a fraction of the nitrogen is converted to N_2O. The relative fraction of N_2O to NO depends on the burning process. Biomass burning produces N_2O, but when wood is burned the proportion of nitrogen converted to nitrous oxide is less than in conventional coal plants. The use of nitrogen-based fertilizer also contributes N_2O to the atmosphere. Hao et al. (1987) estimate that the combustion of fossil fuels releases about 3.2×10^6 tons of N_2O as N per year. Some 86% of this release is due to the burning of coal in large industrial burners; fertilizers contribute significantly less. These figures suggest that anthropogenic sources of N_2O are somewhat less than one-third of natural sources.

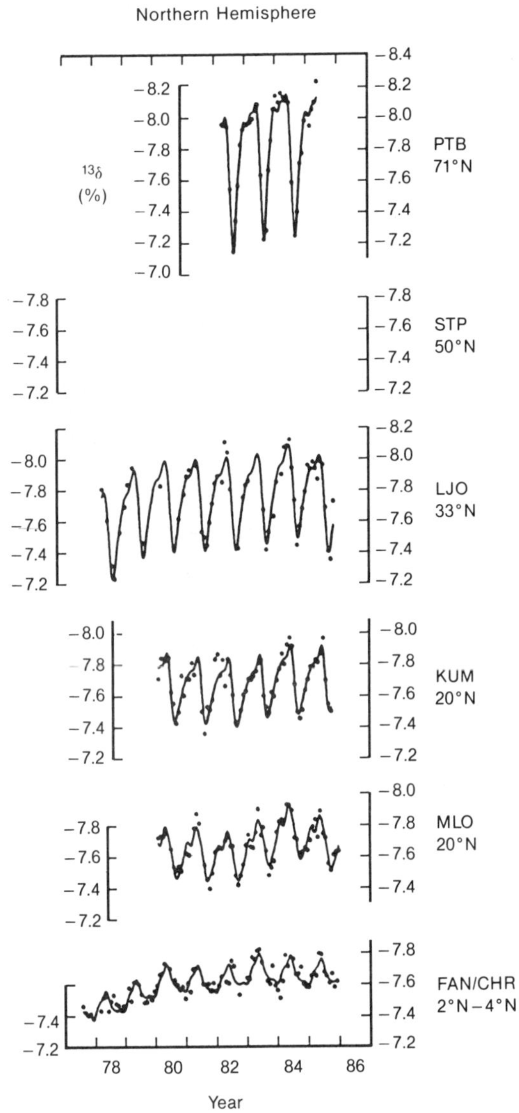

Fig. 18. Time trend of the reduced isotopic ratio, $^{13}\delta$, of atmospheric CO_2 in per mil from standard PDB (source: Keeling et al., 1990).

Tropospheric ozone is produced largely by the catalytic photolysis of nitrogen dioxide in the presence of hydrocarbons and carbon monoxide during urban smog formation. Ozone can also form in non-urban areas, where natural forest provides a major source of hydrocarbons. In rural regions, the abundance of NO_2 limits ozone formation; in urban areas—with their high concentrations of nitrogen oxides—hydrocarbons and carbon monoxide are limiting. Data of the variations of NO, NO_2, and O_3 abundances are insufficient to determine whether urban or rural ozone is the principal contributor to tropospheric ozone.

ICE AGE CLIMATE

Introduction

In order to understand and eventually predict how increasing concentrations of greenhouse gases might alter the earth's climate, both modeling studies and

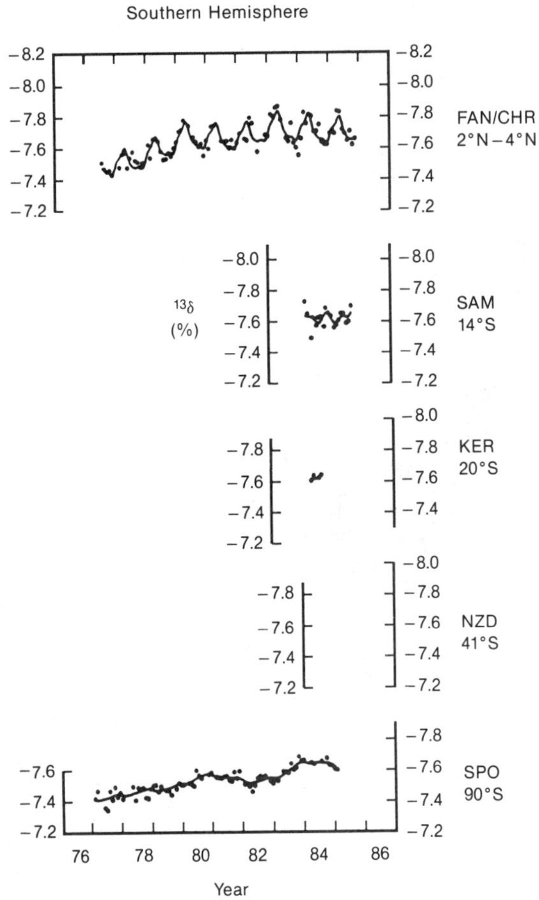

Fig. 19. Same as Fig. 18 but for the Southern Hemisphere (source: Keeling et al., 1990).

empirical analyses of past climates are required. The analysis of past climates is relevant to the present issue of anthropogenic climate change for a number of reasons. Shifts in past climates provide a sample of the natural variability of the climate system. The determination of natural variability is essential if we are confidently to attribute observed climatic changes to changes in atmospheric composition. Detection of a greenhouse signal requires that we identify a trend in nature's "noise." The analysis of past climate allows us insight into the ways in which the atmosphere responds to external forcing, and also into the system's complex internal interactions and feedback mechanisms. In particular, ice core records show that the composition of the atmosphere changed greatly during glacial times, evidence that nature itself has altered the greenhouse gas concentrations of the atmosphere. An understanding of these changes and the climate system's response to them is essential to developing a predictive capability. Past climate also provides test cases for comparison with climate models. Predictions generated from models can be compared with recorded climates as a means of evaluating model performance.

In this section, I will first review the evidence for past climate change, particularly that which pertains to the last glacial advance, and will then consider the mechanisms that could bring about such changes. The emphasis will be on the climatic effects of changing atmospheric composition.

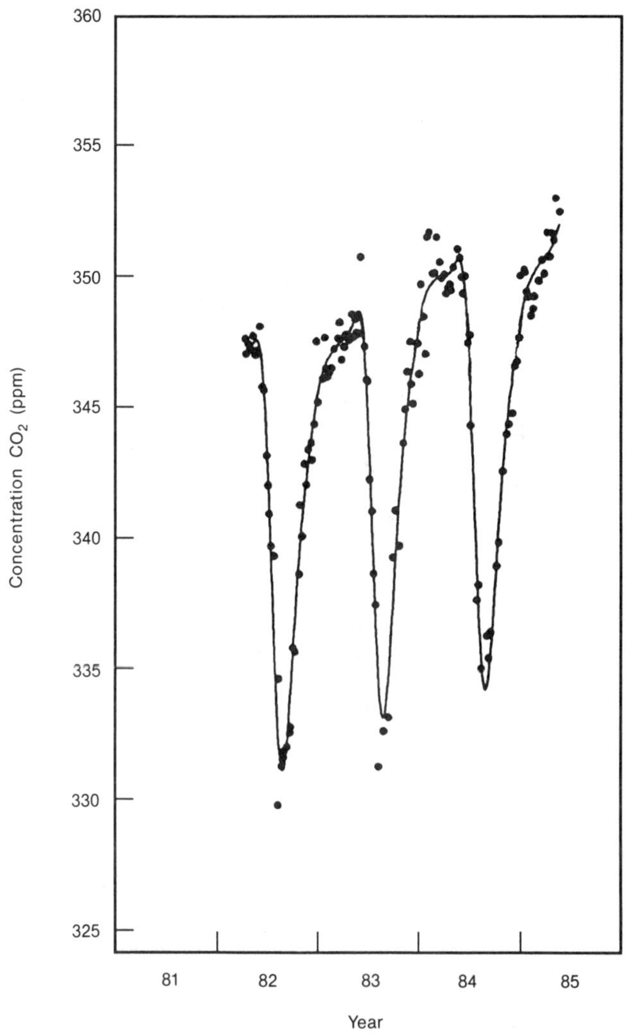

Fig. 20. Time trend of the concentration of atmospheric CO_2, in ppm, at Point Barrow, Alaska (source: Keeling et al., 1990).

Long-Term Climate Changes

Climate records exist on various time scales. Instrumented records contain information on climatic changes with time scales on the order of a hundred years. In order to investigate climate change over longer intervals, geological proxy records are required. These records indicate that climate varies in an irregular fashion at all time scales.

On the very longest time scales of hundreds of millions of years, changes in CO_2 concentration are thought to have been the primary determinant of climate. When the earth was first formed, the sun was much fainter than it is today—probably only 75% as bright—yet geological evidence shows that the earth was not ice-covered at this time. The "Faint Sun Paradox" has been explained by the much higher carbon dioxide and methane content of the early atmosphere, which enhanced the greenhouse effect and thus compensated for the lower level of solar radiation (Owen et al., 1979; Roxburgh, 1980). Since those early days, photosynthesis by plants has reduced CO_2 levels as the sun's radiation has

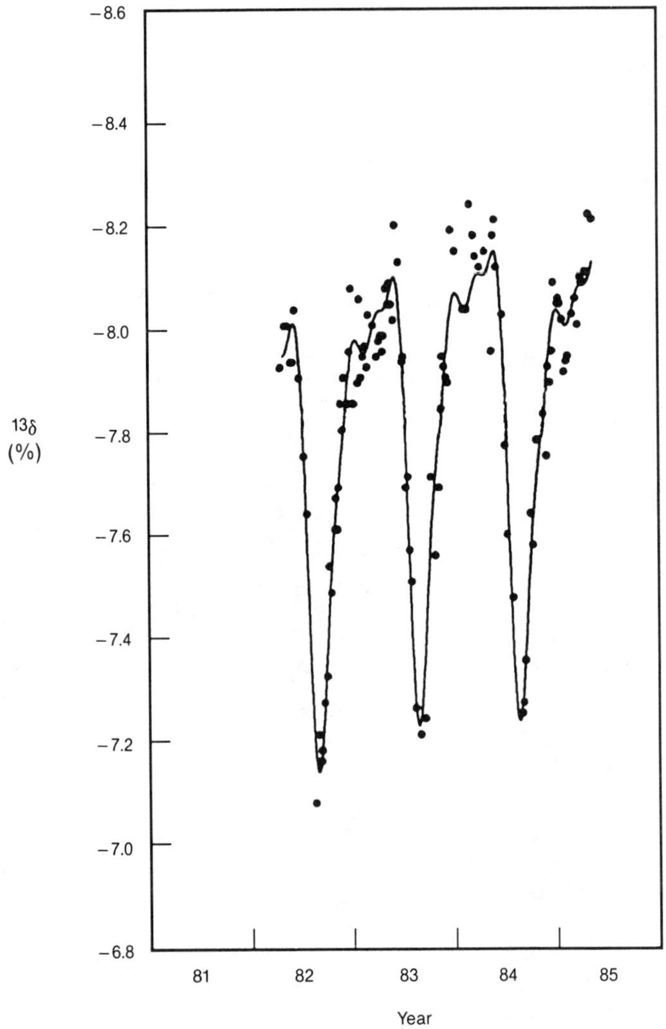

Fig. 21. Time trend of the reduced isotopic ratio, 13δ, of atmospheric CO_2 in per mil difference from standard PDB at Point Barrow, Alaska (after Keeling et al., 1990).

increased. Major fluctuations in climate have been superimposed on the very long time scale of CO_2 decrease; these fluctuations can be seen in the sedimentary record (Berner et al., 1983; Budyko et al., 1987). For example, the warmth of the carboniferous periods (350 million years ago) and the Cretaceous period (100 million years ago) can most easily be explained by elevated atmospheric carbon dioxide (Lloyd, 1984). Budyko et al. (1987) suggest that the average global temperature during these past warm periods was 10°C warmer than at present.

Climate Changes During the Quaternary Period

The climate of the past two million years has been dominated by a sequence of very cold periods. The amplitude of these fluctuations began increasing around 1.7 million years before the present (BP). At that time, Arctic ice appears to have become seasonally permanent (Shackleton et al., 1984), and intense glacial periods began.

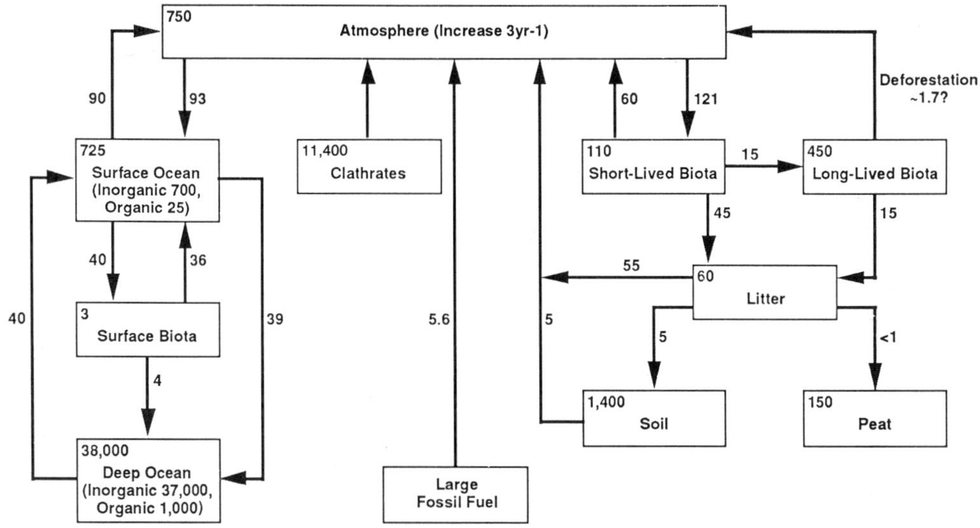

Fig. 22. Land-ocean-atmosphere carbon cycle as of 1988. Units of storage (boxes) are in gigatons (Gt); units of flux (arrows) are in Gt per year (revised and updated from Bolin, 1986).

The last transition between an interglacial and a glacial period took place about 120 kyr BP. The period 120 to 80 kyr BP is thought to have been generally warm, with two cold intervals lasting several thousand years each. Both cold intervals began and ended relatively abruptly, with onset intervals of 1,000 years. The last interglacial period, 120 to about 145 kyr BP, was also generally warm. Global sea levels during this period were some 6 m higher than at present, perhaps due to partial melting of the West Antarctic ice sheet (Mercer, 1978).

From 70 to 15 kyr BP the earth was cool; global mean temperatures dropped 2° to 5°C below today's level (Hansen et al., 1984). Continental glaciation over North America and Eurasia reached its maximum extent around 18 kyr BP, midway through a severe cold period that lasted from 20 to 15 kyr BP. A general outline of the extent of ice coverage in the Northern Hemisphere at the glacial maximum is shown in Fig. 23. During the time of maximum glacial coverage, sea level dropped about 85 m. Between 15 and 10 kyr BP, the recovery from full glacial conditions began. The recovery was spatially and temporally complex. In Western Europe, very rapid warming around 12 kyr BP created conditions almost as warm as they are today. Around 11 kyr BP, a sharp and severe cooling period set in. Rapid warming returned around 10 kyr BP, when winter temperatures warmed by about 15°C and summer temperatures by 8°C (Atkinson et al., 1985) over a period of only a few centuries.

By 10 kyr BP, the earth had warmed to approximately the present mean temperature, although large ice masses persisted in the Northern Hemisphere. The period since 10 kyr BP is known as the Holocene. The early Holocene, from 9 to 6 kyr BP, was a time of regional, and possibly global, warming. In Europe, the ice was gone by 8 kyr BP; in North America, at 6 kyr BP, three residual ice masses remained in Canada. During the warm early Holocene, North American forests extended farther north than they do today. The world's desert regions were substantially wetter, with greater vegetative cover than today, and lake levels everywhere tended to be higher than at present (Kutzbach and Street-Perrott, 1985; Ruddiman and Wright, 1987).

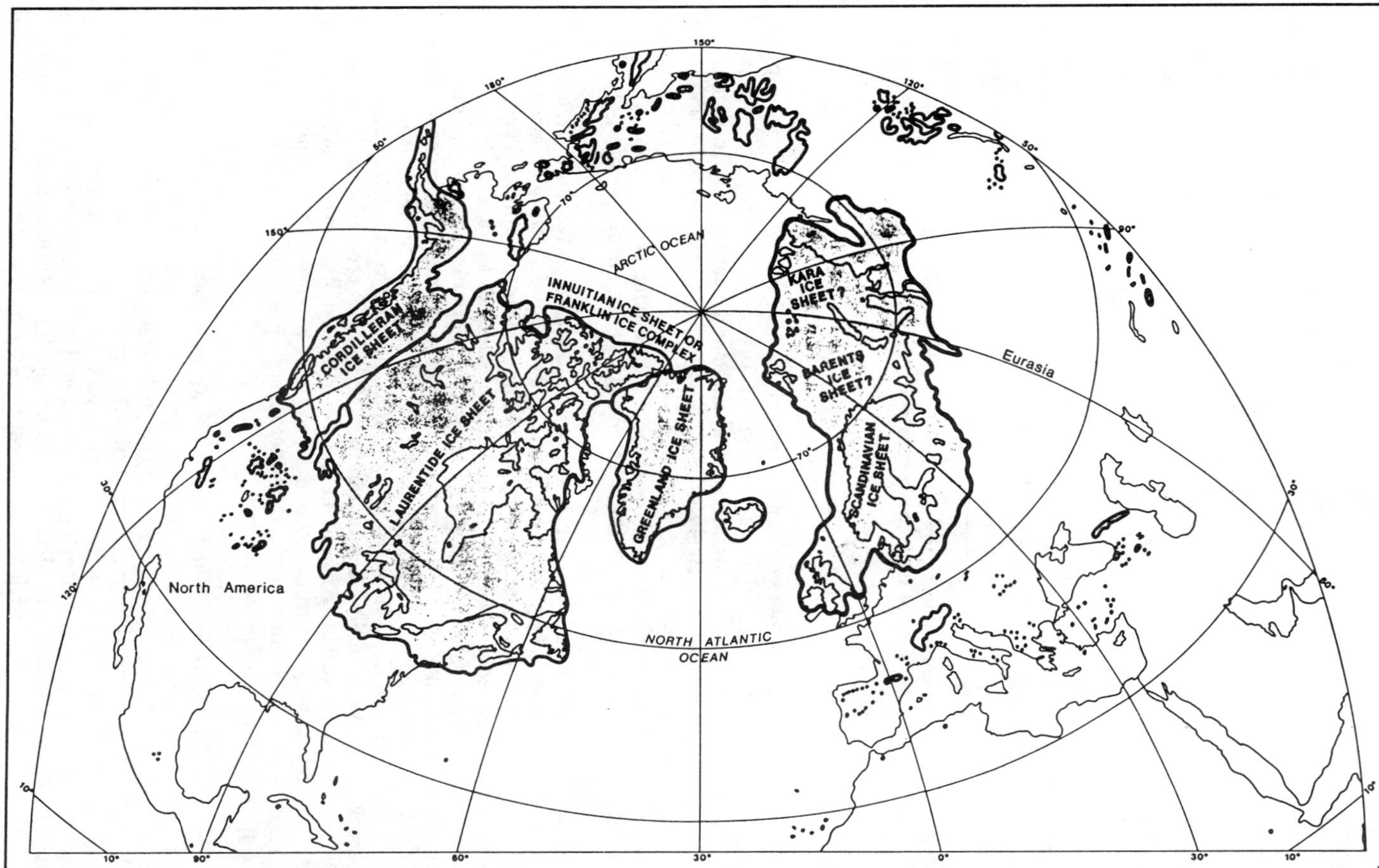

Fig. 23. Schematic diagram of Northern Hemisphere ice sheets during the last glacial maximum (source: Crowlely, 1983). Possible ice shelves are not shown.

The sketchy climatic picture for the early part of the Quaternary Period, the Pleistocene, is based on geologic evidence collected through classical stratigraphic recording methods using glacial till and fine-grained lake deposits, old lake terraces, ancient shorelines, pollen, and other fossil evidence. A chronology was obtained, based largely on ^{14}C, that dates back to the limit of radiocarbon dating some 40,000 years ago. The geologic record has been supplemented over the last 30 years through the use of geochemical evidence, particularly the variation of stable isotopes.

Variation of Carbon and Oxygen Isotopes in Deep Sea Sediments

Detailed evidence relating to ice age glaciations accumulated unexpectedly from the analysis of long piston cores taken from the sea bottom during oceanographic research. The recovered sediments contained fossils of microorganisms called foraminifera. The calcium carbonate of the shells of foraminifera contains oxygen. Oxygen possesses two stable isotopes, ^{16}O and ^{18}O, with an ^{18}O abundance of 0.2%. The relative concentration of ^{18}O to ^{16}O is

$$\delta^{18}O = \left[\frac{\gamma}{\gamma_s} - 1\right] 1,000,$$

where γ is the ratio of ^{18}O to ^{16}O in the sample, and the standard γ_s is the ratio found in a standard sample called the standard mean ocean water (SMOW) (Hoefs, 1987).

The main reason for the fractionation of oxygen isotopes (also separation of hydrogen and deuterium) is vapor pressure differences. The vapor pressure of the heavier isotopic component of water ($H_2\,^{18}O$, HDO) is slightly lower than that of $H_2\,^{16}O$. The lower vapor pressure—about 1% for $H_2\,^{18}O$ and 10% for HDO—results in an atmospheric water vapor content that holds 10% less ^{18}O and 100% less deuterium than does mean ocean water. In the hydrologic cycle, when water evaporated from the ocean cools off and condenses into clouds, the first precipitation has the same composition as the ocean water from which it was derived. This is because the heavy components condense with a preference for $H_2\,^{18}O$ and HDO, of 10% and 100%, respectively. The remaining water vapor is enriched in the light isotopes, and the fraction of these in later precipitation is higher. In the last stages of condensation, the air mass deposits the light water as snow as it moves over an ice sheet. The snow is transformed into ice, where it is stored for times that are very long compared to the times associated with the rest of the hydrologic cycle. As the process is repeated, the ocean is depleted in the light isotope of oxygen and in hydrogen. The changes in the abundance of ^{16}O in the oceans are reflected in the calcareous shells excreted by the foraminifera. Experiments show that small fractionation occurs during formation of the calcium carbonate shell. During times when large volumes of water are stored as ice, foraminifera shells should show high values of $\delta^{18}O$, while shells from interglacial periods should be marked by lower values of $\delta^{18}O$.

Over the past 30 years more than a hundred sedimentary cores have been analyzed for variations in oxygen isotopes. For the purpose of comparing oxygen isotope variations with those of atmospheric carbon dioxide during the last 160,000 years, I have selected an analysis by Shackleton et al. (1983). The East Pacific core (V19-30) records the change in isotopic composition over a time period comparable to that of the Vostok, Antarctica, ice core. Due to its Pacific location, the core contents were probably not as strongly influenced by local changes in currents as those taken in the Atlantic. Core V19-30 also has a higher time resolution than do most cores, because the rate of sedimentation is greater at the site in the Panama basin than in most deep sea sites.

Figure 24 compares the variation of oxygen isotopes from benthic (deep dwelling) foraminifera and from the Vostok ice core. At times of maximum ice

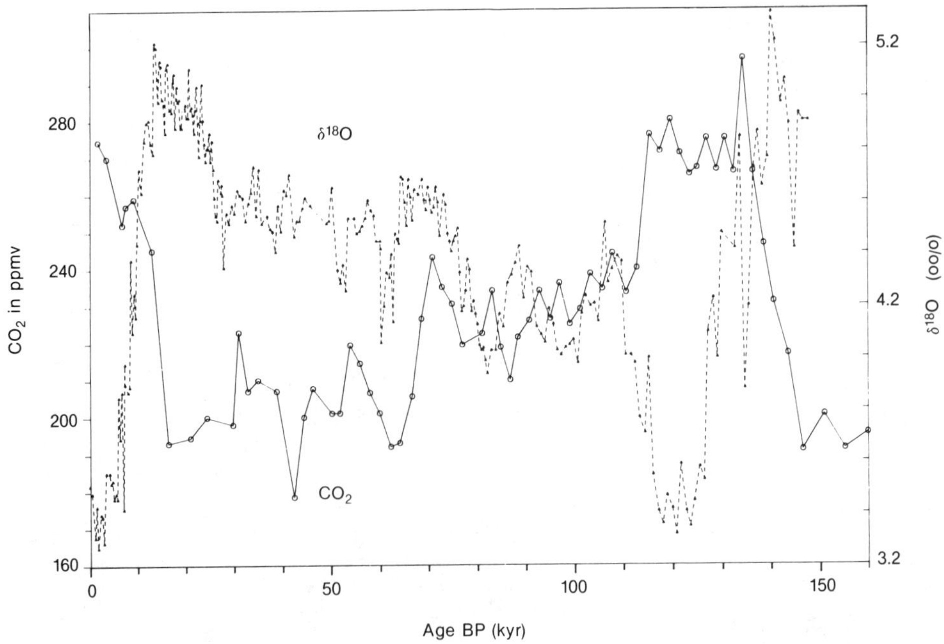

Fig. 24. Comparison of the variation in atmospheric CO_2 concentration derived from analysis of the Vostok ice core and the $\delta^{18}O$ variation from the V19-30 sediment core from the East Pacific.

coverage—as identified by high values of $\delta^{18}O$—atmospheric carbon dioxide (and methane) concentration are low, while during interglacial times the opposite conditions hold. It should be noted that the rapid increase in atmospheric CO_2 at the end of the last glaciation preceded the reduction in ice volume, as measured by $\delta^{18}O$. This would indicate that the increase in atmospheric CO_2 was not driven by the reduction in ice volume.

Shackleton and Pisias (1985) extended the V19-30 core data back to 340 kyr BP, as shown in Fig. 25. The time scale has been derived from ^{14}C dates plus correlations with New Guinea terraces that have been dated by ^{230}Th and ^{231}Pu methods. Minimum $\delta^{18}O$ values, corresponding to minimum ice coverage, occur at about 100,000-year intervals. This apparent recurrence of interglacial intervals is strikingly illustrated in Fig. 26. Imbrie et al. (1984) combined data from five deep-sea cores from the Atlantic, Pacific, and Indian Oceans to extend the time scale and record of $\delta^{18}O$ variation back over 780,000 years (see Fig. 26).

As discussed in the previous section, the carbon isotopic ratio of CO_2 in the atmosphere reflects the amount derived from the biosphere. The surface waters of the ocean also reflect changing carbon isotopic composition in the atmosphere and, taking into account the mixing times for the oceans, so do the deeper waters. During an expansion of the biosphere, plants selectively remove ^{12}C, enriching the atmosphere with ^{13}C. During glacial periods, when forests were destroyed by advancing ice, and colder temperatures inhibited plant growth, light carbon was shifted into the atmosphere and eventually into the ocean (Shackleton, 1977). Figure 27 compares the variation of $\delta^{13}C$ obtained from the analysis of benthic foraminifera from core V19-30 (Shackleton et al., 1983) with the atmospheric CO_2 record. During peaks in glaciation, carbon isotopes tend to be lighter, but the correlation of the atmospheric CO_2 curve with $\delta^{13}C$ is not nearly as clear as its correlation with $\delta^{18}O$. In order to attempt to clarify the origin of the $\delta^{13}C$ variations, Shackleton and Pisias have examined $\delta^{13}C$ variation for both benthic (*U. senticosa*) and near-surface (*Neogloboquadrina dutertrei*) planktonic species.

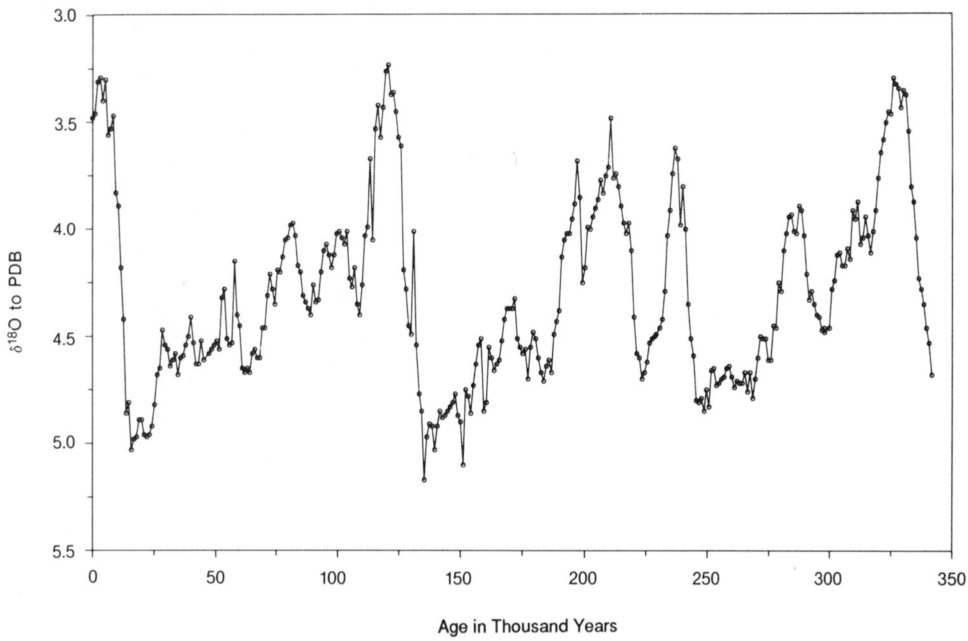

Fig. 25. The $\delta^{18}O$ record derived from the benthic foraminifera *Uvigerina senticosa* from the V19-30 core taken in the East Pacific in the Panama Basin (3°21′S, 83°21′W) in 3,091 m of water (after Shackleton and Pisias, 1985).

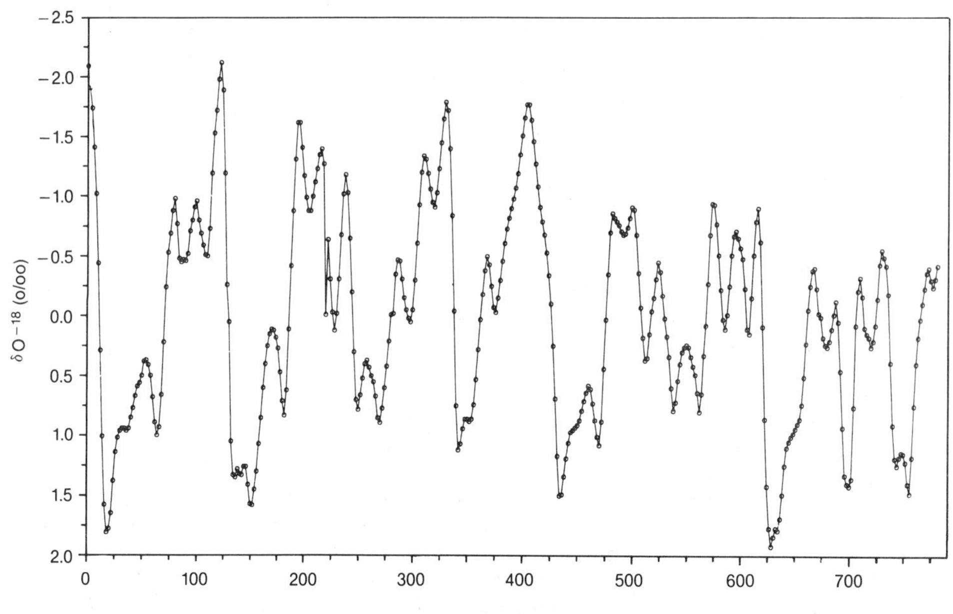

Fig. 26. Variation of $\delta^{18}O$ over 780,000 years as compiled by Imbrie et al. (1984) from five deep ocean cores.

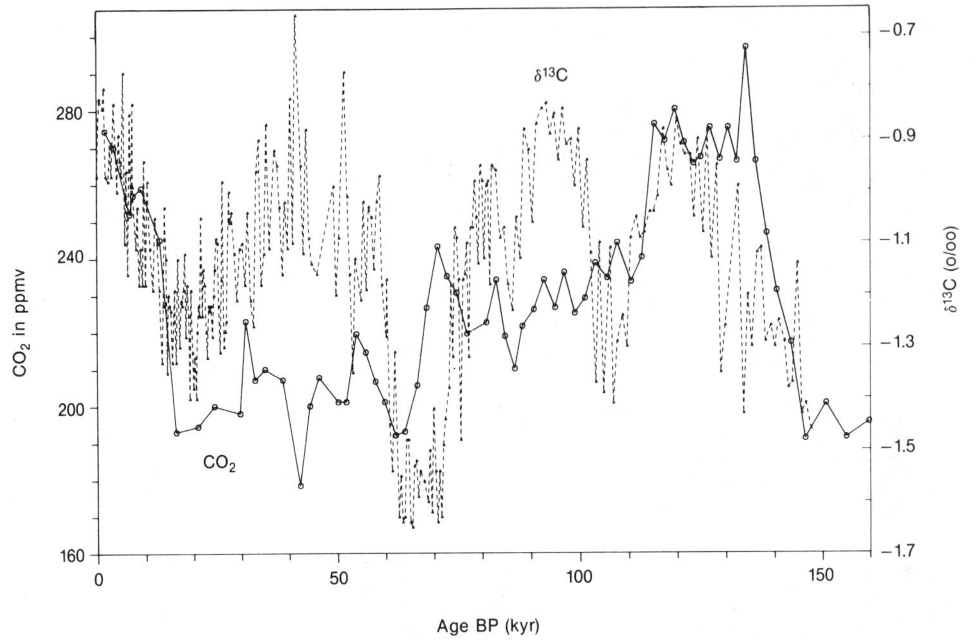

Fig. 27. Comparison of the variation in atmospheric CO_2 concentration derived from analysis of the Vostok ice core and the $\delta^{13}C$ variation from the V19-30 sediment core from the East Pacific (after Shackleton et al., 1983).

The variation in $\delta^{13}C$ for these species is shown in Fig. 28; although both species show variation, the larger fluctuations occur in the deep-ocean species.

Spectral Analysis of Atmospheric Carbon Dioxide and Isotope Data

When records of isotopic variations and atmospheric carbon dioxide concentrations are modeled, they are assumed to contain both a signal, which is determined by some underlying physical process, and noise, which is either of deterministic origin (due to the process), or is caused by stochastic input from sampling and experimental errors. Spectral analysis reveals whether the underlying deterministic signal contains periodic or quasi-periodic components. The analysis of such time series requires the following steps.

1. First select the false alarm probability. The selection is a matter of judgment. If the probability is set too low, physically significant peaks in the spectrum may be missed; if it is set too high, peaks that are due to noise will be mislabeled as true peaks. For the ice age records, I set the probability cutoff at 0.05, so the chance of identifying a noise peak as a true peak is 1/20.

2. Estimate the normalized periodogram. If the record is a noise process without any signal, the periodogram is inappropriate. But if the model includes the possibility of a structured signal, the periodogram provides the best means of signal detection in the maximum likelihood sense. When data are evenly spaced, as they are in the isotope case, a fast Fourier transform (FFT) is used, with the results normalized by the variance. When the data are unequally spaced, as in the case of the CO_2 record, the Lomb periodogram is employed (MacDonald, 1989).

3. Identify the maximum peak in the periodogram and test it against the false alarm probability.

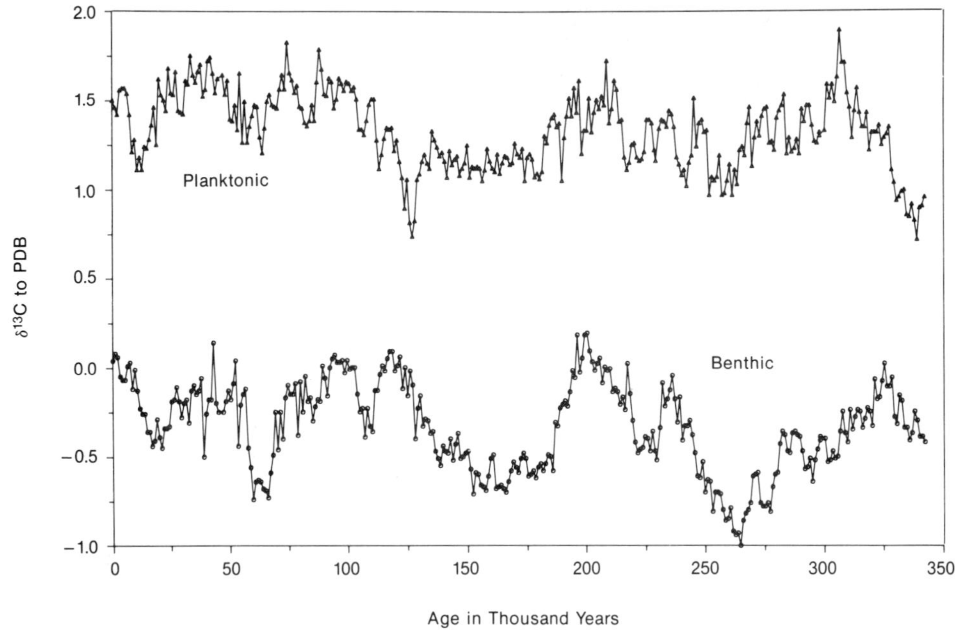

Fig. 28. Carbon isotope data for planktonic species *(N. dutertrei)* and a benthic species *(U. senticosa)* from the V19-30 core of the Eastern Pacific (after Shackleton and Pisias, 1985).

4. If the maximum peak meets the false alarm test, determine by least squares the amplitude and phase of the sinusoid representing the peak.

5. Using the amplitude and phase information, subtract the peak from the record. Because record length is left unchanged, this procedure is superior to band filtering out the peak. Removal of the peak also removes the disturbing side lobes as the analysis continues. After peak removal, the variance in the total record is reduced. The change in variance provides guidance as to the rate of convergence of the peak removal or pre-whitening process.

6. Repeat steps 3 to 5 until no further peaks meet the false alarm criteria. The remaining series is an estimate of the continuum part of the spectrum of the original series. The continuum residual can then be analyzed to determine the nature of the noise or of the deterministic continuum.

The carbon dioxide record for the atmosphere was derived by Barnola et al. (1987) from the analysis of air samples trapped in a 2,083-meter-long ice core, which contained samples ranging in age from 1,700 to 159,690 years BP. The ice core was sampled at 66 depths. Although the record is short and noisy, it is of great interest for ascertaining whether or not the record contains a deterministic component. The Lomb periodogram for the carbon dioxide data is shown in Fig. 29. Initially, only two peaks exceed the probability threshold, but after applying the procedure outlined above, the record yields five significant peaks, as indicated in Table 8. These five peaks contain 88.2% of the total variance in the record. The fit of the five-sinusoidal term to the data is shown in Fig. 30, with Fig. 31 showing the spectrum after peak removal. The spectrum still appears to show structure, but none of it is significant at the 0.05 level.

The normalized periodogram for δ^{18}O data obtained from benthic foraminifera is shown in Fig. 32. Four peaks are significant at the 0.001 level.

Table 8. Frequencies and amplitudes of peaks in the periodogram of variation of atmospheric carbon dioxide.

Frequency (cpkyr)	Period (kyr)	Amplitude ppmv of CO_2	Percentage Contribution to Total Record Variance
0.00772	129.5	44.1	48.7
0.01429	70.0	18.2	17.5
0.00561	178.2	14.8	11.0
0.02273	44.0	9.4	5.6
0.04768	20.9	9.2	5.4

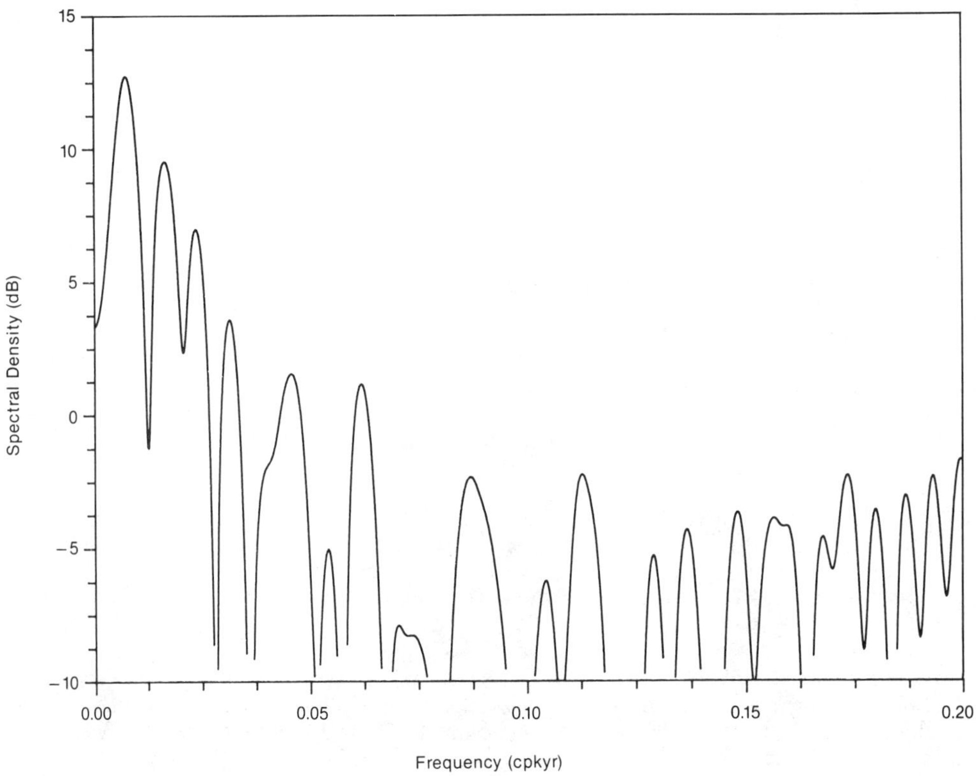

Fig. 29. Periodogram of the carbon dioxide time series shown in Fig. 7.

These peaks correspond to periods of 109.9 kyr, 40.5 kyr, 23.1 kyr, and 54.6 kyr. The spectral analysis process outlined above leads to the identification of nine significant peaks at the 0.05 level (see Table 9). The significant peaks carry 91.2% of the total variance in the record. The nine-sinusoidal fit to the data is shown in Fig. 33, with the residuals plotted in Fig. 34. Figure 35 gives the periodogram for benthic $\delta^{18}O$ after removal of the nine significant peaks.

Table 9. Significant peaks in the Benthic $\delta^{18}O$ record from core V19-30.

Frequency (cpkyr)	Period (kyr)	Combination Frequencies	Percent of Variance in Peak
0.0091	109.9	f_1	42.2
0.0247	40.5	f_2	21.7
0.0433	23.1	$2f_1 + f_2$ (0.0429)	10.7
0.0183	54.6	$2f_1$ (0.0182)	5.0
0.0147	67.9	$f_2 - f_1$ (0.0156)	3.4
0.0080	125.5	?	3.4
0.0340	29.4	$f_1 + f_2$ (0.0338)	2.9
0.0520	19.2	$3f_1 + f_2$ (0.520)	1.3
0.0114	87.9	$4f_1 - f_2$ (0.0117)	0.6

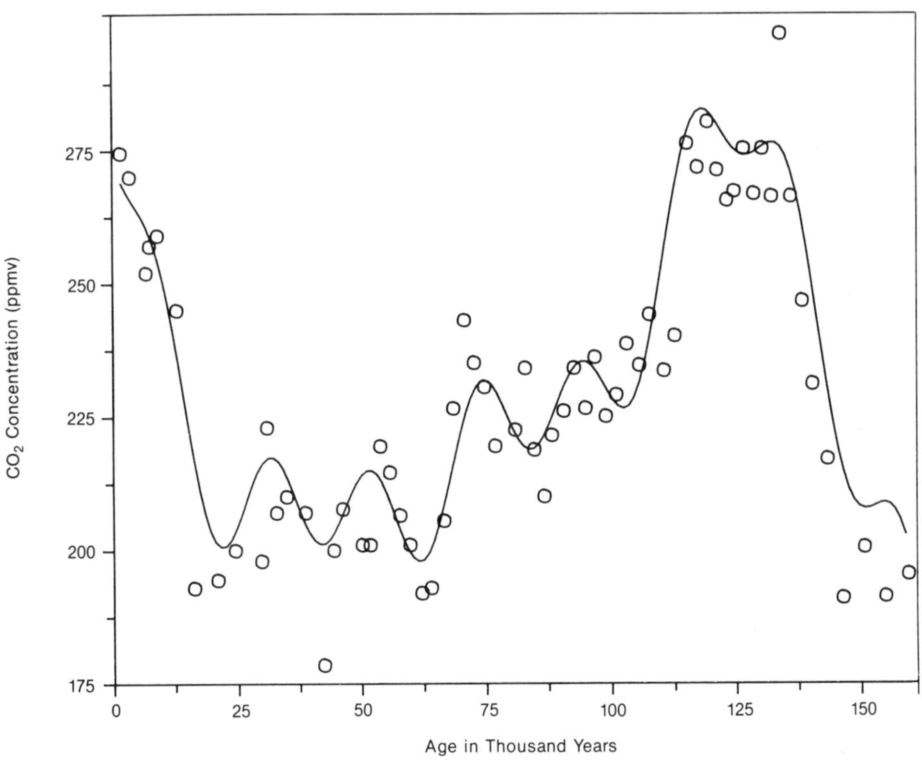

Fig. 30. Fitting of the Vostok ice core carbon dioxide data with five sinusoids significant at the 0.05 level.

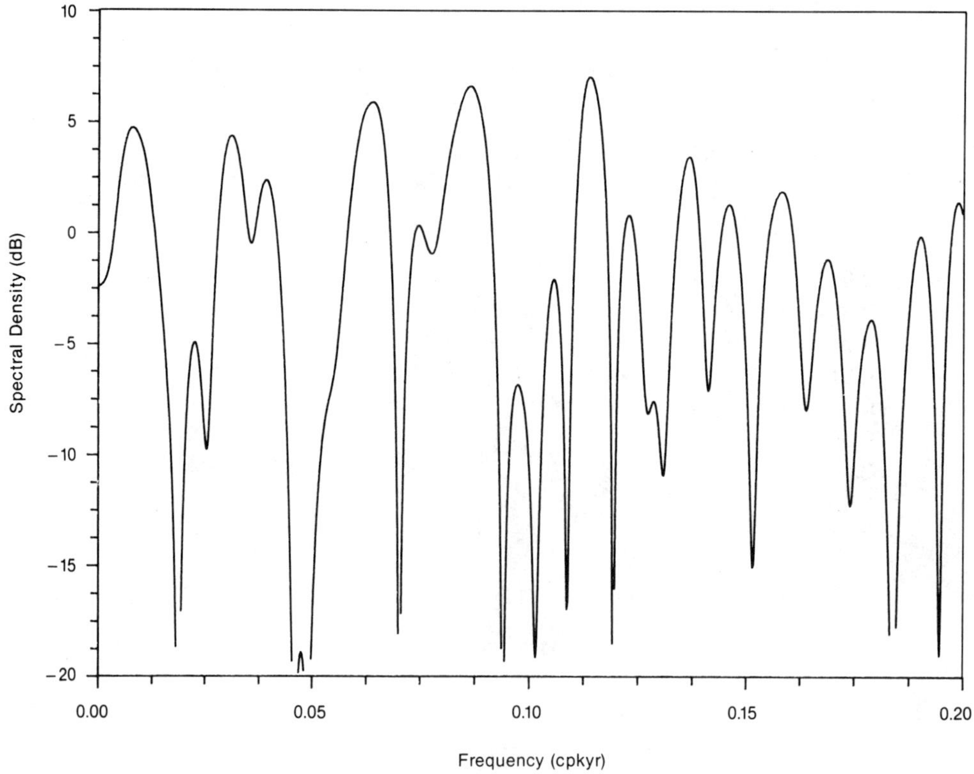

Fig. 31. Periodogram of atmospheric carbon dioxide time series after pre-whitening spectrum (see Fig. 29) by removing five peaks significant at 0.05 level.

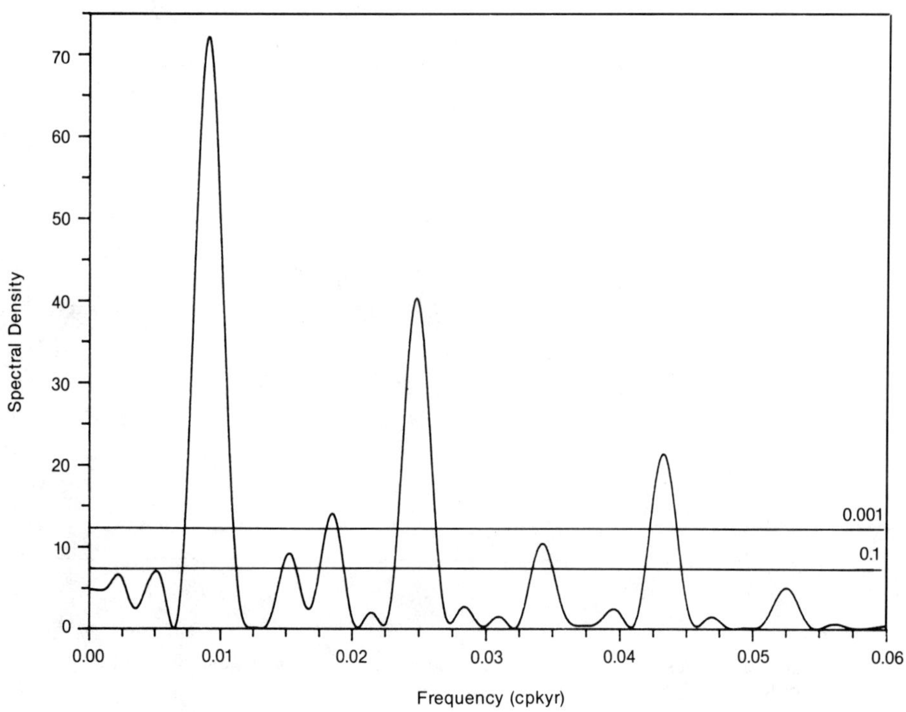

Fig. 32. Normalized periodogram for $\delta^{18}O$ obtained from *U. senticosa* for core V19-30 from Eastern Pacific.

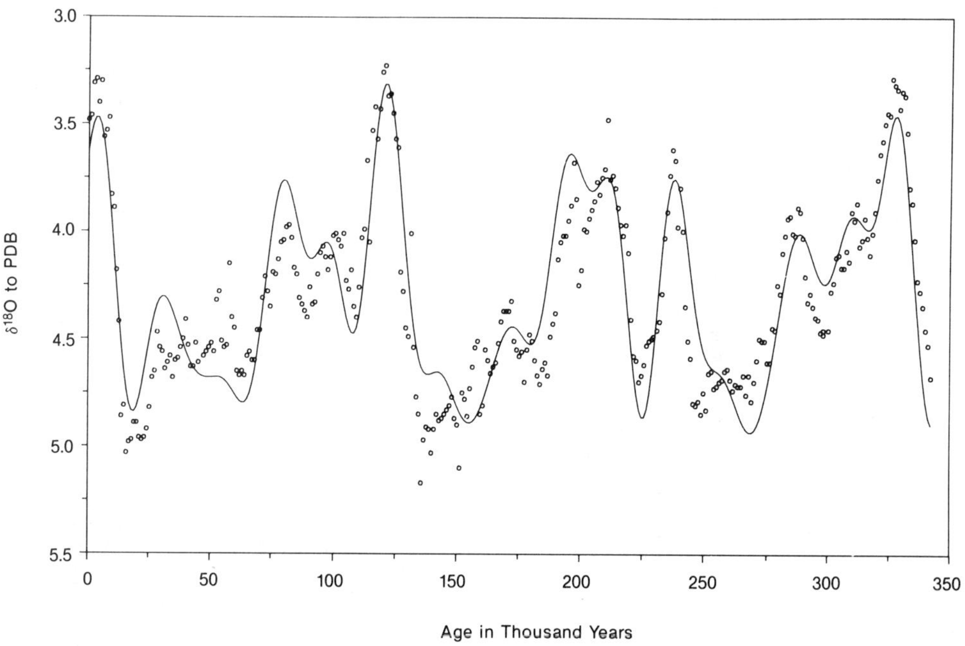

Fig. 33. A nine-sinusoid fit to the $\delta^{18}O$ data obtained from benthic foraminifera from core V19-30

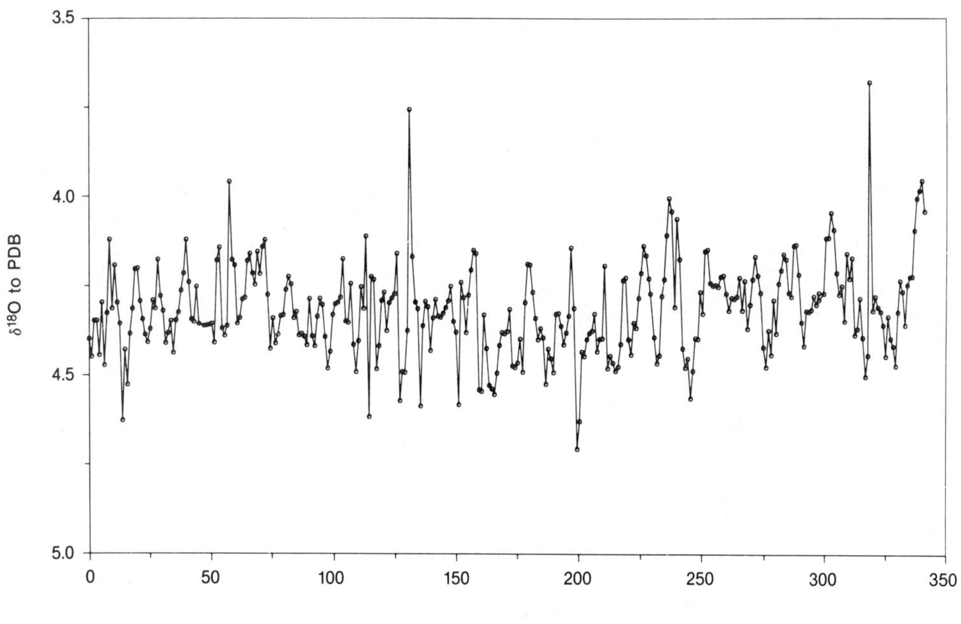

Fig. 34. Residuals remaining after removal of the nine significant peaks in the $\delta^{18}O$ data.

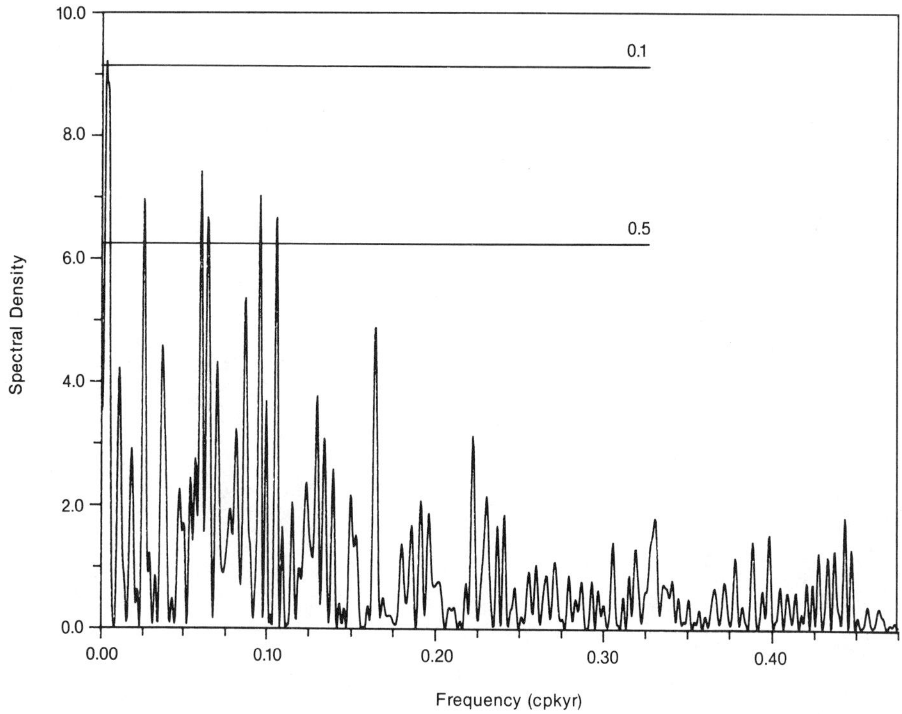

Fig. 35. The residual spectrum for $\delta^{18}O$ after removal of the nine significant peaks.

Figure 36 shows that for the variation of $\delta^{13}C$ in benthic foraminifera from Core V19-30, two peaks are initially significant at the 0.001 level. These peaks, with periods of 110.2 kyr and 40.4 kyr, are closely similar to the main peaks in the $\delta^{18}O$ record. Pre-whitening spectral procedure yields a total of ten peaks that meet the 0.05 significance criteria. These peaks contain 85.1% of the variance of the total record.

The periodogram for near-ocean-surface variations of $\delta^{13}C$ shows a much more complicated spectrum than the one for benthic $\delta^{13}C$, and shows abundant energy at very low frequencies (see Fig. 37). The strongest peak, at a period of 112.5 years, is followed by a cluster of peaks at longer periods. In all, the pre-whitening procedure leads to the identification of eleven significant peaks, but these contain only 67.7% of the total variance—less than the percentages carried by the peaks for the $\delta^{18}O$ and benthic $\delta^{13}C$ records. Clearly, the planktonic record for $\delta^{13}C$ contains much higher noise levels than do the other records.

Figure 38 shows a comparison between the peak positions in the spectrum of atmospheric carbon dioxide and in the spectrum of $\delta^{18}O$, which is a measure of the volume of water held in ice. In this and the following figures, the highest peak is assigned a unit amplitude, and the other peaks are scaled proportionally, thus providing a measure of relative amplitudes. The spectra clearly show a number of similarities, but there are also differences. The structure of the benthic $\delta^{18}O$ is richer, as might be suspected, because the record is 342.3 kyr long, with 326 points spaced at 1.05 years. The Vostok CO_2 record is shorter, with only 66 irregularly spaced points. The main peaks in both records correspond to long-period oscillations of 129.5 kyr and 109.9 kyr for CO_2 and $\delta^{18}O$, respectively. But because the benthic $\delta^{18}O$ record also contains a peak at 125.5 kyr, the CO_2 peak may represent some combination of the low-frequency peaks in the benthic $\delta^{18}O$ record. At high frequencies, the $\delta^{18}O$ record contains peaks at 23.1 and 19.2 kyr, but the CO_2 record has only a single peak between the two high-frequency

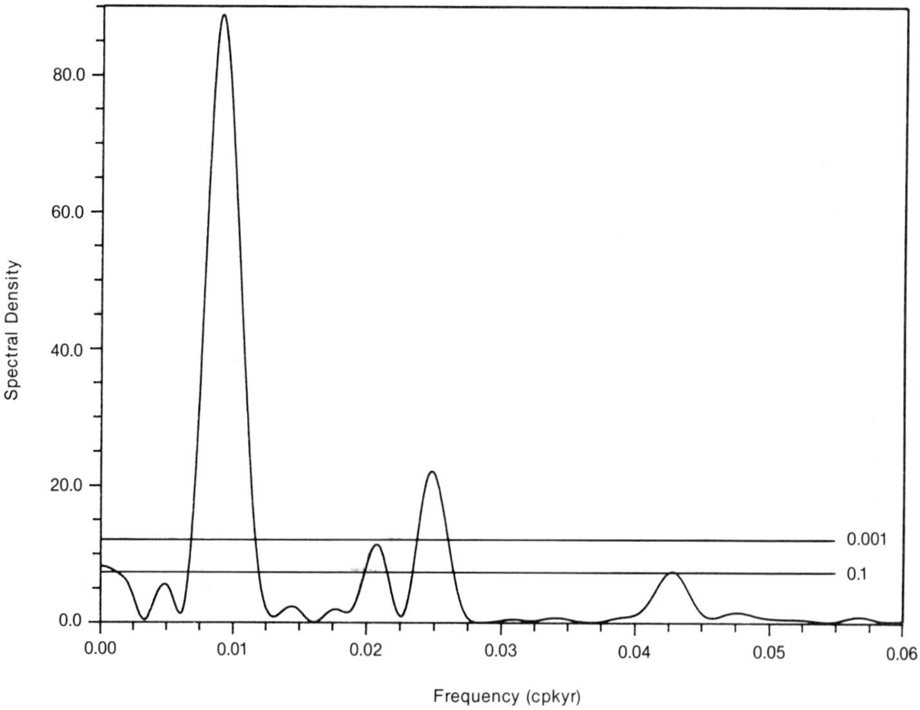

Fig. 36. Periodogram for variations of $\delta^{13}C$ derived from benthic foraminifera from core V19-30 of the East Pacific.

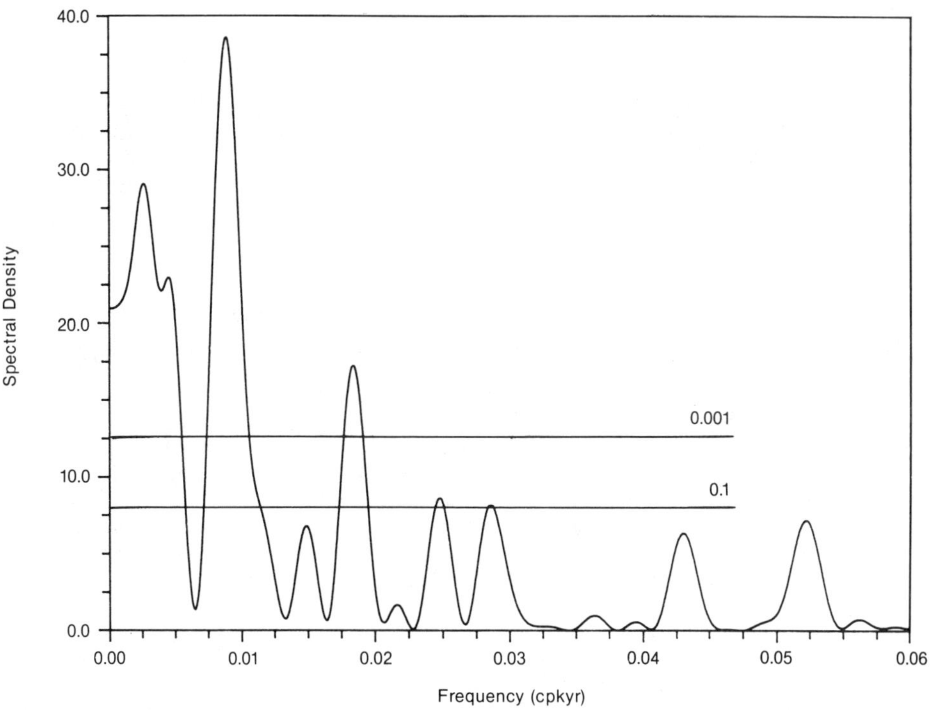

Fig. 37. Periodogram for variations of $\delta^{13}C$ obtained from planktonic foraminifera from East Pacific core V19-30.

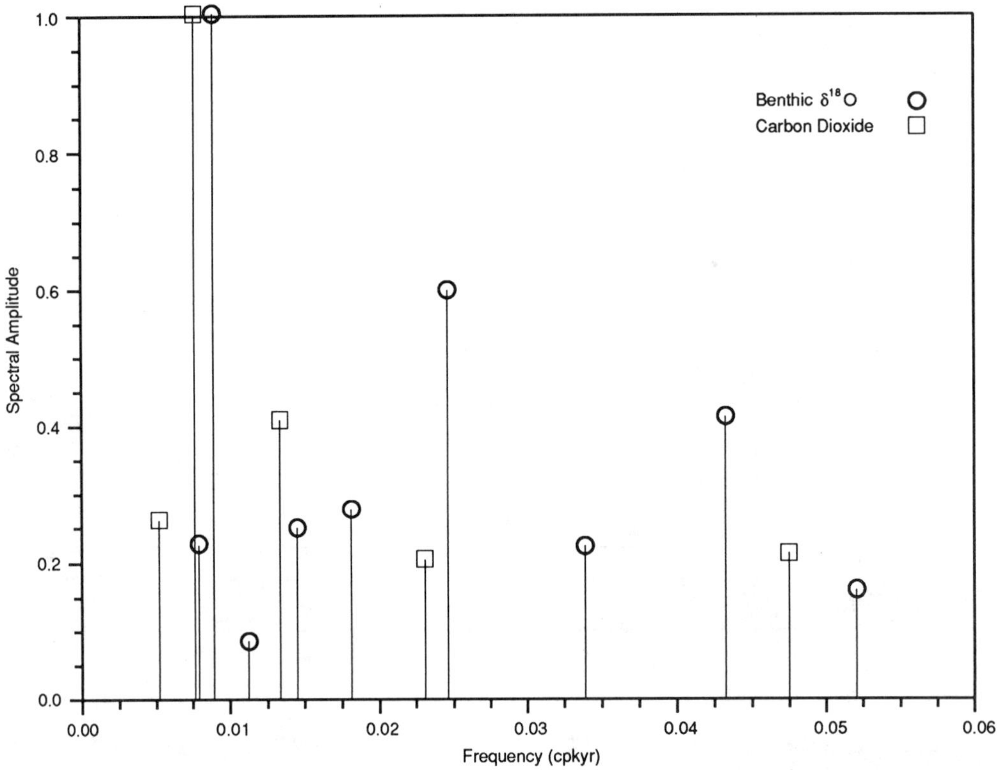

Fig. 38. Comparison of the amplitude spectrum for atmospheric carbon dioxide and benthic $\delta^{18}O$. The peak amplitude in any record is assigned unit amplitude and the other peaks are scaled proportionally.

$\delta^{18}O$ peaks. A summary comparison of the amplitude spectra for the V19-30 core data and the atmospheric carbon dioxide record is shown in Fig. 39. All the isotope data have their strongest peak at about 110 kyr, with further peaks at 40 kyr and 23 kyr. The best correlation between the atmospheric CO_2 and isotope data is with the difference of planktonic and benthic $\delta^{13}C$.

Development of the Orbital Theory of Climate Change

After the recognition by Louis Agazzi in 1837 that the earth had experienced a great ice age, a number of hypotheses were advanced to explain its origin. The global scale of the ice ages prompted a search for global or larger causes. The discovery of the ice age phenomena came soon after the great advances in celestial mechanics made by Gauss, Laplace, Lagrange, and Leverrier. Speculation about causes turned to the possibility that changes in the path the earth traveled about the sun would give rise to changes in insolation, and that these changes in incident radiation could drive the ice ages. Changes in the earth's orbit, such as the 22,000-year precession of the equinoxes, were consistent with the then-prevalent notion of the time scales for ice ages. This coincidence in time scale further strengthened the belief that orbital changes drove the ice ages.

In 1920, Milankovitch published his first analysis of the climatic effects of changes in insolation due to orbital variation. Milankovitch (1920) argued that at high latitudes in the Northern Hemisphere, the intensity of incoming radiation during the summer was a decisive climatic factor. Milankovitch's theory dominated discussions of the ice ages until the late 1950s, particularly in Europe. His insolation curves—calculated approximately from celestial mechanics

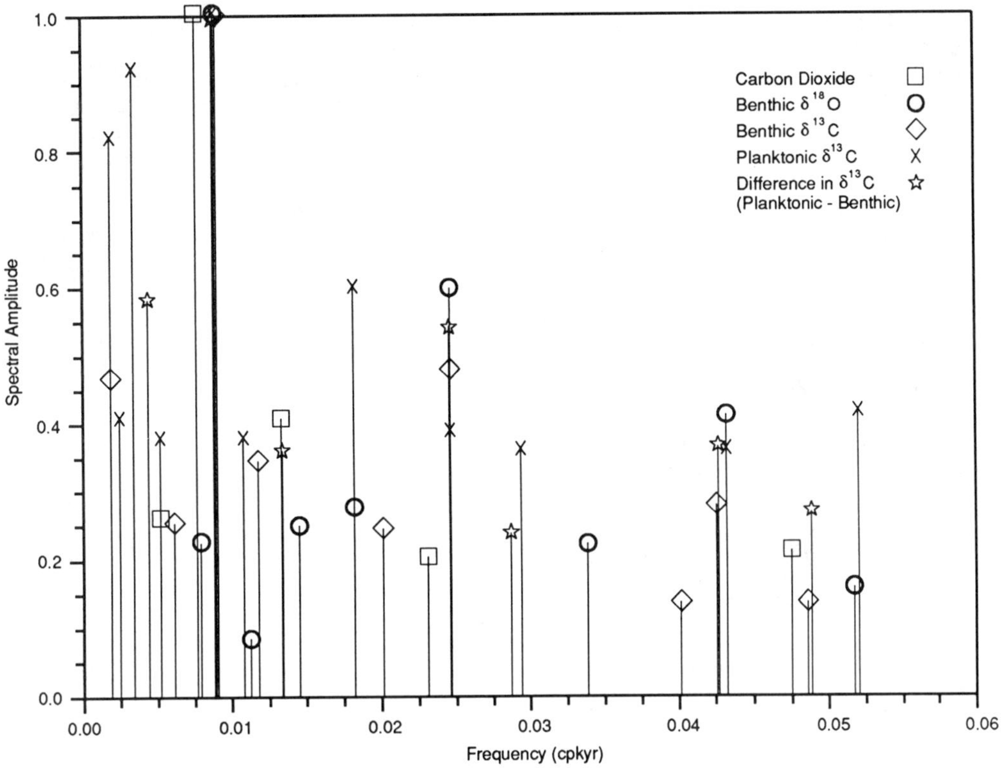

Fig. 39. Summary comparison of amplitude spectra for isotope data from core V19-30 and atmospheric carbon dioxide.

theory—were in general agreement with reconstructions of past climates. In the 1950s, the advent of radiocarbon dating began to undermine Milankovitch's arguments. Ice age time scales were produced that differed from those derived from purely geological evidence, and the generally accepted picture of glacial age climate change was altered significantly as further field work unraveled its complex history.

In the 1960s, evidence from ocean cores started to shift interest back to orbital variation. Astronomical theory predicted variation with 40-, 23-, and 19-kyr periods, and these periods were dramatically evident in the spectra of oxygen isotope variations (see Fig. 32) (Hays et al., 1976).

Theory of Orbital Changes

In the absence of any other celestial bodies, the earth would describe an elliptic orbit, with the sun at one focus. The shape and position in space of this ellipse, which is called the ecliptic, would be fixed in time, provided the small dissipative tides raised by the sun on the earth are neglected. Again neglecting dissipation, if the earth were perfectly spherical and homogeneous in density, the earth's axis of rotation would be fixed in space. The axis of rotation would be at a constant angle normal to the plane ecliptic, which is fixed in space if only the sun and earth are present.

However, the gravitational attraction of the other planets perturbs the elliptic motion about the sun. Because the masses of the planets are small compared to that of the sun, the perturbations are small. It is thus possible to describe the motion of the earth about the sun as an ellipse whose defining parameters, shape, and position in space vary slowly in time.

The choice of orbital elements used to describe the motion of the earth is somewhat arbitrary and depends on the problem under consideration. For short-term predictions needed to support spacecraft exploration of the planets, the elements and methods used would differ from those needed to calculate earth's past orbit over hundreds of thousands of years. For problems of past values of the insolation, the eccentricity (e) of the orbit is generally chosen to describe the shape of the ellipse. In defining the position of the ellipse in space, use is made of the fact that the position of the earth's axis of rotation is determined almost entirely by the gravitational pull of the sun and moon on earth's equatorial bulge. The effect of the planets on the position of earth's axis of rotation, which describes a circular core in space, is entirely negligible. The effect of the planets on the position of the ecliptic can thus be measured as the angle of tilt of the axis to the ecliptic, independent of its conical precession about the normal to the ecliptic. The angle of tilt is termed the obliquity (ε). The obliquity is chosen as the measure of the ellipse's position in space.

In order to describe the slowly varying position of the earth's ecliptic orbit in the plane of the ecliptic, it is customary to use the precession angle $\tilde{\omega}$ of the perihelion—the point on the ellipse that is closest to the sun—with respect to the line of intersection of the ecliptic with a reference plane that can be taken as the mean ecliptic of all the planets. In fact, the total precession is made up of a combination of the lunar-solar effect on the earth's mass distribution and the planetary effect on the total mass.

The position of the earth in its orbit about the sun is of no interest; time scales of a year are negligible for problems involving changes in 10,000-year periods. The orbital elements of the yearly motion and variations in the earth's axis of rotation alter secularly on time scales that are long when compared with the time scales of change of eccentricity, obliquity, and precession.

The difficulties encountered in practical, long-term calculation of the N-body problem of celestial mechanics brought Poincaré (1892) and Birkhoff (1927) to consider some of the underlying theoretical issues. Their considerations led to the emergence of the qualitative theory of differential equations, algebraic topology, and other branches of abstract mathematics. More recently, the theory has evolved rapidly in the analysis of nonlinear dynamical systems (Lorenz, 1963).

The question of stability was studied by Poincaré and others in the search for quasi-periodic solutions. Their solutions consisted of a superposition of trigonometric functions with rationally unrelated frequencies, but a difficulty arises in that, as the parameters slowly change, resonance appears between various frequencies. It was only in the 1950s and 1960s that Kolmogorov, Arnold, and Moser were able to show that resonances can be avoided if certain technical conclusions are met (Arnold, 1978).

In order to calculate variations in insolation resulting from orbital changes, the orbital parameters are first expanded in a trigonometric series. The obliquity is given by

$$\epsilon = \epsilon_o + \sum_i A_i \cos(f_1 t + \delta_i),$$

the eccentricity by

$$e = e_o + \sum_i E_i \cos(\lambda_i t + \Phi_i),$$

and the precessional parameter entering into the expression for insolation is

$$e \sin \tilde{\omega} = \sum P_i \, sin(\alpha_i t + \varsigma_i).$$

The coefficients for the trigonometric terms are then calculated from the equations of motion, with the number of terms used being dependent on the order of terms kept in order of mass of the planets, the order of terms in the inclination, and the eccentricities of the orbits (Berger, 1978a). The largest amplitude terms for obliquity, precessional parameter, and eccentricity are shown in Table 10.

The computation of variation of earth's orbital elements over time proceeds if the coefficients of the terms in the trigonometric expansion are accepted. Figure 40 shows the variation in earth's eccentricity, taking the 19 terms in the expansion given by Berger (1978b) for the period covered by the SPECMAP isotope record (Imbrie et al., 1984). The variation of the obliquity over this same time scale is given in Fig. 41, which illustrates the time variation of obliquity, with 24 terms kept in the expansion, while the variation of the precessional parameter, with nine terms, is illustrated in Fig. 42. The fundamental frequencies of about 100 kyr for the eccentricity, about 20 kyr for the precessional parameter, and 40 kyr for the obliquity are clearly evident.

MODELING ICE AGE CLIMATE

Introduction

The proxy record for ice volume, the $\delta^{18}O$ record, contains power at three of the fundamental frequencies associated with orbital fluctuations corresponding to periods of 23 kyr, 40 kyr, and 100 kyr. In this section, I will examine questions concerning the ways in which orbital changes interact with climate. For example:

Table 10. Coefficients in series expansion of obliquity (ε), precessional parameter (es in ω), and eccentricity (e) (after Berger, 1978a).

Amplitude	Frequency (cpkyr)	Period (kyr)
Obliquity (")		
−2,462.22	0.02439	41.0
−857.32	0.02517	39.7
−629.32	0.01865	53.6
−414.28	0.02468	40.5
308.94	0.02390	41.8
Precessional Parameter es in $\tilde{\omega}$		
0.018608	0.042166	23.7
0.016275	0.044587	22.4
−0.013007	0.052698	19.0
0.009888	0.052206	19.2
Eccentricity		
0.0110294	0.00242	412.9
−0.00873296	0.010532	95.0
−0.0074955	0.0081105	123.3
0.00672394	0.010041	100.0
0.00581229	0.0076192	132.5

Fig. 40. Variation of eccentricity (e) of orbit over the last 780 kyr.

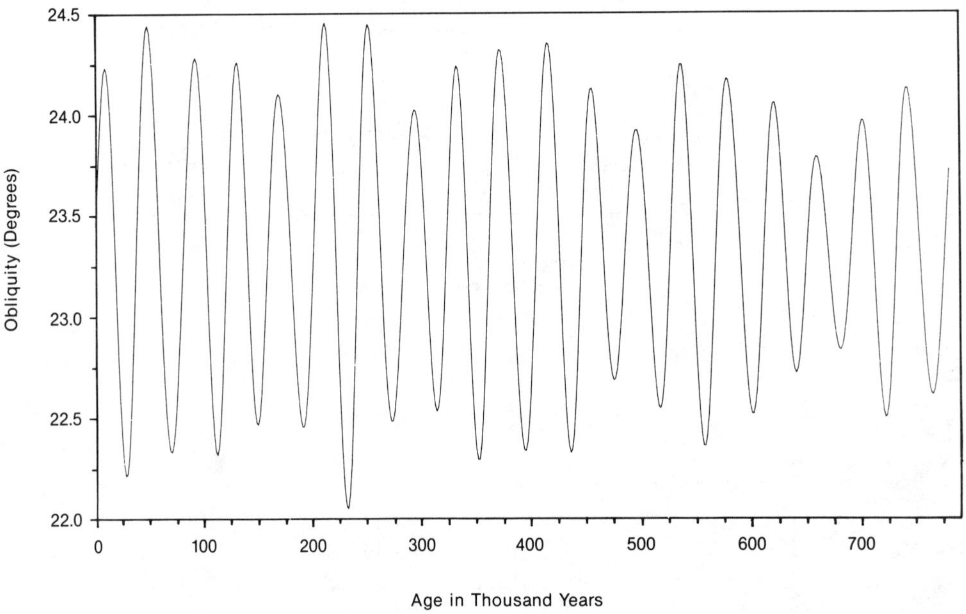

Fig. 41. Variation of obliquity over time.

What are the physical mechanisms by which these orbital changes affect climate? To what extent does the variation in climate follow the relative magnitude of the orbital fluctuation? And how do variations in atmospheric composition interact with changes due to orbital perturbation?

After first describing how insolation varies as a function of the orbital elements, I will turn to a comparison of the spectra of various proxies for climate

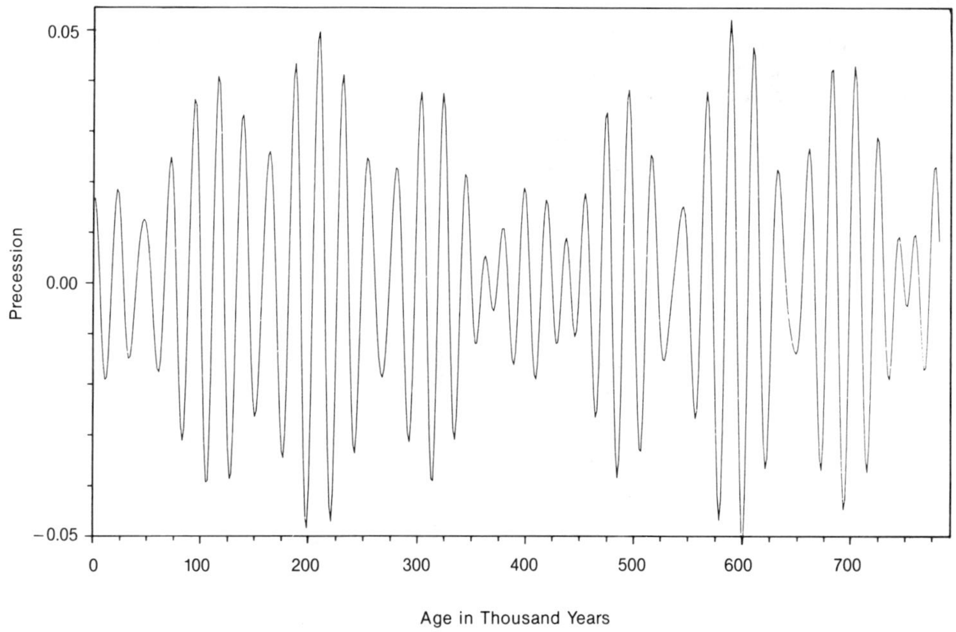

Fig. 42. Variation of the precessional parameter $e \sin \widetilde{\omega}$ with time.

change and the astronomical forcing function. The comparison shows that orbital forcing alone is inadequate to describe the observed fluctuations in climate proxies. The process through which orbital changes are transformed into climate changes is complex. Unraveling this process requires an understanding of the various feedback mechanisms that may be at work, particularly feedbacks from the biosphere, which appear to be of great importance. Because feedback mechanisms are inadequately portrayed in currently available climate models, predictions of future climate change must be examined with great caution.

Insolation

The flux of solar heat at the earth's mean distance from the sun r_0 is known as the solar constant S_0 :

$$S_0 = P/4\pi r_0^2,$$

where P is the radiated power (irradiance) of the sun, which by measurements of S_0 is

$$P = 4 \times 10^{20} \, MW.$$

The value of the solar "constant" is taken as 1,366 Wm^{-2}. Variations of the solar constant of perhaps as much as a few tenths of a percent have recently been observed by satellite, but its value is taken as constant over the past million years.

The flux of solar heat incident upon a unit horizontal area at the limit of the atmosphere over a given point on the earth and at a given time is

$$I' = S_0 \left(\frac{r_0}{r}\right)^2 \cos \varsigma,$$

where

$$\cos \varsigma = \sin \phi \sin \varsigma + \cos \phi \cos \delta \cos \psi,$$

and r is the distance of the earth from the sun at a given time; ς is the sun's zenith angle at a point on the earth; ϕ is the geographical latitude of the point; δ

is the geocentric declination of the sun—that is, the angle between the sun's direction and the plane of the earth's equator; $\psi = 2\pi t/T$ is the hour angle—that is, the angle between the plane of the meridian of the given place and the plane passing through the earth's axis of rotation and the sun, reckoned to west of the meridian; t is the time; and T = 86,400 s is the length of the solar day.

The daily insolation is found by integrating I' over the daylight hours. Integration is facilitated because r and δ can be approximated as constant over the 24 hours:

$$I = \frac{1}{T}\int_{t_0}^{t_0} I'\,dt = \frac{S_0}{\pi}\left(\frac{r_0}{r}\right)^2 (\psi_0 \sin\phi \sin\delta + \cos\psi \cos\delta \sin\psi_0),$$

where the hour angle of the sun is $\psi_0 = 2\pi t_0/T$ at sunset, $\psi_0 = \pi$ on polar nights when the sun never sets, and $\psi_0 = 0$ when the sun does not rise.

The daily insolation is independent of longitude, making it latitudinally zoned. At the equinoxes, insolation is a maximum at the equator and zero at the poles. At summer solstice in the Northern Hemisphere, the daily insolation reaches a maximum 45.1 $\mathrm{MJm^{-2}}$ at the North Pole, where there are 24 hours of daylight, while at the equator it is only 33.2 $\mathrm{MJm^{-2}}$. However, the insolation does not vary monotonically with latitude at this time; there is a secondary maximum between 40° and 45°N (41.7 $\mathrm{MJm^{-2}}$) and a secondary minimum between 60° and 65°N (41.0 $\mathrm{MJm^{-2}}$). But because the annual insolation in the polar region is much less than that at the equator, climate tends to be zonally distributed and thus reflects the annual average insolation.

If the eccentricity were zero, the seasonal oscillation of the insolation would be in phase at all latitudes and would differ in phase only by π radians in different hemispheres. Because the eccentricity is not zero, the insolation of the Northern and Southern Hemispheres is asymmetric. For example, the earth is near aphelion ($r = 1.52 \times 10^8$ km) in the Northern Hemisphere in summer, and near perihelion ($r = 1.47 \times 10^8$) in winter, making summer insolation less in the Northern Hemisphere than the Southern Hemisphere, while the winter insolation is greater.

In the theory of planetary orbits, Laplace demonstrated that, accurate to the first power in e, the semi-major axis r_0 of an elliptic orbit is not perturbed. To the same accuracy, we can say that the solar constant S_0 and length of year to T_0 do not vary in time, and thus the mean annual insolation over the earth's surface $S_0T_0/4\sqrt{1-e^2}$ is constant as well. To second order in eccentricity, changes in insolation are proportional to $(\Delta e)^2/2$, where Δe is the change in eccentricity. The longest change in eccentricity—over 780 kyrs—is about 0.05 (see Fig. 40); the total change in average flux is about one part in a thousand, which is comparable to changes in the solar constant. Changes in the orbital elements—eccentricity, obliquity, and precessional parameter—can only change the distribution of radiation over the earth's surface. An increase in the obliquity will enhance the difference between seasons, making climate more continental-like. If it can be assumed that warm winters and cool summers promote glacial growth, then one must look for cycles of cool summers and warm winters at high latitudes when searching for the onset of ice ages.

Figure 43 illustrates the calculated variation of the insolation at 60°N during the summer solstice. There are minima in the summer insolation (cool summers) at 120 kyr, 80 kyr, and 25 kyr, corresponding to the time of onset of cold conditions. The spectra of insolation variations show vanishingly small power at periods of 110 kyr (see Figs. 44 and 45), where the isotope proxy spectra show maximum power. At low latitudes (lower than 45°N), the maximum in the insolation spectrum comes at periods of 23 kyr (precession), and the variation at 40 kyr is small. At 60°N, the spectral energy in the 40-kyr period builds up (see Fig. 44), and the importance of the obliquity variations increases with latitude (see Fig. 45).

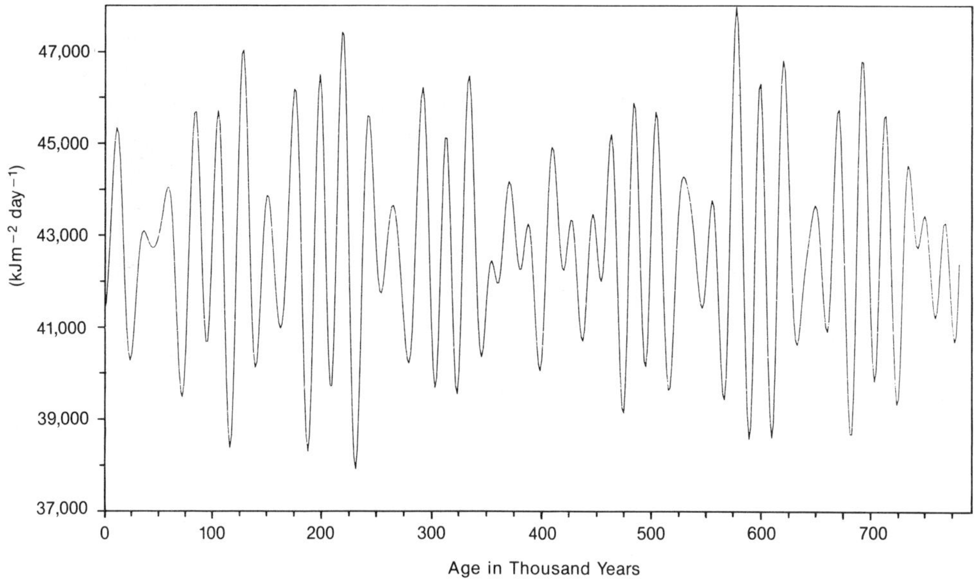

Fig. 43. Variation of insolation at a 60° N latitude for summer solstice, about June 21.

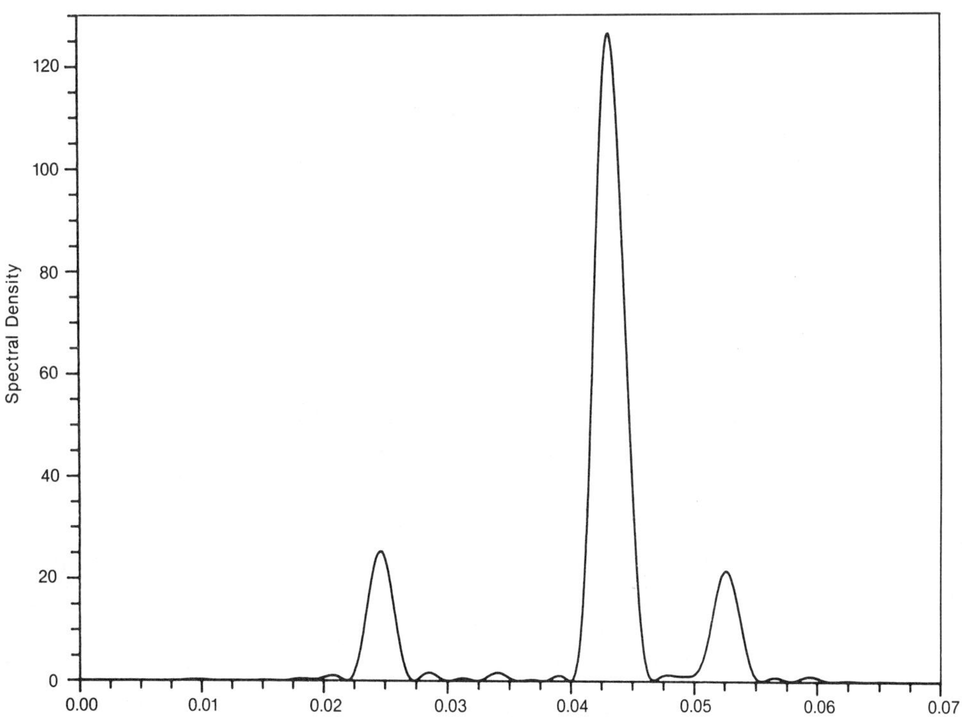

Fig. 44. Spectrum of insolation at 60° N at the summer solstice, about June 21.

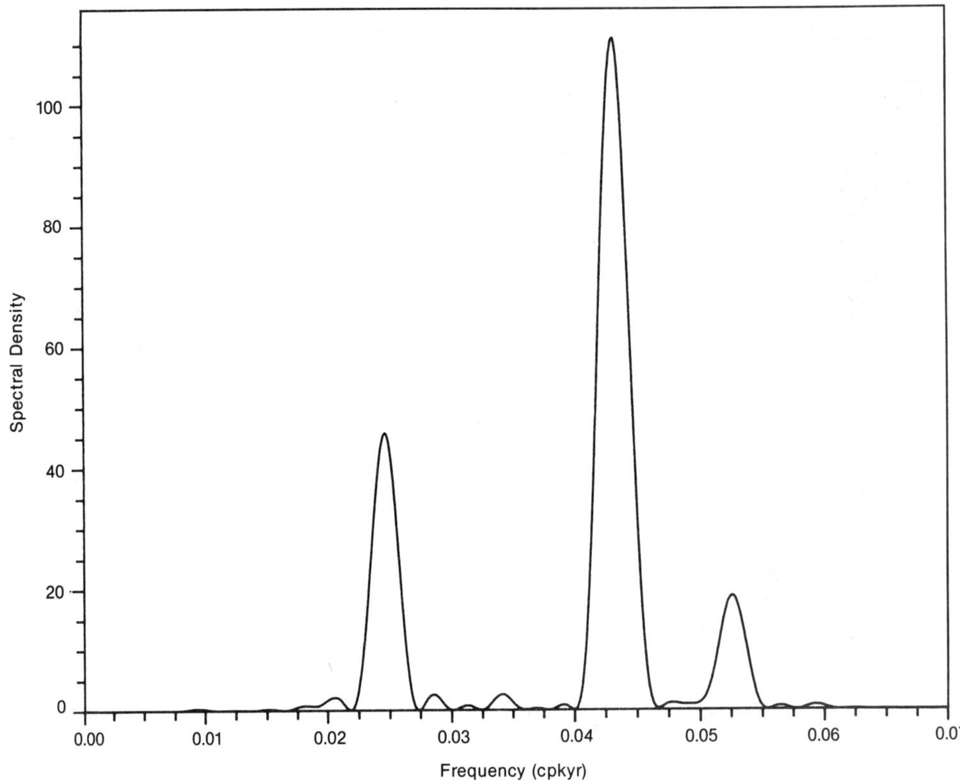

Fig. 45. Spectrum of insolation at 80° N at the summer solstice, about June 21.

In terms of insolation, the earth is currently at a typical interglacial value. The next minimum in insolation at 60°N lies about 10 kyr in the future, and its minimum value corresponds approximately to the insolation value of 25 kyr BP. If orbital changes were solely responsible for climate variability, the forecast for the future would be for slow cooling, with global average temperatures dropping by 2° to 5°C over the next 10,000 years.

Comparison of Spectra of Orbital Elements and Proxy Isotope Data

The spectra of insolation vary with both geographic position and latitude. Because of this variation, it may be more meaningful to compare the spectra of the individual orbital elements with the proxy isotope data. One must remember, however, that in terms of direct effects on insolation, the impact of a variation in eccentricity is vanishingly small, and that of the precessional parameter is larger than that of obliquity, at all but the very highest latitude. The spectra of the orbital elements were obtained from the calculated time variation of the elements, using the periodogram method and the theory described in the previous section. All peaks significant at the 0.05 level are shown. The spectra were calculated on the basis of a 364-kyr record so as to be compatible with the proxy isotope records but not with the CO_2 record of 160 kyr. The amplitude of the maximum peaks in each of the orbital elements was normalized to unity, so that the relative amplitudes for each element, as well as those of the proxy data, are represented.

Figure 46 shows a comparison of the astronomical and carbon dioxide amplitude spectra. The main peak in the carbon dioxide spectrum coincides with one of the lines for the eccentricity, but lies at a lower frequency than most of the power in the variation of eccentricity. The spectrum of $\delta^{18}O$ variations derived

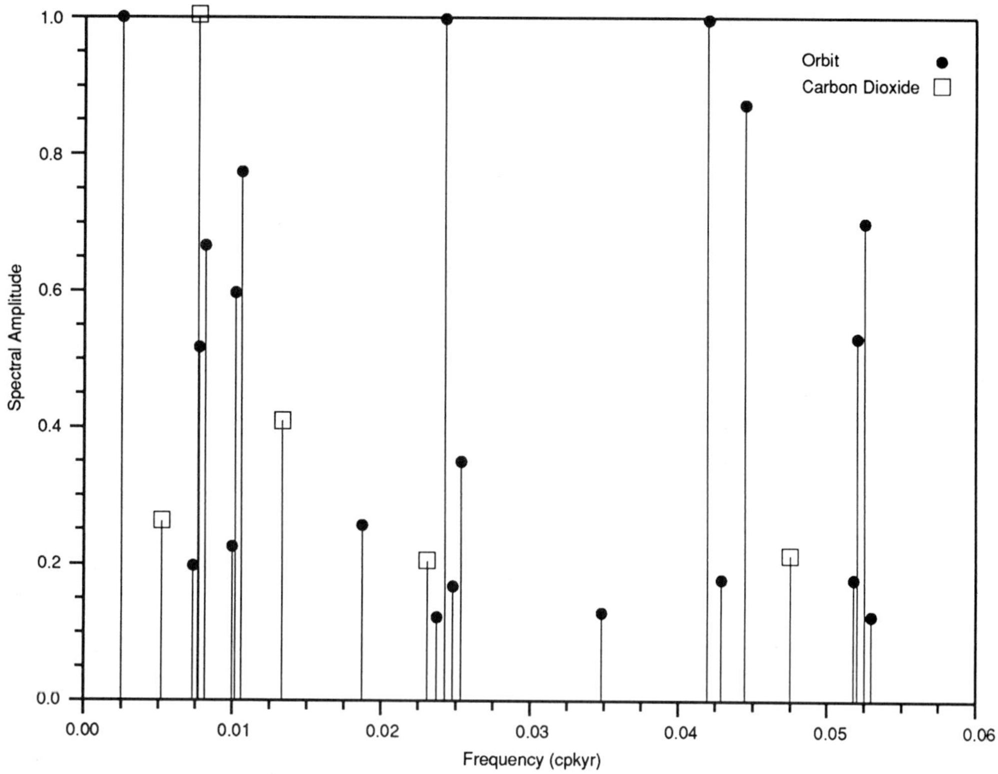

Fig. 46. Comparison of amplitude spectra for the orbital elements and the spectrum for the variation of atmospheric carbon dioxide, as determined from the Vostok core.

from the V19-30 ocean core and the spectra of the orbital elements are shown in Fig. 47. The agreement with the orbital elements is spectacular: the $\delta^{18}O$ peak at 0.043 cpkyr lies within the peaks for the strongest variation in the precessional parameters; the peak at 0.052 cpkyr in the $\delta^{18}O$ record lies close to the cluster of lines for the precessional parameters; and even the ratio of the amplitude of the lines at 0.043 and 0.052 is in general accord with expectations based on the orbital theory. Equally striking is the match of lines for the obliquity and $\delta^{18}O$. The $\delta^{18}O$ record not only picks up the main cluster of lines at 0.02439 cpkyr, the 41-kyr period, but also the lines at higher and lower frequencies: 0.01865 cpkyr (53.6 kyr) and 0.03459 cpkyr (28.9 kyr). The line structure in the eccentricity frequency range also shows the similarity of the $\delta^{18}O$ record to the astronomical elements, but with some differences. The principal peak in the $\delta^{18}O$ record does not coincide with any of the peaks in variation of eccentricity, though these peaks are overlapped by the triplet lines in $\delta^{18}O$ around the 110-kyr period. In terms of direct comparison, the only analogous line in the $\delta^{18}O$ spectrum lies at 0.014 cpkyr (67.9 kyr), well removed from lines associated with any of the orbital elements.

The longer record for $\delta^{18}O$—obtained by patching together five ocean cores (Imbrie et al., 1984)—might be thought to provide a better basis for comparison with the orbital elements (see the previous section). However, because the dating of the records used in constructing the SPECMAP record was based on assumed astronomical forcing at the basic 41-, 23-, and 19-kyr periods, agreement between the proxy record and the spectra of the major peaks in the orbital elements should be assumed. Table 11 provides a list of peaks in the SPECMAP record that contribute at least 1% to the total variance of the record and are significant at the 0.05 level. Major peaks listed in Table 11 contain 86.7% of the total variance. Figure 48 shows the comparison between the spectrum of the 780-kyr SPECMAP

Table 11. Spectral peaks of the 780-kyr $\delta^{18}O$ record*.

Frequency (cpkyr)	Period (kyr)	Percent of Variance
0.01010	99.0	35.9
0.02462	40.6	9.71
0.04239	23.6	6.78
0.01176	85.0	6.04
0.01477	67.7	5.26
0.00141	709.2	4.71
0.00914	109.4	4.54
0.0172	58.1	2.21
0.00793	126.1	2.19
0.00627	159.5	2.12
0.05256	19.0	2.03
0.01662	60.2	1.60
0.04444	22.5	1.30
0.02347	42.6	1.26
0.01924	51.2	1.00

*SPECMAP time scale, Imbrie et al., 1984.

composite record and the spectra of the orbital elements. The coincidence of peaks at the three principal frequencies is expected, but the subsidiary peak at 0.04444 represents real agreement that is not superimposed by assumption. The 780-kyr record permits the separation of peaks at 0.04239 and 0.04444 cpkyr, but the agreement between peaks in the variation in obliquity at 28.9 kyr and 56.6 kyr is not as good as in the V19-30 record. Based solely on this comparison, the dating of SPECMAP is probably not as accurate as that for the single, long V19-30 record. The SPECMAP record also shows a complex spectrum at low frequencies, as well as a prominent line at 0.0148 cpkyr (67.7 kyr), which is also found in the V19-30 record but is not of orbital origin.

A summary comparison of the three proxy elements obtained from the V19-30 core—benthic $\delta^{18}O$ and $\delta^{13}C$, and planktonic $\delta^{13}C$—is shown in Fig. 49. In general, variations in both the deep and shallow carbon isotopes follow the variation in orbital elements. It was noted in the previous section that the difference in benthic and planktonic carbon isotope ratios most closely approximated the spectra for the variation in atmospheric CO_2 concentration. Figure 50 shows that although the spectrum of the difference in deep and shallow $\delta^{13}C$ has some common features with the spectra of the orbital elements, there are significant differences.

Summary of Effects of Orbital Variations of Insolation on Climate

Variations in obliquity and precession have no net global or annual effects on the insolation received at the outer limits of the atmosphere. Variation in these elements of the earth's orbit alters only the contrast between seasons. For obliquity, the change in contrast is the same in both hemispheres. For precession, the contrast is larger in one hemisphere than in the other because of the eccentricity term.

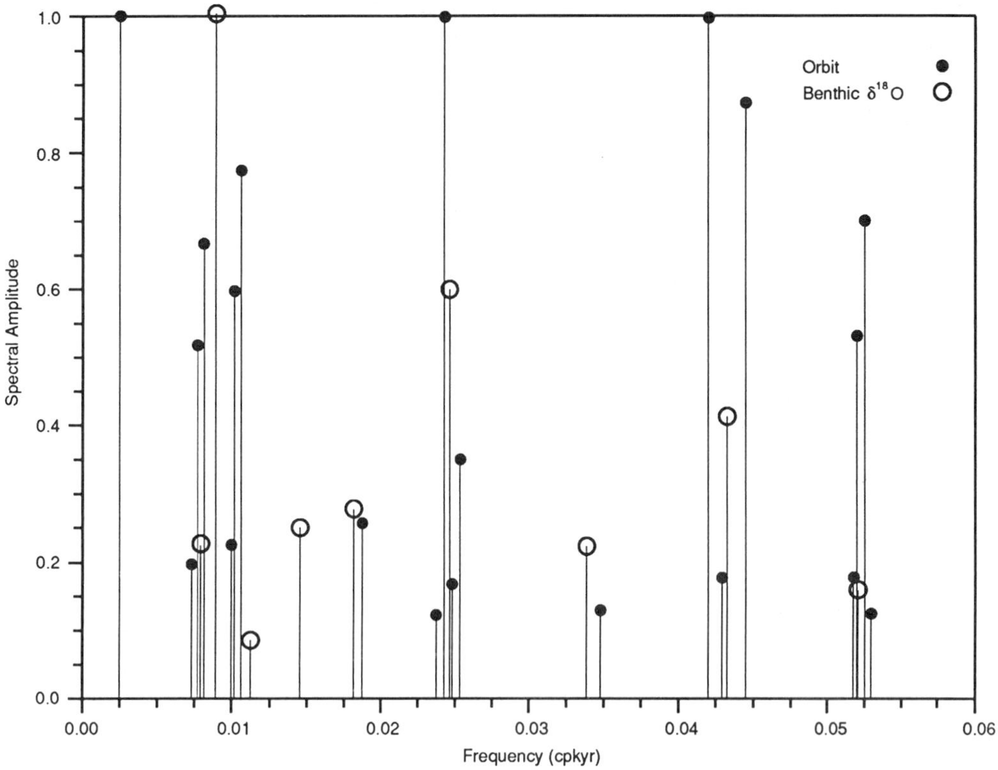

Fig. 47. Comparison of the spectrum of $\delta^{18}O$ from core V19-30 from the Eastern Pacific and the spectra of the orbital elements.

The precessional effect $e \sin \tilde{\omega}$ remains the same within one hemisphere, while the obliquity term increases with latitude. As a result, the obliquity term becomes more important at high latitudes. Changes in insolation due to either precession or obliquity are small, but are still much larger than the effect of eccentricity.

Variations in eccentricity bring about changes in the global average insolation proportional to $(Ae)^2/2$. The resulting alterations in average global temperature are quite small. The most elementary energy budget for the atmosphere requires that the outgoing infrared radiation be balanced by the incoming insolation corrected for the earth's albedo:

$$\sigma T_e^4 = \frac{S_0}{4}(1 - \alpha),$$

where

σ = Stefan Boltzmann constant (5.6×10^{-8}) Wm^{-2} $(°K)^{-4}$
$\overline{T}_e$ = global mean effective temperature of the earth radiating into space
S_0 = solar constant (1,366 Wm^2)
$\overline{\alpha}$ = global mean albedo (0.31)

The relation between the effective temperature of the earth and the global-mean surface temperature T_S is given by

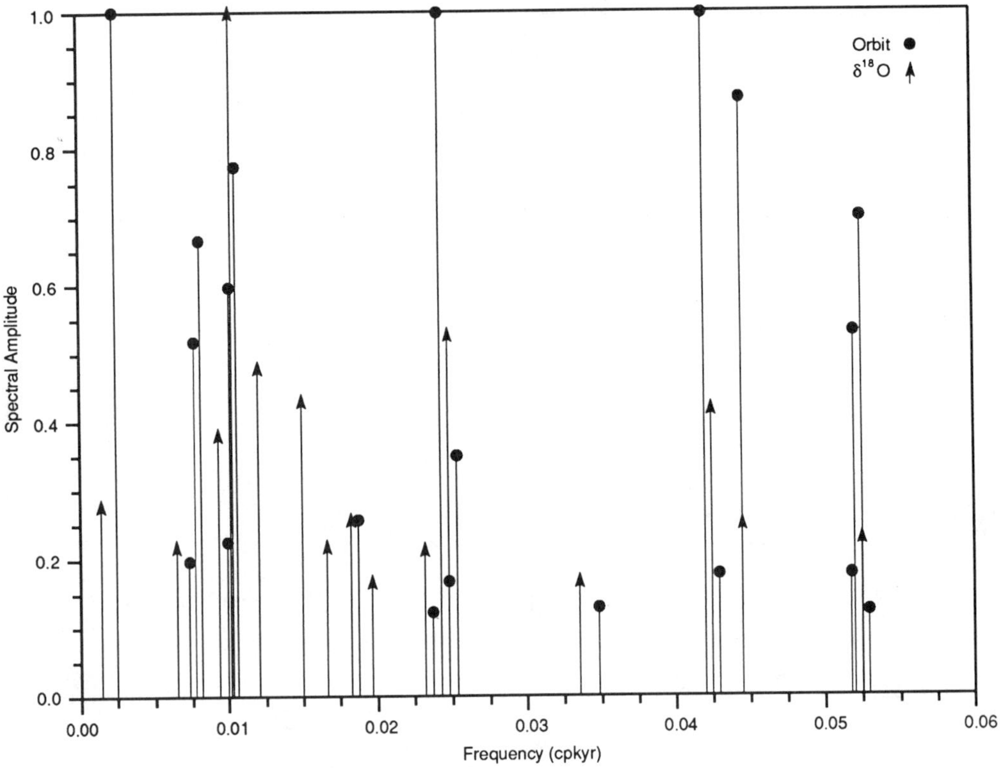

Fig. 48. Comparison of the spectra of the orbital elements with the variation of $\delta^{18}O$ derived from the composite SPECMAP record.

$$\overline{T}_S = \left(1 + \frac{3}{4}\tau\right)^{1/4} T_e,$$

where τ is a measure of the opacity of the atmosphere. For the current atmosphere, τ is about 0.75 (MacDonald, 1982). The radiative balance equation leads to an effective temperature $T_e = 255$ K and a surface temperature $T_S = 285$ K.

A change of 0.05 in eccentricity *(e)* leads to a change in insolation of about one thousandth, with an accompanying variation in global average temperature of less than a tenth of a degree. This observation highlights the paradox in the proxy data for ice volume. The highest power in the spectrum for $\delta^{18}O$ is at a frequency where the effect of orbital change is minimal. The maximum power in variation in insolation is at the precessional period of 23 kyr, yet this peak in the proxy data contains significantly less energy than those at 100 or 40 kyr.

The quest for a mechanism that translates minute annual and small seasonal insolation effects into major shifts in climate usually begins by considering the atmosphere as a highly nonlinear system that is subject to quasi-periodic forcing. The hydrologic cycle is assumed to play a significant role in amplifying small external perturbations into a major climate shift. Conditions of large snow accumulations in winter, coupled with lower snow ablation in summer, work toward increasing the mass of water trapped in ice. Based on meteorological observations, warmer winters with higher atmospheric water-carrying capacity, coupled with cooler summers, would meet these requirements. But is there some mechanism that amplifies the regular minute changes in insolation caused by eccentricity variations into major changes in ice volume? In other words, how are alterations in atmospheric composition tied to orbital variations?

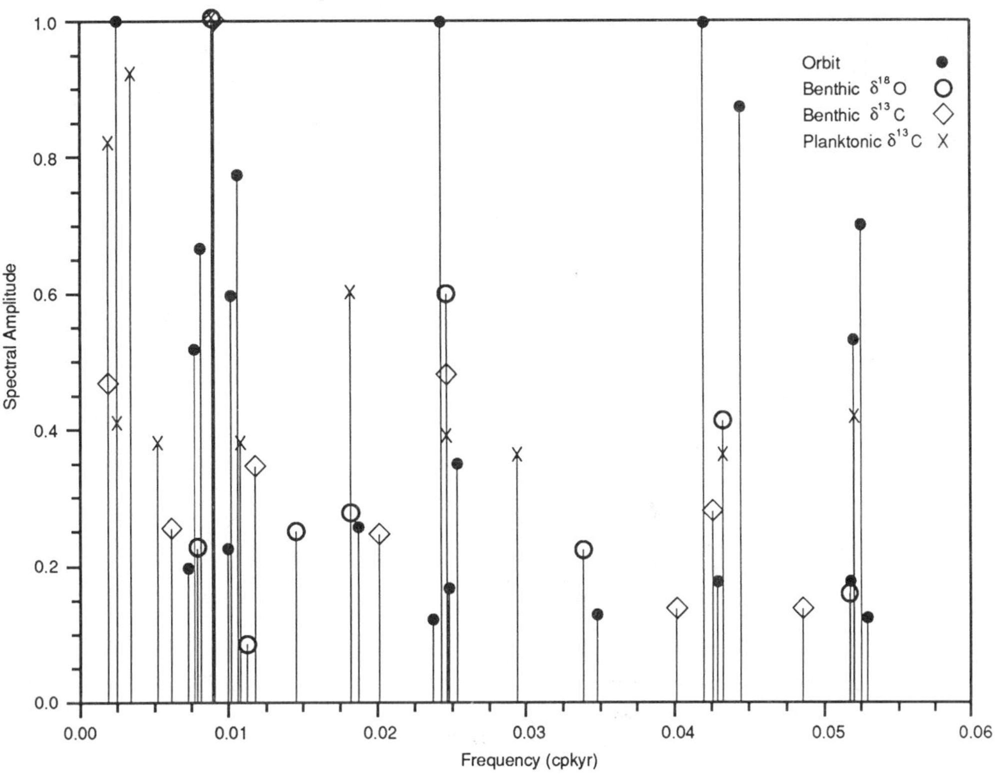

Fig. 49. Comparison of spectra of orbital elements with spectra for benthic and planktonic $\delta^{13}C$, and benthic $\delta^{18}O$ from core V19-30 taken in the Eastern Pacific.

Ice Age Feedback Mechanisms

At the end of the last glacial maximum, approximately 200 billion tons (Gt) of carbon moved into the atmosphere on a time scale of a few thousand years (see Fig. 7). A similar movement of carbon took place at the beginning of the interglacial period some 150 kyr BP. During periods of glaciation the reverse occurred: the atmosphere gave up a comparable quantity of carbon. Because methane and carbon dioxide strongly affect the atmosphere's radiative budget, these shifts could be associated with the nonlinear mechanism that amplifies orbital changes in insolation.

During glaciation, destruction of northern forest by the advancing ice sheets, coupled with deteriorating climatic conditions worldwide, would be expected to release biospheric carbon to the atmosphere, thus causing a drop in $\delta^{13}C$. The ocean core proxy data does show lower values for $\delta^{13}C$ during glacial times (see Fig. 28), but the release of biospheric carbon should cause an increase in atmospheric carbon dioxide (Shackleton, 1977) rather than the observed decrease. The reduction of standing biomass during glaciation may have been more than compensated for by extensive development of peat bogs and wetlands in the large periglacial regions (MacDonald, 1988). But because tills and glacial outwash sediments tend to be poor in carbon, the existence of extended regions of carbon-rich soils in land areas bordering the ice sheets remains conjectural.

A large, known reservoir of carbon is the ocean (see Fig. 22). Broecker (1981, 1982) postulated that changing photosynthetic rates in the near-surface ocean may have influenced atmospheric CO_2 concentration. During periods of glaciation, nutrient-rich continental shelf areas were exposed by falling sea level and subjected to erosion. The addition of these nutrients to near-surface ocean

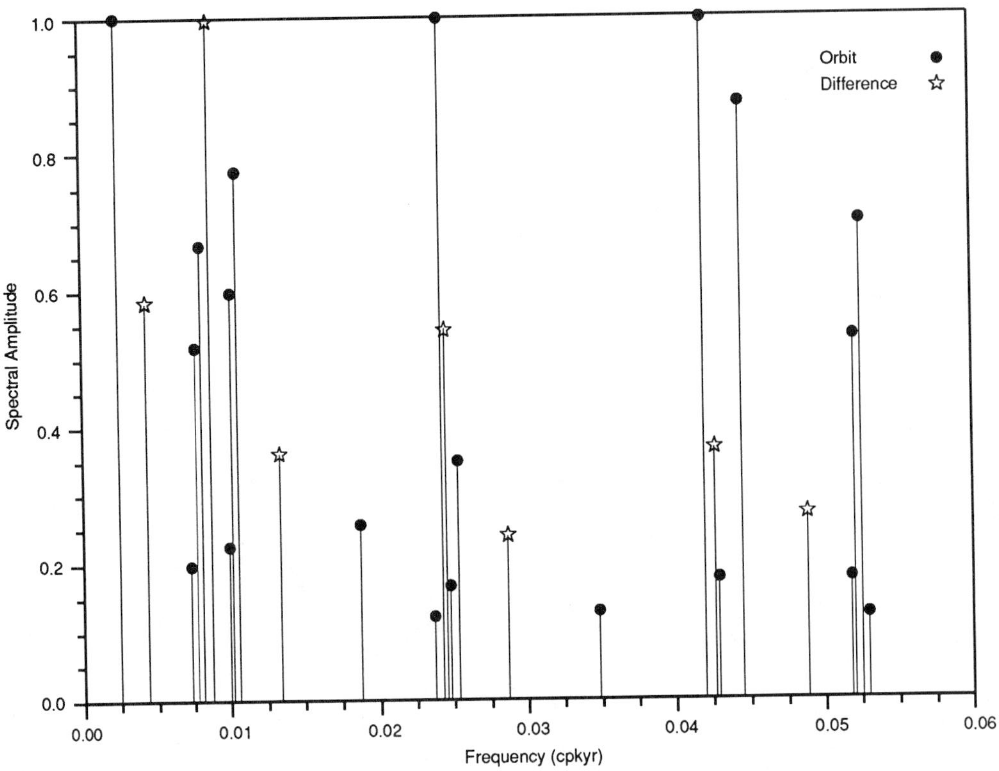

Fig. 50. Comparison of the spectra of the orbital elements with the spectrum of the difference between benthic and planktonic $\delta^{13}C$, as recorded in core V19-30 from the Eastern Pacific.

waters stimulated photosynthetic activity, thus reducing atmospheric carbon dioxide. During interglacial times, inundation of the shelf areas lessened the flow of nutrients to the open ocean, reducing photosynthesis and increasing the atmospheric CO_2 concentration. However, Broecker's hypothesis requires a greater amount of sedimentation than is observed (Walsh, 1984). Furthermore, the amount of carbon released from glacial destruction of forests must be minimized with this hypothesis, and the implied changes to ocean chemistry (low bottom-water oxygen content) are not observed (Broecker and Peng, 1986).

Still another hypothesis for glacial age changes in atmospheric carbon dioxide content makes use of a different large carbon source of biological origin: methane stored in near-surface rocks as clathrates (MacDonald, 1990).

Role of Methane Clathrates in Climate Change

Methane clathrates are ice-like compounds in which cages of water molecules hold methane molecules, $CH_4 \cdot 5.75\ H_2O$. Not all the methane sites within the clathrate lattice need be filled for the crystalline structure to be stable. The laboratory density of the fully filled clathrate is 0.91 g cm^{-3}; one cubic meter of methane clathrate contains 170 m^3 of methane gas at standard conditions. Alternatively, a cubic meter of clathrate contains 91.4 kg of carbon. The field of stability for methane is shown in Fig. 51. An important fact is that methane clathrate can exist at a range of temperatures and pressures at which pure ice would not be stable. Beginning at pressures greater than 26 bars, methane clathrate is stable at temperatures greater than 0°C.

In order for methane clathrate to form, not only must the temperature and pressure be in the field of stability, but methane must also be present in sufficient

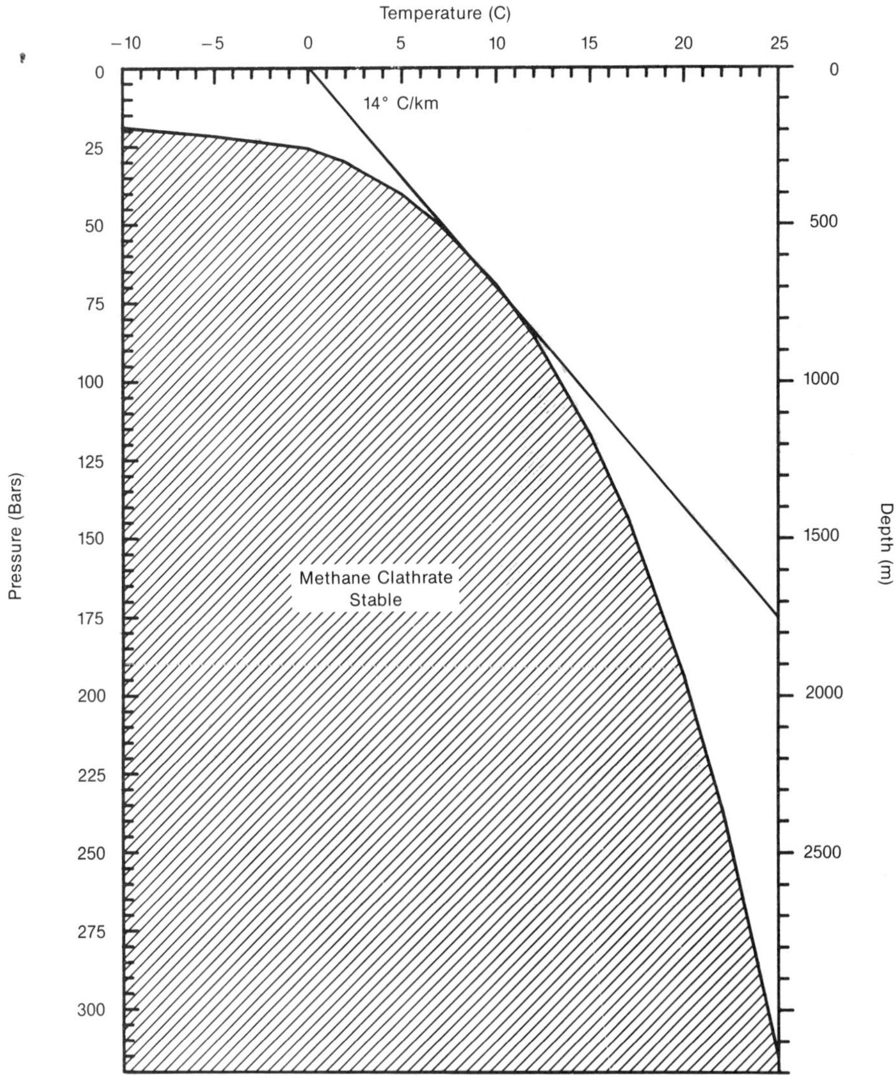

Fig. 51. Phase diagram showing the boundary between free methane gas and methane clathrate for the H_2O-CH_4 system. The ice-water phase boundary would be a vertical line through 0°C. The depth scale is drawn assuming hydrostatic equilibrium for pore water. The limiting geothermal gradient of 14°C km^{-1} for the stability of methane clathrate outside permafrost regions is also indicated.

quantities. The amount of methane within sediments depends on the source of methane and the porosity and permeability of the rocks. The availability of methane is often the limiting factor in clathrate formation. Methane can form in place in sediments as a result of microbial methanogenesis, or by thermal decomposition of matter at high temperatures (Hunt, 1979). Because the clathrate stability field is limited to near-surface conditions, a microbial origin is of greatest interest.

A general picture of methane generation by bacteria in ocean sediments has evolved, largely as a result of cores taken in the Deep Sea Drilling Program (Claypool and Kvenvolden, 1983). The bacterial decomposition of fermentable organic matter in marine sediments begins at deposition below the sediment-water boundary. Organic matter is first attacked by aerobic bacteria that produce carbon dioxide, which dissolves in the water or escapes to the atmosphere. If the rate of sedimentation is low (< 1cm/kyr), the organic content is small (<< 1% C), and oxygen is abundant, aerobic bacteria can remove most or all of the organic material. When sedimentation is rapid—several centimeters per thousand years—the aerobic environment is replaced by anaerobic conditions at depths within the sediments as shallow as a few centimeters. In the upper part of the aerobic region, bacterial action reduces sulfates and generates carbon dioxide and hydrogen sulfide. Deeper down, where sulfate has been removed, methane becomes the principal product of bacterial degradation of organic matter.

The anaerobic production of methane under sedimentary conditions is a complex biological process requiring synergism among a variety of microorganisms (Oremland, 1988). Bacterial action on the organic source matter leads to the production of carbon dioxide and hydrogen. The dissolved carbon dioxide forms bicarbonate, which is combined by the methanogenic bacteria with hydrogen ions to form methane and water. Methanogenesis requires a highly reducing environment (En below -400 mV) and a pH in the range of 6 to 8. The temperature range for methanogenesis is broad; in lake sediments, methanogens function from at least 4°C to about 55°C, with an optimum between 35° and 42°C (Zeikus and Winfrey, 1976). Compounds other than carbon dioxide, such as acetate, can serve as substrates, but these are probably less important to the production of methane within sediments.

At a great enough depth, biological reactions cease, owing to increased temperature and pressure, buildup of toxic materials, or exhaustion of organic material. The depth at which biologic activity ceases is not known but is almost certainly in excess of a kilometer.

Clathrates have been sampled at a number of localities, both on land and in ocean sediments, and have also been seismically identified. The locations of methane clathrate occurrences are given in Fig. 52. Using data on geothermal gradients and carbon content of sediments, MacDonald (1990) finds that approximately 400 Gt of carbon are stored as clathrates in cold regions, and 11,000 Gt in subocean sediments. These estimates agree with an independent estimate by Kvenvolden (1988) of about 10,000 Gt of carbon in subocean sediments.

Surface disturbances associated with glaciation perturb subsurface conditions and alter the volume of sediments lying within the methane clathrate stability field. A drop in temperature in cold periglacial regions stabilizes methane clathrates at depth, as is schematically illustrated in Fig. 53. Under an ice sheet, added pressure from the overburden of ice increases the stability field volume, thus decreasing the amount of carbon that would otherwise leak into the atmosphere as methane. As an ice sheet melts, the pressure drop beneath it leads to destabilization of methane clathrates. In ocean sediments, methane clathrates are stable where water is in the 200-m to 3-km depth range. The lowering of sea level reduces the stability volume in continental shelf regions, provided there is no accompanying drop in temperature. In ocean sediments on high-latitude continental shelves, the inundation of cold land by relatively warm waters destabilizes underlying clathrates, while the increase of pressure due to a rise in sea level leads to stabilization.

The effect that periodic fluctuations in ice volume and surface temperature have on the volume of clathrates in the stability region can be analyzed in terms of simple heat conduction in a semi-infinite solid (Carslaw and Jaeger, 1959). If surface temperature undergoes a harmonic variation with frequency f of the form

$$T_S = A\cos(2\pi f t - \epsilon),$$

then the variation of temperature at any depth z is

$$T = A\exp(-\mathbf{k}z)\cos(2\pi f t - \mathbf{k}z + \epsilon),$$

Fig. 52. Geographical distribution of methane clathrate occurrences.

Table 12. Temperature variation at depth due to harmonic disturbances at glacial periods.

	Conductivity 1.76 $Wm^{-1}\,{}^{\circ}C^{-1}$			Conductivity 3.39 $Wm^{-1}\,{}^{\circ}C^{-1}$		
Period (Years)	20,000	40,000	100,000	20,000	40,000	100,000
Wavelength (m)	2,580	3,640	5,600	3,570	5,050	7,990
Attenuation at 500 m	0.30	0.42	0.57	0.41	0.54	0.67
Attentuation at 1,000 m	0.09	0.18	0.33	0.17	0.29	0.45
Phase Lag at 1,000 m in kyr	7.8	11.0	17.4	5.6	7.9	12.5

where **k**, the wave number, is given by

$$\mathbf{k} = (\pi f/\alpha)^{1/2},$$

and where the thermal diffusivity σ is the quotient of the thermal conductivity K and the specific heat per unit volume

$$\sigma = K/\rho c.$$

The wavelength λ of the temperature wave is

$$\lambda = \frac{2\pi}{\mathbf{k}} = (4\pi\alpha/f)^{1/2}.$$

Table 12 lists the wavelengths of disturbances corresponding to periods observed in the $\delta^{18}O$ record and for differing conductivities (see Fig. 32). The conductivity of frozen, water-rich sediments is taken as 3.39 $Wm^{-1}\,{}^{\circ}C^{-1}$, and that of thawed sediments as 1.76 $Wm^{-1}\,{}^{\circ}C^{-1}$. These two values bracket the observed values in thermal conductivity. The specific heat per unit volume is roughly 2.1×10^6 J/m^3 and does not vary with materials. Table 12 gives the amplitudes of the disturbances, which fall off exponentially with depth and are greatest at the lower periods. The temperature curve at depth lags behind the surface changes. The phase lag Δe is

$$\Delta\epsilon = z(\pi f/\alpha)^{1/2}$$

and decreases with the period of the surface disturbance. In interpreting the values given in Table 12, it is useful to note that the current depth of the clathrate zone of stability in Siberia lies at 1,500 m in the Tugunska low in the east, and at depths of 1,200 m over much of the Viluy Basin, both of which are large reservoirs of clathrate carbon. At these depths, the higher-frequency variation periods of 23 kyr are highly attenuated as compared to the long-term changes. The large depth of penetration of the 100-kyr oscillation may provide a mechanism by which the variation in orbital eccentricity is amplified. A greater volume of clathrate stability zone is affected by the 100-kyr variation than by the higher-frequency variations.

Phase relations among the orbital forcing function, $\delta^{18}O$ variations, and atmospheric carbon dioxide records are consistent with the hypothesis that clathrate stability plays a role in glaciation. Cross-spectral analysis between the

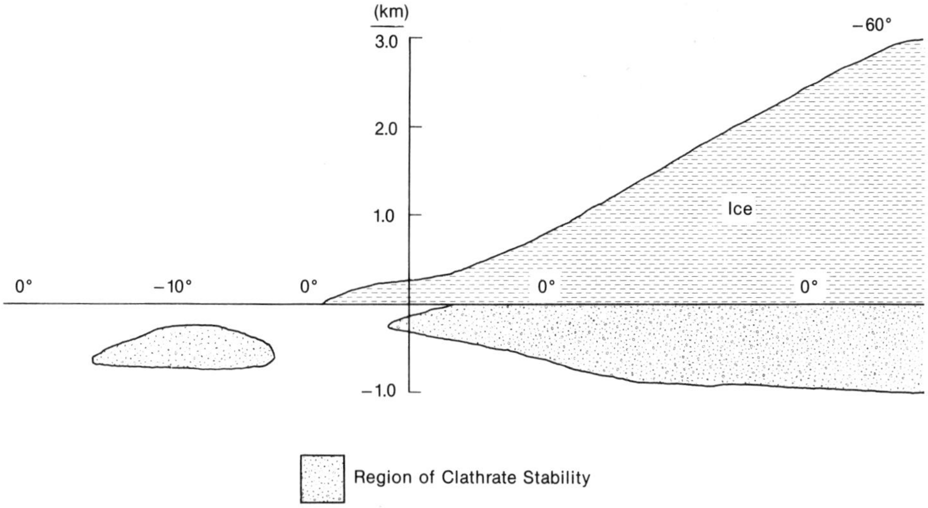

Fig. 53. Schematic representation of the region of methane clathrate stability in the vicinity of an advancing ice sheet.

proxy $\delta^{18}O$ record and the calculated astronomical forcing function shows that the phase of the ice volume ($\delta^{18}O$) lags the astronomical forcing function by 5 kyr and 9 kyr at the precessional and obliquity frequencies, respectively. The ice volume record and the atmospheric CO_2 record are much more nearly in phase, at least over the last 40 kyr (see Fig. 24). As indicated in Table 12, the response at depth lags behind the surface forcing function as a temperature wave penetrates. The stability limits of clathrate fix the time scale for carbon releases. The similarity of observed time scales with those calculated on simple conduction theory is striking (see Table 12).

As mentioned earlier, the rising of sea level during interglacial periods results in the warming of exposed, low-lying shelf. In high latitudes, a sharp change in surface temperature from periglacial cold to the relative warmth of the ocean bottom leads to a downward-traveling temperature wave that can release methane from underlying clathrates. Simple thermal conduction calculations provide insight into the time scales involved in the release of the trapped methane. The temperature at a depth z can be written as

$$T(z,t) = T_0 + \gamma_0 z + T'(z,t),$$

where the initial temperature before ocean flooding has reached equilibrium at a surface temperature T_0 and constant geothermal gradient γ_0. The time-dependent part of the temperature field is given by the one-dimensional heat conduction equation:

$$\frac{\delta^2 T'(z,t)}{\delta Z^2} = \frac{1}{\alpha}\frac{\delta T'}{\delta t}\,.$$

If the surface temperature is assumed to vary as

$$T'(0,t) = Dt^{n/2}\,, \qquad 0 < t < t_0,$$

where D is a constant and n is zero or a positive integer, then the temperature $T(z,t)$ is

$$T(z,t) = T_0 + \gamma_0 z + 2^n\,\Gamma\left(\frac{1}{2}\,n + 1\right)\,i^n\,erfc(z/\sqrt{4\alpha t})\,.$$

In this expression, Γ is the gamma function and $i^n\,erfc$ is the nth repeated integral of the error function (Carslaw and Jaeger, 1959). The case for $n = 0$ represents a step increase in temperature.

The history of the evolution of subsurface temperatures for a flooded permafrost section that has an initial surface temperature of −10°C is shown in Fig. 54. The initial equilibrium temperature gradient is 31.25°C/km, with a thermal conductivity of 1.76 $Wm^{-1}\,°C^{-1}$. These conditions correspond approximately to those that hold today around Point Barrow, Alaska. The initial top of the clathrate stability zone, 230 m down, is reached after 150 years, but the stability zone is reduced by only 10 m after 500 years. However, after 2,000 years the entire 400-m-thick stability zone is removed. The most rapid rate of methane release occurs during the interval between 1,000 and 2,000 years, as the geothermal gradient and the phase boundary become parallel.

The case of high-conductivity, ice-saturated sediments that overlie lower-conductivity, water-saturated sediments in a permafrost area is illustrated in Fig. 55. The solution that is valid for the upper layer is extended to the second homogeneous layer by requiring continuity in temperature and heat flow at the boundary of the frozen and thawed sediments. Assuming an initial surface temperature of −15°C, the initial clathrate zone is 1,000 m thick. Erosion of the top of the clathrate zone proceeds slowly downwards until about 10,000 years. Fourteen thousand years after the step warming at the surface, the clathrate stability zone vanishes.

The examples given in Figs. 54 and 55 show that a sudden warming, resulting either from inundation by the sea or the formation of a large lake, causes the elimination of the clathrate stability zone on a time scale of 2 to 14 kyr. Low initial surface temperatures and high-conductivity sediments favor the longer time scale.

Figure 56 illustrates the long time scale involved in the formation of a thick clathrate stability zone. The initial surface is taken to have a temperature of −1°C (maintained by the overlying sea or lake) and is underlain by low-conductivity sediments. A stepwise drop in temperature to −15°C is assumed to accompany a drop in sea level, which, at high latitudes, exposes the surface to cold periglacial conditions. The geothermal gradient drops below the clathrate stability zone in 2,000 years, but after 10,000 years the thickness of the clathrate stability zone has grown to just 230 m, and after 60 kyr the zone is still only 480 m thick.

The difference between time scales for heating and cooling can be understood in terms of heat flux. As indicated in Fig. 57, cooling at the surface proceeds as heat is drawn from deeper levels, and thus works against the low conductivity of the deep sediments. Conversely, in the case of surface warming, heat flows into the earth at the surface (see Fig. 58); the warming illustrated in Fig. 54 proceeds from the outer surface inwards. The asymmetry introduced into the problem by the surface results in the very different time scales for warming and cooling.

Sea level changes on the order of 100 m accompanied the formation and melting of continental ice sheets. The resulting change in pressure at depth had a major impact on the volume of the clathrate stability zone in ocean-bottom sediments. The effect of changing sea level on the thickness of the clathrate stability zone is a function both of ocean depth and subsurface temperature gradient, as is illustrated in Fig. 59. At present, some 11,000 Gt of carbon are stored as clathrate in a stability zone having an average thickness of about 500 m. A decrease in thickness of 10 to 20 m resulting from a 100-m drop in sea level (see Fig. 59) would destabilize 200 to 400 Gt of carbon under current conditions. The time scale for the release of carbon from clathrates depends on the rate of sea level drop and on the thermal properties of the overlying sediments.

As pressure drops, clathrates dissociate, absorbing energy. At 0°C, the heat of dissociation of methane clathrate is 53.5 kJ per mole of CH_4. The dissociation

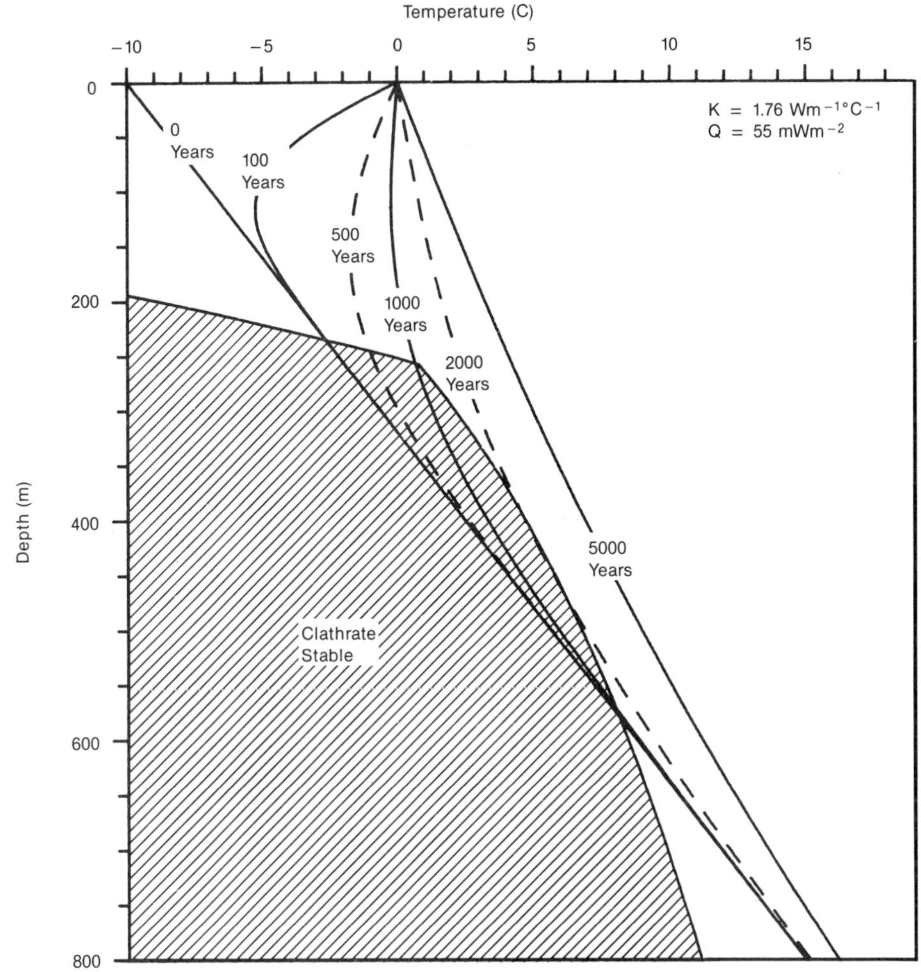

Fig. 54. Development of subsurface temperature for a stepwise warming of 10°C following an ocean incursion over sediments with a low thermal conductivity.

of a cubic meter of fully filled clathrate requires 4.1×10^8 Jm^{-3}, a quantity of heat considerably higher than the volumetric specific heat of 2.1×10^6 Jm^{-3} $°C^{-1}$. The cooling of the dissociating clathrate sets up a thermal gradient between the clathrate and the surrounding sediments, initially drawing heat from sediments above and below. The time scale Δt for the dissociation of a layer of thickness Δt is approximately

$$\Delta t = (\Delta z)^{2/4}\alpha^*,$$

where $\alpha^* = K/(\rho c + L)$, and h is the heat of dissociation of clathrate per unit volume. For a thermal conductivity of 1.0 Wm^{-1} C^{-1} in ocean sediments with 0.5 porosity,

$$\alpha^* = 7.7 \times 10^{-2} \text{ m}^2/\text{yr}.$$

The dissociation of a 20-m layer of clathrate would take roughly 1,300 years. Because this time is short compared to the time scale for sea level drop, the latter would control the rate of methane release. The clathrate zone would erode more rapidly in the upper layer; at the bottom of the clathrate zone, the large thermal inertia of the underlying sediments would cause the time scale for dissociation to be longer. Further, it is likely that released gases from the bottom erosion would be trapped in place by the seal of the overlying clathrates.

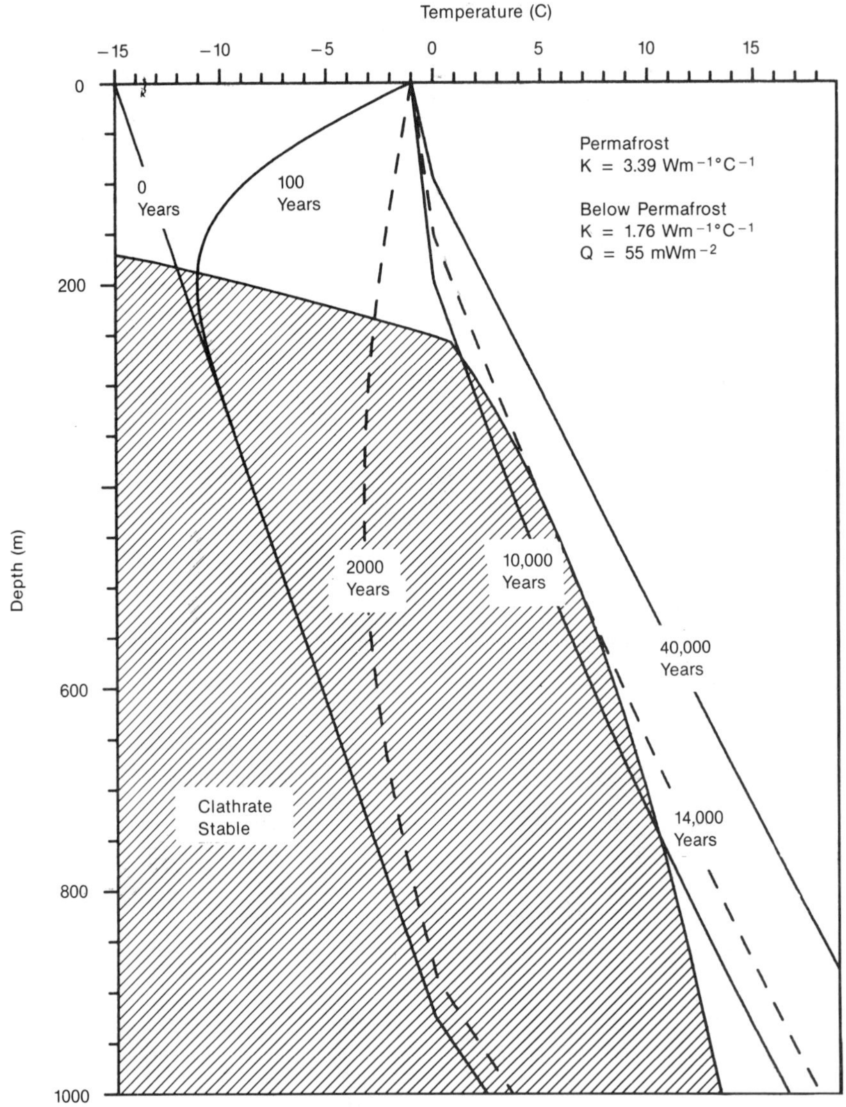

Fig. 55. Development of subsurface tempertaure field following an initial stepwise warming of 14°C. The initial permafrost region extends to a depth of 920 m; the initial clathrate stability zone extends to a depth of 1,200 m.

In summary, major shifts in surface conditions connected with glaciation had a considerable impact on subsurface clathrates and the mass of carbon stored in them. The mass of carbon involved in the release of methane is clearly large enough—hundreds of gigatons—to have had significant impacts on atmospheric composition, and thus on the radiative properties of the atmosphere. In the atmosphere, released methane converts to carbon dioxide over a time scale of tens of years. The time scales for the movement of methane out of and into clathrate are thousands and tens of thousands of years, respectively. These time scales are determined solely by the thermal properties of sediments, which are known to within a factor of two, and the depth of the clathrate stability zone, which is known to a few percent. The amount of carbon stored within the clathrate stability zone is far more speculative, though observations taken over the last two decades in both ocean sediments and permafrost regions suggest that clathrates are ubiquitous. If this conclusion is correct, clathrates must have played a major

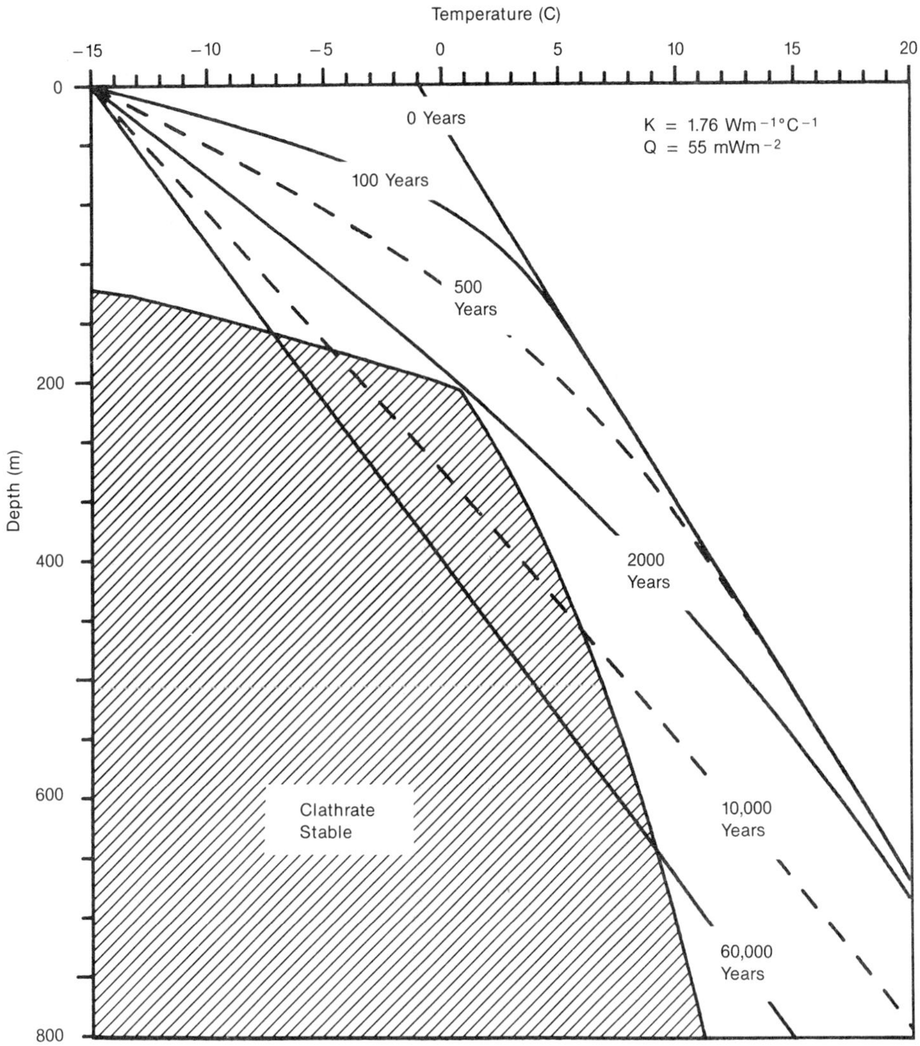

Fig. 56. Development of clathrate stability zone as a result of surface cooling. At the initial time the surface temperature drops by 14°C.

role in regulating the composition and climate-controlling radiative properties of the atmosphere during the ice ages.

DETECTING GREENHOUSE WARMING

Introduction

The previous section dealt in part with the need to include major feedbacks into the atmospheric system when modeling climate if ice age climate change is to be explained. All the postulated feedbacks involve important biospheric functions such as photosynthesis in the upper ocean, or the action of methanogenic bacteria on buried organic matter in sediments. The ability to predict future climate will depend on developing an understanding of the interaction of the biosphere with the atmosphere, a relatively neglected aspect of current climate research.

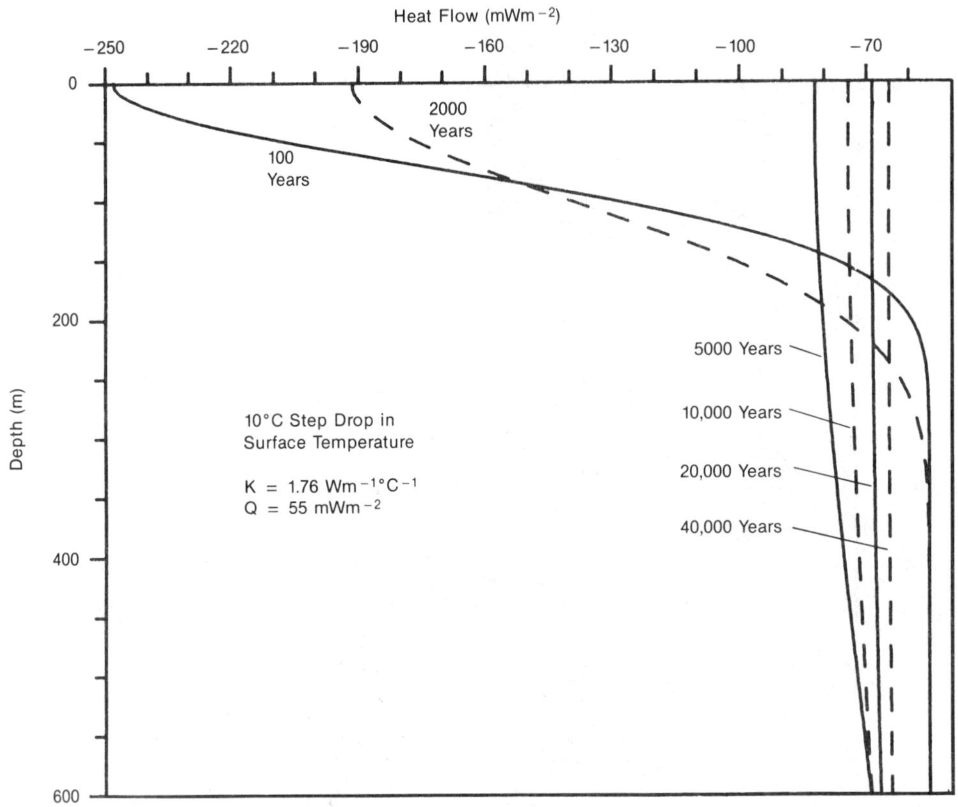

Fig. 57. Depth variation of heat flow as a function of time for a stepwise decrease in surface temperature after 40 kyr. The surface heat flow is 65 mWm $^{-2}$ as contrasted with the equilibrium heat flow of 55 mWm $^{-2}$.

In order to predict future climate trends, we need first to determine positively that the climate is currently changing. If the change is in the warming direction, is the subsequent warming a result of the shifting composition of the earth's atmosphere? In the section on Changing Atmospheric Composition, I described the evidence for warming. The present section will focus on the statistical significance of the observed climate changes, the degree to which observed changes can be confidently attributed to the changing composition of the atmosphere, and the types of future observations that would permit a robust assessment of trends.

Attributing shifts in atmospheric composition to industrial and agricultural activities is a much simpler problem than attempting to attribute climate change to compositional changes. In the former case, the record of air trapped in glaciers shows that the concentrations of methane and carbon dioxide remained constant over thousands of years. The most recent observed shift in atmospheric composition is correlated temporally with the beginning of the industrial revolution. Budgets for various constituents of the atmosphere can be created (see the section on Sources and Sinks for Greenhouse Gases), and these agree relatively well with observation. Thus, for compositional change, a reasonably secure model can be created. The situation is far different with respect to warming. The effects of compositional change on the atmosphere's radiation budget are well known, but feedback effects are only rudimentarily modeled, with some feedbacks—particularly those associated with the biosphere—remaining very much a mystery.

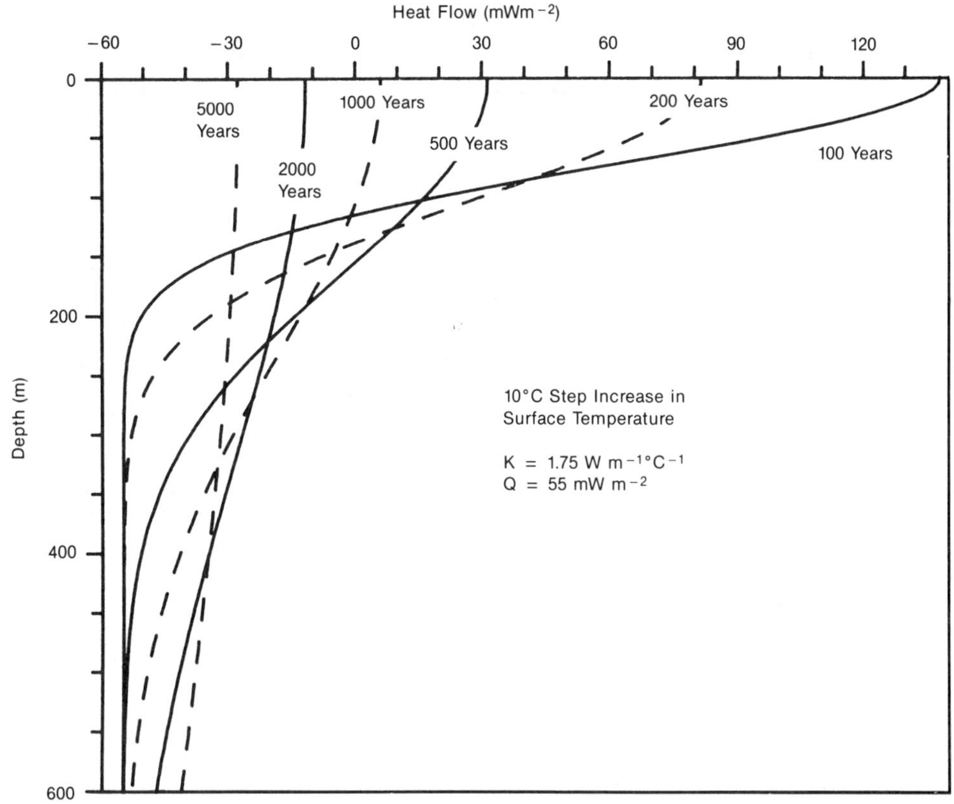

Fig. 58. Depth variation of heat flow as a function of time resulting from stepwise increase of temperature of 10°C. For the first 1,200 years, the flow of heat is downward from the surface.

Statistics of Global Temperature Change

The 108-year temperature record compiled by Hansen and Lebedeff (1987, 1988) (see Fig. 9) will be used to illustrate various considerations arising from questions concerning statistics and attribution. The correlation function for the global average temperature is shown in Fig. 60, where the correlation function is

$$C(\tau) = \frac{1}{\sigma^2} E[T(t + \tau)\, T(t)],$$

and σ^2 is the variance for the record. The correlation function determines the elements of the 108-by-108 covariance matrix **R**

$$C = \sigma^2 \begin{bmatrix} 1 & C(1) & \cdots & C(N-1) \\ C(1) & 1 & & \vdots \\ \vdots & & & C(1) \\ C(N-1) & \cdots & & 1 \end{bmatrix},$$

$$C = \sigma^2 \mathbf{R}$$

which enters into discussions of the statistical validity of an observed shift in temperature. Although the correlation function drops away from unity, correlation is still present at high lags, indicating that the temperatures at intervals separated

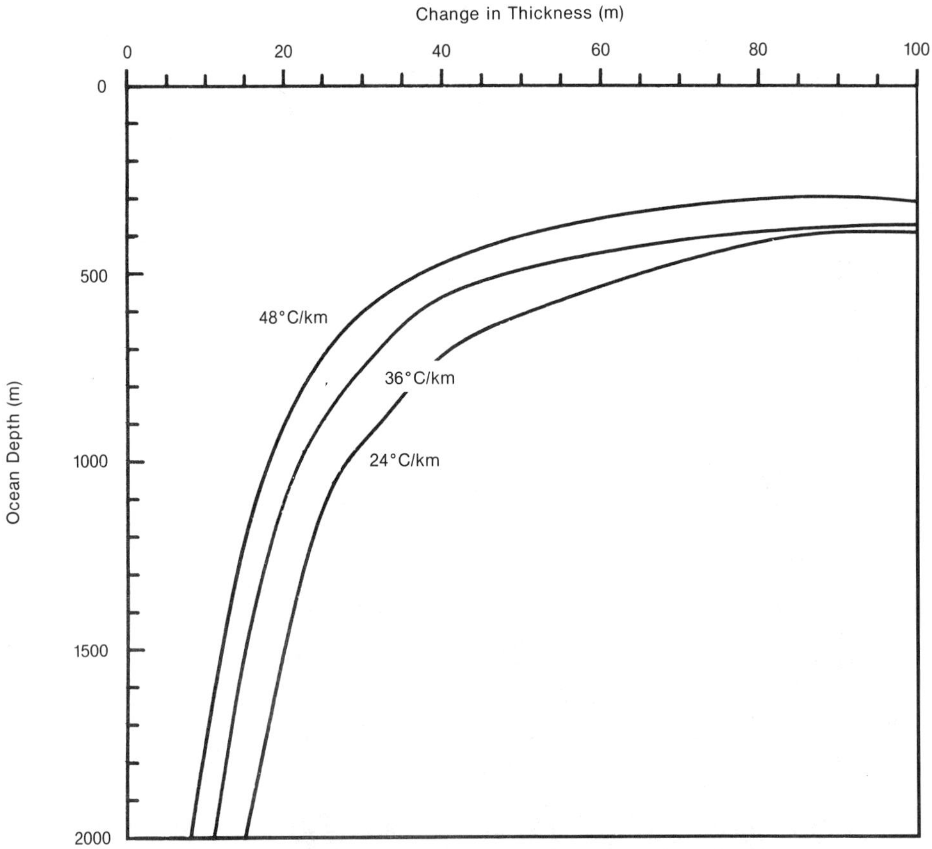

Fig. 59. Change in thickness of clathrate stability zone change results from a 100-m drop in sea level and is a function of ocean depth and temperature gradient.

by several years possess a degree of correlation or persistence. At long enough lags, the correlation function is negative, indicating anticorrelation. If the temperatures were uncorrelated, the covariance matrix **R** would be diagonal, and the terms along the diagonal would be the variance.

The power spectrum for global average temperature is shown in Fig. 61. The spectrum shows most of the power at low frequencies, a feature that is typical of many geophysical records. Power at low frequency implies long periods of change. Distinguishing these long-term changes from shifts in the mean is problematic, though these shifts will contribute to the low-frequency power.

Given the general upward trend of the temperature curve, there is a temptation to fit a least squares line to the data. However, the correlation matrix contains large off-diagonal terms, making classical least squares inappropriate. The correct procedure for dealing with this case is described later in this section. For the present, we adopt a more robust procedure than least squares by minimizing the absolute deviation of the observed temperatures from a straight line. The model is a simple straight line:

$$T(t, a, b) = a + bt,$$

where t is the time measured in years. The merit function to be minimized is

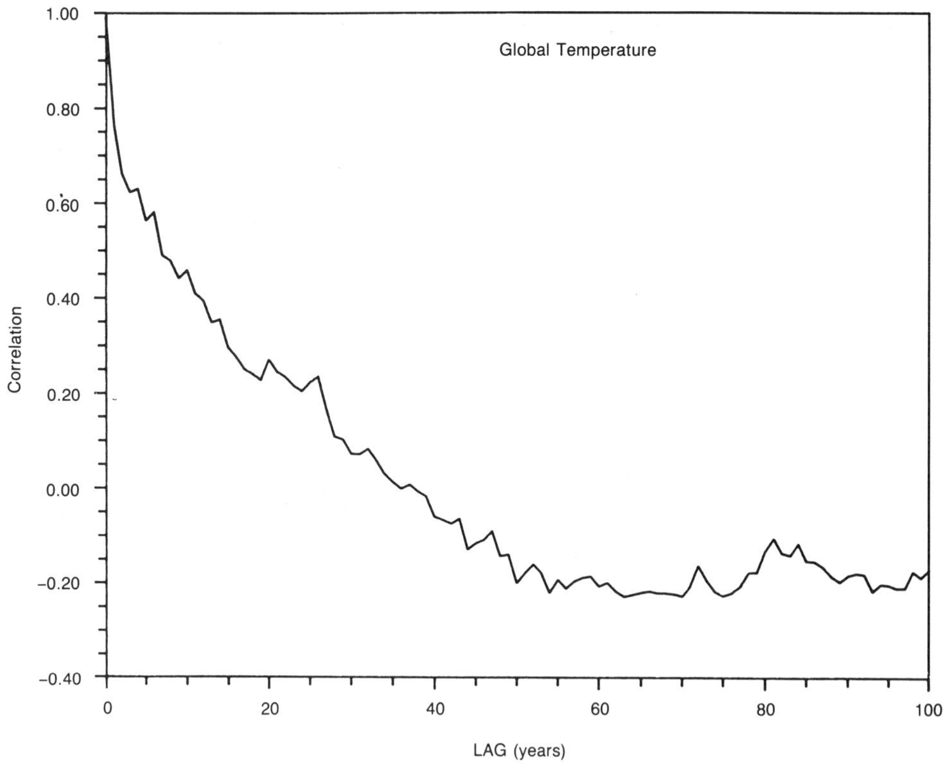

Fig. 60. Autocorrelation function for the variation of global average temperature shown in Fig. 9.

$$\sum_{i=1}^{N} | T_i - a - bt_i | .$$

The calculation is simplified by noting that the median C_M of a set of numbers C_i is also the value that minimizes

$$\sum_{i=1}^{N} | C_i - C_M | .$$

For a fixed slope b, the value of a, which minimizes the merit function, is

$$a = \text{median } (T_i - bt_i).$$

The slope b is determined by

$$\sum_{i=1}^{N} t_i \, Sgn(T_i - a - bt_i).$$

Figure 62 shows the result of fitting a straight line that minimizes the absolute deviation to the average global temperature data. For the complete period of 108 years, the best straight-line fit, obtained by using the absolute deviation criterion, is 0.6°C per century. However, inspection of the curve reveals that the increase in temperature has not been steady over time. Indeed, a marked decrease in temperature appears to begin in the 1940s. Figure 62 also shows the fit of three segments of straight lines that minimize the absolute deviation; the issue becomes whether a model with one, three, or n lines is most appropriate.

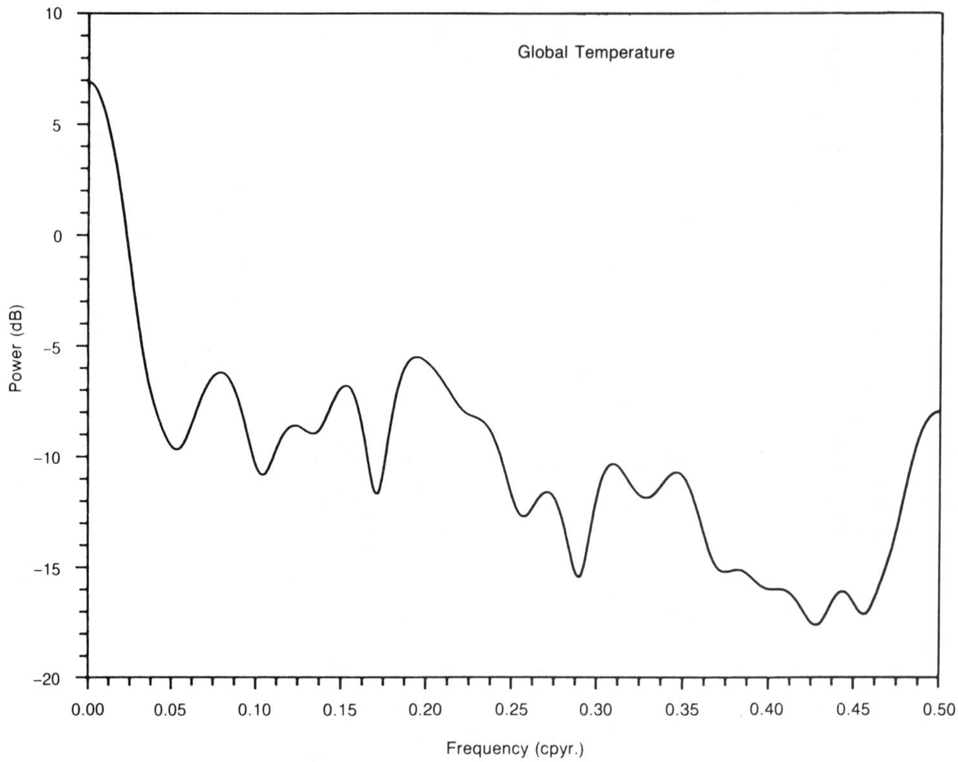

Fig. 61. Power spectrum of global average temperature normalized to unit variance (see Fig. 9).

The issue of detecting a change in mean in a time series is equivalent to determining whether a time series is stationary or non-stationary. One approach is to examine cumulative sums in order to detect shifts in the mean. For a given time series T_i, the mean $\overline{T}_i$ is defined by

$$\overline{T} = \frac{1}{N} \sum_{i=1}^{N} T_i;$$

the cumulative sum is

$$T_p = \sum_{i=1}^{p} T_i - p\overline{T} \; ; \; T_N = 0.$$

If there is no change in mean, the quantity T_p fluctuates about zero. Alternatively, if the time series is a pure linear trend,

$$T_1 = 1, T_2 = 2, \ldots T_N = N,$$

then the cumulative sum is

$$T_p = \frac{p^2 - pN}{2},$$

a concave, upwards function of p.

The cumulative sum for the global mean temperature record is shown in Fig. 63. As the graph shows, the mean of this record for the first 45 years was less than the mean for the entire record, while that of the second half was greater than the global mean. This intrinsic division suggests breaking the record into two

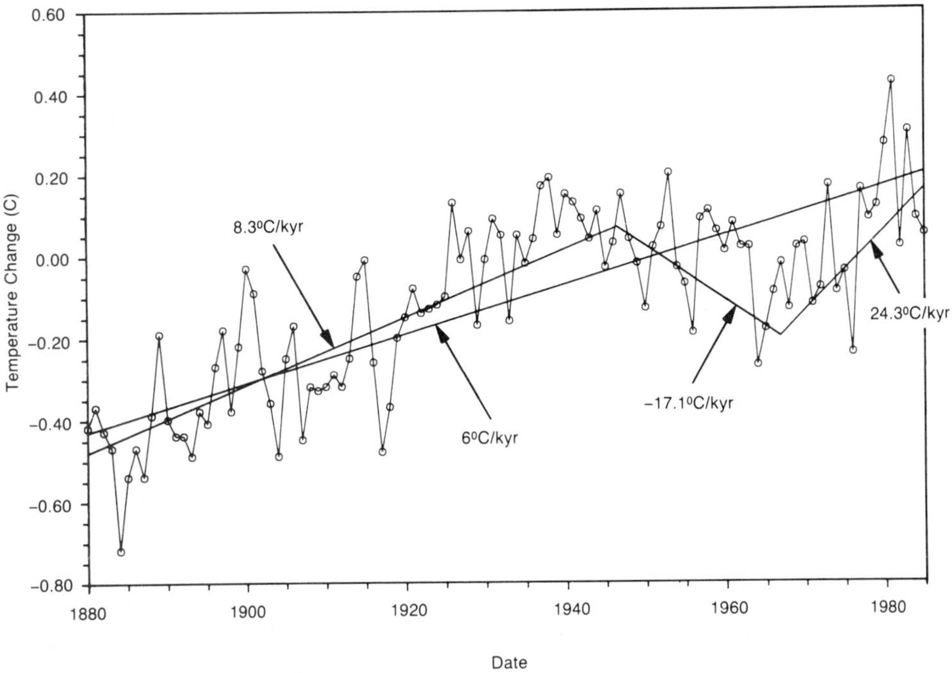

Fig. 62. Fitting of global average temperature data with straight lines using minimization of the absolute deviation.

parts at the minimum of the cumulative-sum curve, and then using the Student-t test to determine if the means differ significantly in the two segments of the record. The procedure can be continued by using stretches of the record identified through cumulative sums. When the stretches no longer have significantly different means at the 0.001 level (according to the Student-t test), the procedure is terminated. The results of the procedure are shown in Table 13. Clearly, the mean-average global temperature has shifted temporally but not monotonically. The mean during the 1943 to 1970 period was significantly lower than the mean before or afterwards. The remaining issue is whether or not the observed changes can be attributed to changing atmospheric composition.

Between 1880 and 1942, the mean increased by about 0.5°C. The fossil fuel contribution to atmospheric carbon dioxide grew at about 4.3% per year between 1860 and 1913, but at only 1.5% per year from 1913 to 1950. After 1950, up to 1973, growth in fossil fuel use was rapid (4.5% per year). Yet it was during this time that global average temperature actually decreased. These discrepancies show that the response of the atmosphere to shifting atmospheric composition is complex, and does not relate directly to changes in greenhouse gas concentrations.

Detecting a Trend in Noise

The identification of a trend in the global temperature record is a problem of detecting a signal in "noise." The solution hinges on devising a suitable statistic. The method outlined below is perfectly general and will work for any specified trend, but for simplicity only a linear trend will be examined. A linear trend can also be justified on physical grounds. The concentration of the principal infrared absorber, carbon dioxide, is increasing exponentially. However, the increase of absorption is only logarithmic in concentration, because the significant CO_2 absorption bands are saturated. Thus, to a rough approximation, the temperature is expected to increase linearly in time as CO_2 increases exponentially. This approximation does not hold for gases such as CFCs, for which the absorption

Table 13. Changes in global mean temperature in the period 1880 to 1988.

Period	Mean Temperature (°C)	Standard Deviation (°C)
1880-1883	-0.495	0.11
1884-1892	-0.392	0.09
1893-1909	- 0.254	0.13
1910-1942	0.022	0.10
1943-1970	-0.057	0.12
1971-1988	0.183	0.12

bands lie in clear windows. The time series for the temperature is taken as a stationary Gaussian process. As noted above, the autocorrelation function shown in Fig. 60 is associated with the covariance matrix $\mathbf{C} = \sigma^2\mathbf{R}$; the diagonal terms in the matrix $\mathbf{C}$ represent the variance σ^2 of the total record. The estimation begins with a sample that is $N = 108$ points long (1880 to 1987). The statistical question is whether the sample is a noise process alone, or one in which a linear trend has been added to the noise. Suppose that the trend has a slope b, and that $b\mathbf{k}$ was added to the record at the time k, where $\mathbf{k}$ is the N-long column vector whose transpose is

$$\mathbf{k}' = (0, 1, 2 \ldots, N - 1),$$

and the observed temperature record is the column vector $\mathbf{T}$, whose transpose is

$$\mathbf{T}' = (T_0, T_1, T_2, \ldots T_k, \ldots T_{N-1}).$$

For simplicity in notation, the inverse of $\mathbf{C}$ is denoted by $\mathbf{S}$ and the inverse of $\mathbf{R}$ by dimensionless matrix $\mathbf{B}$:

$$\begin{aligned} \mathbf{S} &= \mathbf{C}^{-1} \\ \mathbf{B} &= \mathbf{R}^{-1} = \sigma^2\,\mathbf{C}^{-1} = \sigma^2\mathbf{S}. \end{aligned}$$

The null hypothesis H_o states that the sample was selected from a noise process with no linear trend; H_1, the causal hypothesis, asserts that the sample was selected from a process that included noise plus a linear trend. By standard theory, the probability density for the sample—given the null hypothesis—is the multivariate Gaussian

$$\frac{(det\ \mathbf{S})^{\frac{1}{2}}}{(2\pi)^{\frac{N}{2}}} \exp\left(-\frac{1}{2}\,\mathbf{T}'\,\mathbf{S}\mathbf{T}\right)$$

(Cramér, 1946).

The probability density under the causal hypothesis is

$$\frac{(\det\ \mathbf{S})^{\frac{1}{2}}}{(2\pi)^{\frac{N}{2}}} \exp\left[-\frac{1}{2}(\mathbf{T}' - \mathbf{T}'_*)\,\mathbf{S}\,(\mathbf{T} - \mathbf{T}_*)\right],$$

where $\mathbf{T}_*$ is the column vector $b\mathbf{k}$. Thus, the log odds or log likelihood of H_1 over H_0 is

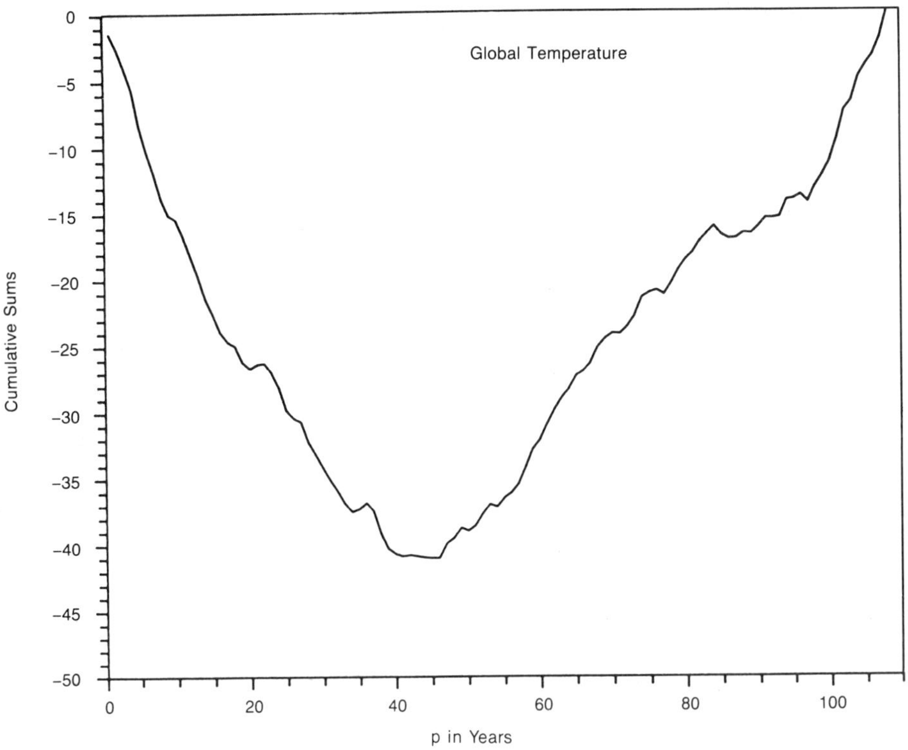

Fig. 63. Cumulative sum for global average temperature record.

$$L = \mathbf{T}'\,\mathbf{S}\,\mathbf{T}_* - \frac{1}{2}\,\mathbf{T}_*'\mathbf{S}\mathbf{T}_*$$
$$. = \frac{b}{\sigma^2}\,\mathbf{T}'\,\mathbf{B}\,\mathbf{k} - \frac{1}{2}\,\frac{b^2}{\sigma^2}\,\mathbf{k}'\,\mathbf{B}\,\mathbf{k}.$$

The slope b of the trend is determined in the maximum likelihood sense by selecting the value of b that maximizes L; i.e.,

$$b = \frac{\mathbf{T}'\,\mathbf{B}\,\mathbf{k}}{\mathbf{k}'\,\mathbf{B}\,\mathbf{k}}.$$

If the process is uncorrelated Gaussian noise, then $\mathbf{B}$ is the identity matrix $\mathbf{I}$ and

$$\mathbf{S} = \frac{1}{\sigma^2}\mathbf{I}.$$

For uncorrelated Gaussian noise, the expression for the slope is the same as that for ordinary linear regression

$$b = \frac{\mathbf{T}'\,\mathbf{k}}{\mathbf{k}'\,\mathbf{k}}.$$

The normalization term $\mathbf{k}'\,\mathbf{k}$ is

$$\mathbf{k}'\mathbf{k} = \sum_{k=1}^{N-1} k^2 = \frac{(N-1)N(2N-1)}{6}$$
$$\cong \frac{N^3}{3}.$$

Routine calculations then give the expected value of L for both the causal and null case:

$$E[L \mid \mathbf{H}_1] = \frac{1}{2} \mathbf{T}'_* \mathbf{S} \mathbf{T}_* = \frac{1}{2} \frac{b^2}{\sigma^2} \mathbf{k}' \mathbf{B} \mathbf{k}$$

$$E[L \mid \mathbf{H}_o] = -\frac{1}{2} \mathbf{T}'_* \mathbf{S} \mathbf{T}_* = -\frac{1}{2} \frac{b^2}{\sigma^2} \mathbf{k}' \mathbf{B} \mathbf{k}$$

$$\text{var}(L \mid \mathbf{H}_1) = \text{var}(L \mid \mathbf{H}_o) = \mathbf{T}'_* \mathbf{S} \mathbf{T}_* = \frac{b^2}{\sigma^2} \mathbf{k}' \mathbf{B} \mathbf{k}.$$

For Gaussian, uncorrelated noise the expected value of the log likelihood for the hypothesis of a linear trend is

$$E[L \mid H_1] = \frac{1}{6} \frac{N^3 b^2}{\sigma^2}.$$

The log odds of a trend with an estimated slope increase with the cube of the record length and are inversely proportional to the variance. If the noise is correlated, as is the case for the global temperature record (see Figs. 9 and 60), then account must be taken of the non-diagonal covariance matrix. This point is illustrated by contrasting $\mathbf{k}' \mathbf{B} \mathbf{k}$ and $\mathbf{k}' \mathbf{k}$ for the 108-point temperature record:

$$\frac{\mathbf{k}' \mathbf{B} \mathbf{k}}{\mathbf{k}' \mathbf{k}} = 0.09.$$

Thus, for an estimated slope and variance, the assumption of uncorrelated Gaussian noise would lead to an expected value for the log likelihood that is ten times larger than the estimate derived using the observed covariance matrix.

In order to use the methodology described above, we should ideally have a long temperature record that precedes the CO_2 era, from which good estimates of the covariance matrix could be found. But because the best temperature record available is that of the last 100 years or so, we are forced to use data that might be tainted. It seems sensible, however, to simply assume that the available temperature record does not reflect a trend due to CO_2 and other gases, and so gives us a reliable covariance estimate. It is this covariance matrix that was used to compute the log likelihood of a trend, and it is from this standpoint that Table 14 should be interpreted. Even if no trend is supposed, large log-factors favor the presence of a trend.

Table 14 gives the variation of both the estimated slope of the trend and the log likelihood, as longer intervals of the temperature record are used to obtain the estimates. The covariance matrix was estimated on a circle by taking a Fourier transform (FFT) of the 108-year temperature record, forming the power spectrum, taking the cosine transform of the power spectrum to obtain the covariance function, and then casting the covariance function into a Toeplitz matrix. Using only the first 30 years of the record, the log likelihood is only 1.2, corresponding to odds of 3.3 to 1 that the record consists of noise plus a trend rather than noise alone. By 1969, even after a cooling interval of almost 30 years (see Fig. 9), the odds in favor of the hypothesis of a trend plus noise grow to 1,250 to 1. After 108 years, the evidence in favor of a trend plus noise increases the odds to a highly convincing 1.74×10^5 to 1.

Hansen and Lebedeff (1988) indicate that there is a small urban heat-island effect (about 0.1° to 0.2°C) in the data used to construct Fig. 9 (see also Karl and Jones, 1989, and Karl et al., 1989). The existence of such an urban heat-island effect would lower the estimate of the appropriate slope for the hypothesized trend. However, we can estimate the robustness of the conclusion that there is a linear trend in the temperature record by supposing that the estimated slope should be 0.3°C/century rather than 0.565°C/century. Because the likelihood scales as the square of the slope, the likelihood for a slope of 0.3°C is 3.4, with

Table 14. Detection of a linear trend in global average temperature.

Period	Record Length (years)	Variance $(°C)^2$	Slope (°C/ century)	Log Likelihood	S.D.* of Log Likelihood	Odds	S.D.* of Odds
1880–1909	30	0.0212	0.339	1.20	1.55	3.32	4.7
1880–1939	60	0.0428	0.673	5.95	3.45	384	31
1880–1969	90	0.0451	0.502	7.13	3.77	1250	44
1880–1987	108	0.0498	0.565	12.07	4.91	1.74×10^5	136

*S.D. = Standard Deviation

odds in favor of the trend hypothesis of 30:1. The conclusion that a trend exists in the global average atmospheric temperature seems inescapable, even with imperfections in the temperature record.

Table 14 shows that the estimated slope of the record varies as the observational period increases. Although the next section presents an argument that such behavior is expected for a highly nonlinear atmospheric system with long periods of persistance, the hypothesis testing procedure outlined above follows the classical Bayesian pattern. Confidence in the true hypothesis increases as more observations become available, and the parameters, in this case the slope b, are adjusted. Note that the variance increases monotonically with the length of the record and is fairly close to the expected rate. Indeed, if one assumes that the time series consists of a random process with variance τ^2, to which a trend of slope b has been added, then a simple calculation shows that the expected value of the variance of the time series is

$$\mathbf{E}\,[\text{var}] \approx \sigma^2 + \frac{N^2\, b^2}{12}.$$

From the values in Table 14 we obtain

Period	E[var]	τ^2
1880–1909	$\tau^2 + 0.0008$	.0204
1880–1939	$\tau^2 + 0.0135$	.0293
1880–1969	$\tau^2 + 0.0170$	.0281
1880–1987	$\tau^2 + 0.0310$	.0188

with a weighted mean for τ^2 of 0.0241 $(°C^2)$.

Question of Attribution

The apparent upward trend in global temperature and the statistically significant shifts in global mean temperature have led to speculation that greenhouse warming is here, and that the observed increases are not due to internal fluctuations within the system. The apparent irregularity of the atmosphere is a well-known, fundamental atmospheric characteristic (Lorenz, 1984a). Further, the atmosphere is a highly nonlinear system. Recent

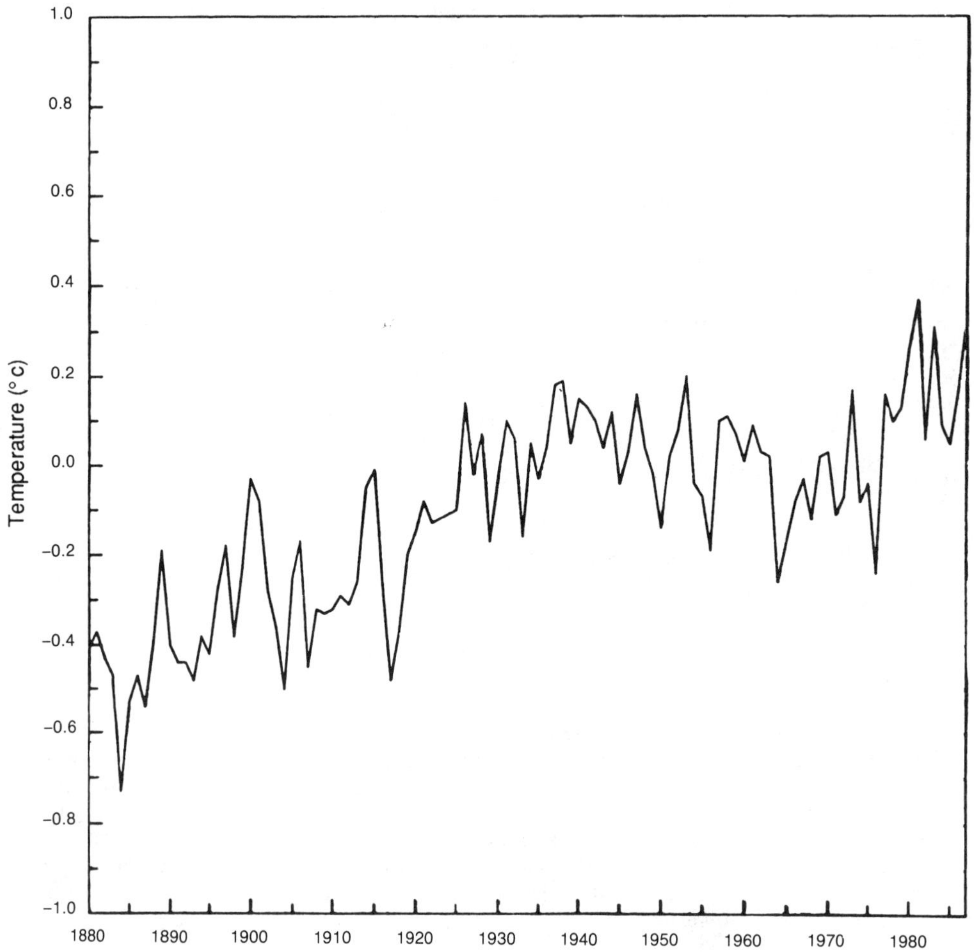

Fig. 64. Variation of the global-mean, annual-mean, surface air temperature for 100 years in a numerical solution of the GISS model II. The temperatures shown are departures in degrees C from a standard value (source: Hansen et al., 1988).

developments in the theory of nonlinear dynamic systems emphasize the atmosphere's variability over a very wide range of time scales. These fluctuations are a manifestation of what has lately been termed chaos. Chaotic behavior is characterized by an aperiodic or continuous spectrum. In chaos, deterministic, governed behavior leads to an apparently random motion.

General circulation models (GCMs) with long-term integrations clearly exhibit chaotic irregularity. Figure 64 shows a 100-year series of annual, global-mean temperatures produced by the GISS Model II in a control run in which there are no variable external influences (Hansen et al., 1988). The model consists of nearly 100,000 coupled nonlinear equations. Besides showing a highly irregular form, it bears a close resemblance to the global-mean temperature curve (Fig. 9).

The irregularity that characterizes the atmosphere also shows clearly in simpler models. Lorenz (1984b) has developed a low-order model in 27 variables. The physical variables include the mixing ratios of water vapor and liquid water, as well as temperature, pressure, and the usual dynamic variables. In Lorenz's model, the atmosphere and ocean exchange heat and water through evaporation and precipitation. The model also produces clouds, which reflect incoming solar

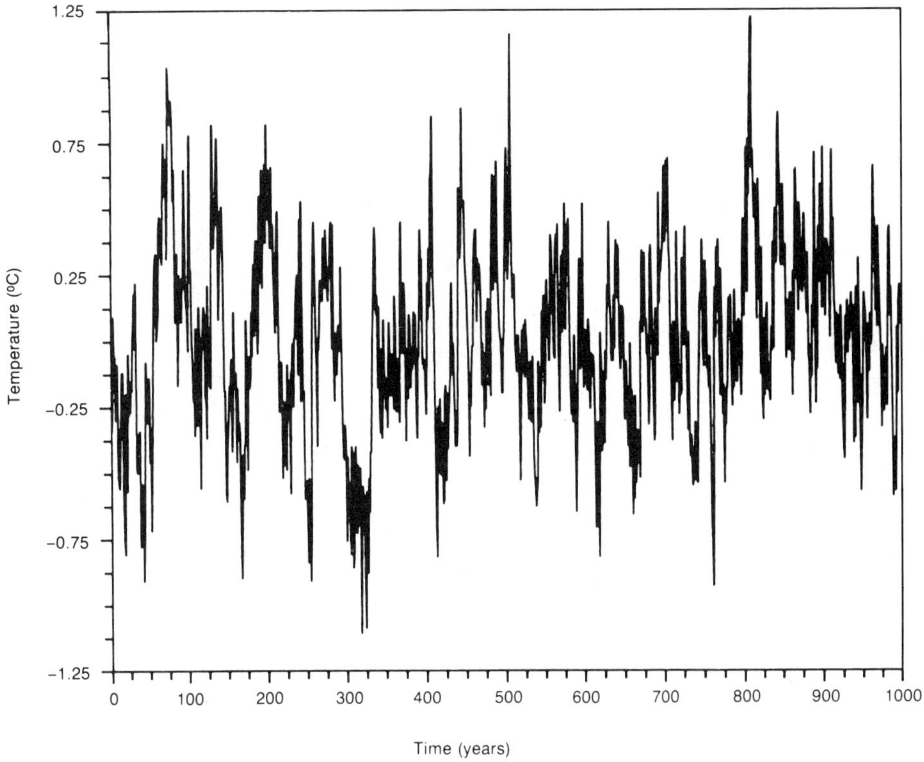

Fig. 65. Variations of the global-mean, annual-mean, sea level air temperature for 1,000 years in a numerical solution to Lorenz's 27-variable model of the atmosphere (Lorenz, 1984b).

radiation, while both phases of water absorb and re-emit infrared radiation. Figure 65 shows the model's sea level air temperatures, averaged over the globe and over one-year intervals, for 1,000 consecutive years. The model produces upward and downward trends lasting for several decades, yet there is no variability of external features or of atmospheric composition other than the ever-changing mixing ratios of liquid water and water vapor. The total range of upward and downward trends is 2.5°K, which can be contrasted with the warming of about 0.6°C over the last 108 years.

The autocorrelation function for the 1,000-year record is shown in Fig. 66; it shows similarities to the same function for the global-mean temperature record (compare with Fig. 60). Figure 67 gives the power spectrum for the temperature record shown in Fig. 65. In the power spectrum, the characteristic buildup of power at low frequency can be seen, together with statistically significant peaks at about 20 and 11.5 years. These peaks are reminiscent of the often-claimed detection of solar cycles in observed temperature records.

The similarities between the model and the observed global temperature curve are illustrated by examining shorter sections of the 1,000-year record. In Fig. 68, the temperature between 400 and 600 years is shown. During the 35-year interval between years 415 and 450, a 1.5°C temperature increase occurred, followed by a sharp drop. Similar variability is shown for the 150-year stretch beginning at year 150 (see Fig. 69). The cumulative sums are used to determine the non-stationary character of the record shown in Fig. 70. Dividing the record into the three intervals indicated in Fig. 70 leads to the results shown in Table 15.

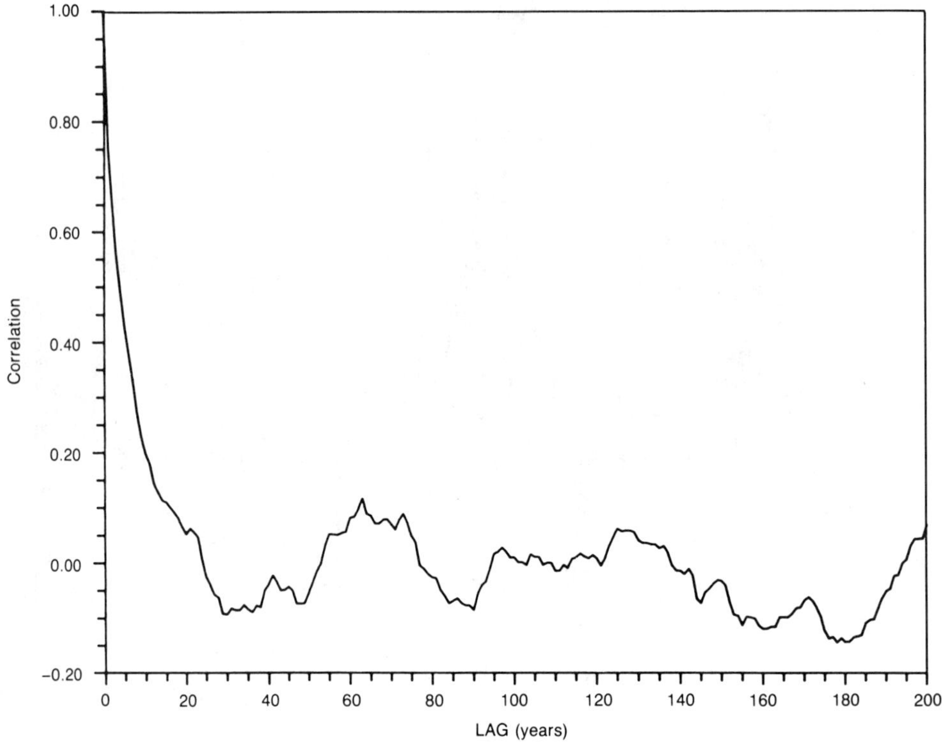

Fig. 66. Autocorrelation function for the 1,000-year record derived from the Lorenz (1984b) 27-variable model.

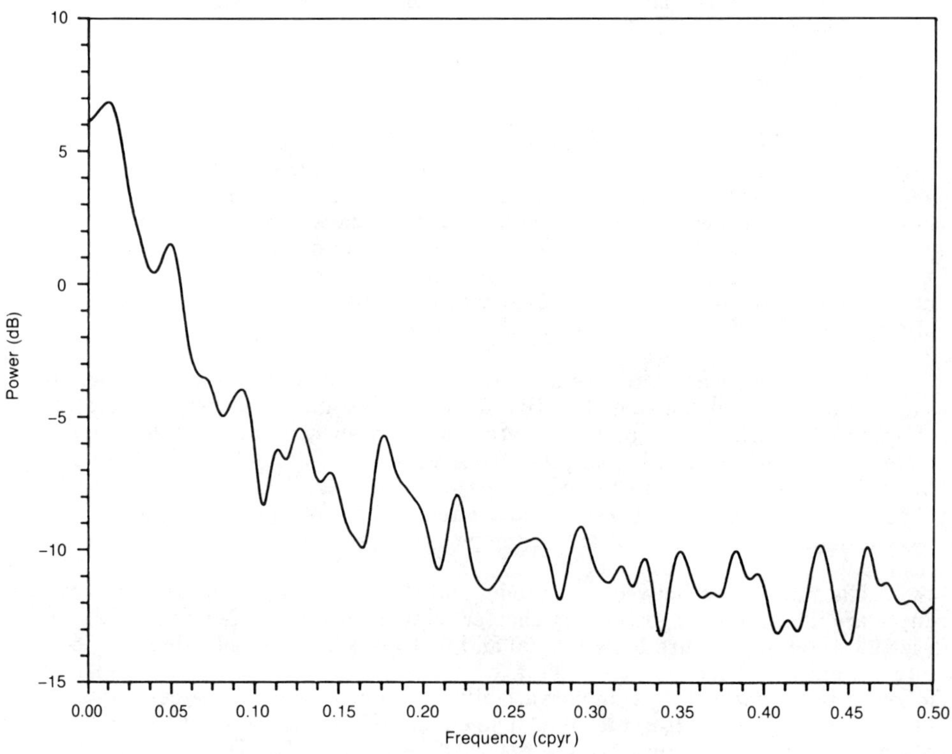

Fig. 67. Power spectrum for the 1,000-year temperature record shown in Fig. 65.

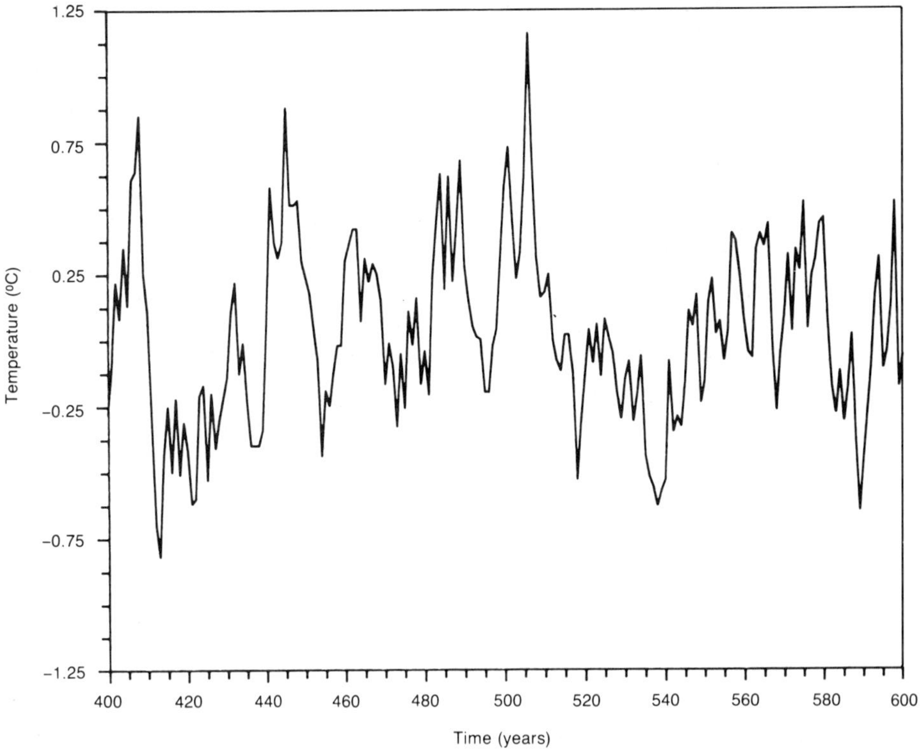

Fig. 68. A 200-year sector of the temperature record shown in Fig. 69.

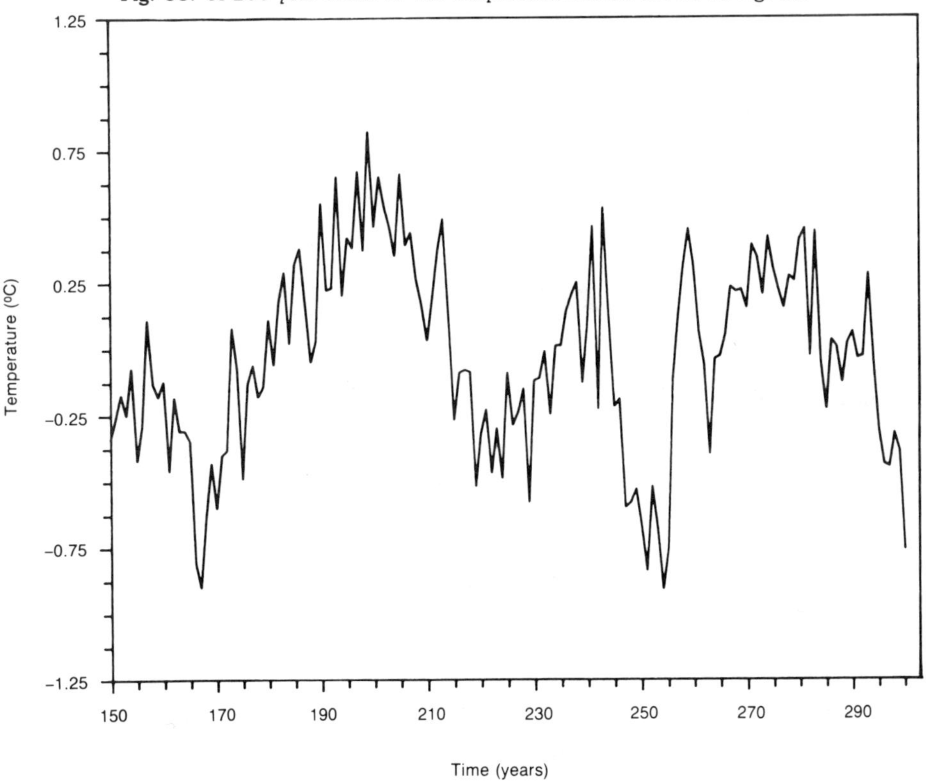

Fig. 69. A 150-year segment of the 1,000-year record shown in Fig. 65.

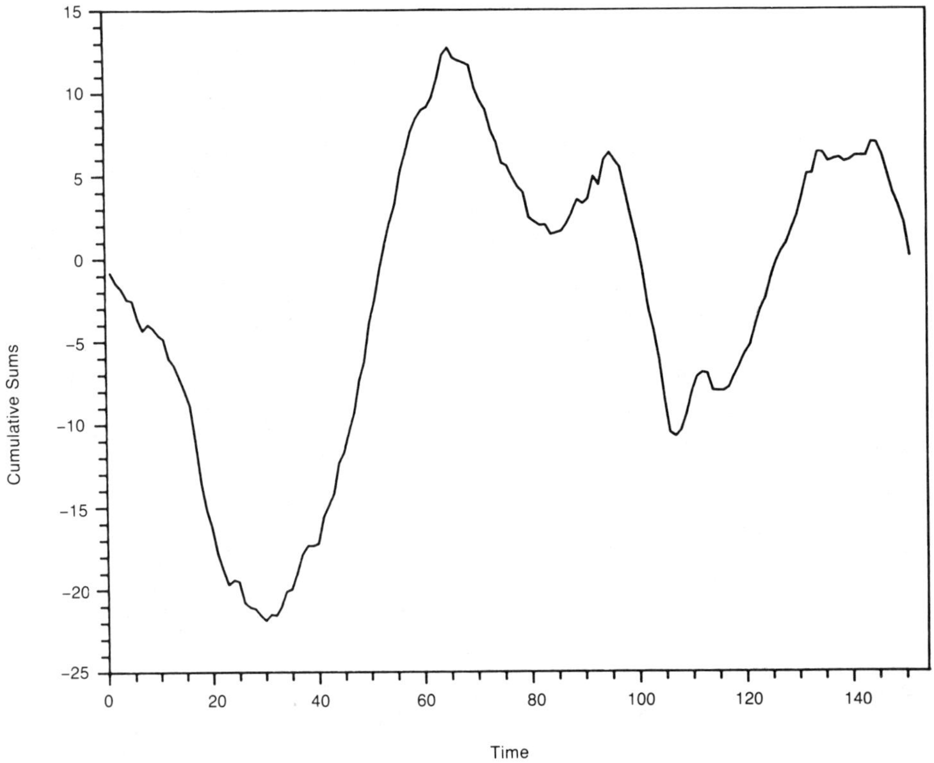

Fig. 70. Cumulative sums for the model global-mean temperature record shown in Fig. 69.

Table 15. Variation of mean temperature in Lorenz model atmosphere*

Period	Mean (°C)	Student-t Test
150-180	−0.28	
		−10.1 (prob = 1×10^{-16})
181-215	0.32	
		7.3 (prob = 1×10^{-10})
216-300	−0.08	

*Shown in Fig. 69

The intervals from 150 to 180 years and 181 to 215 years have means that differ with almost statistical certainty. Although there is a significant variation of the mean, the total variance of the record remains constant. For the record shown in Fig. 65, a 30-year period would be required to establish a trend of one degree per 20 years.

The long period variations are due to the positive cloud albedo feedback process. If the albedo were set to a constant, the observed long-term trends in the model would be eliminated. In the model, higher temperatures reduce the cloudiness and, consequently, the albedo, allowing the temperature to increase further. The heating does not cause the temperature to shoot up or to level off at some extreme value. The moving synoptic systems, which fluctuate chaotically over periods of weeks or months, act as random forcing functions and draw the system back to globally averaged conditions. Although the model is far from an accurate representation of the real atmosphere, it illustrates that chaotic dynamical systems generally show very long period fluctuations. These fluctuations are much longer than any obvious time constants appearing in the governing laws and will develop without external influence.

The difficulties of detecting a warming trend are further illustrated in Fig. 71, using the 27-variable model constructed by Lorenz. The parameter that determines solar insolation has been increased linearly in time, which raises global mean temperature 10°C in 1,000 years. A 200-year segment of this temperature record is shown in Fig. 72. Statistical analysis of the 100-year interval from 440 to 540 shows no change in mean. Thus, in the presence of a 0.01°C/yr warming, the chaotic behavior of the model forces the appearance of a constant mean over a 100-year period.

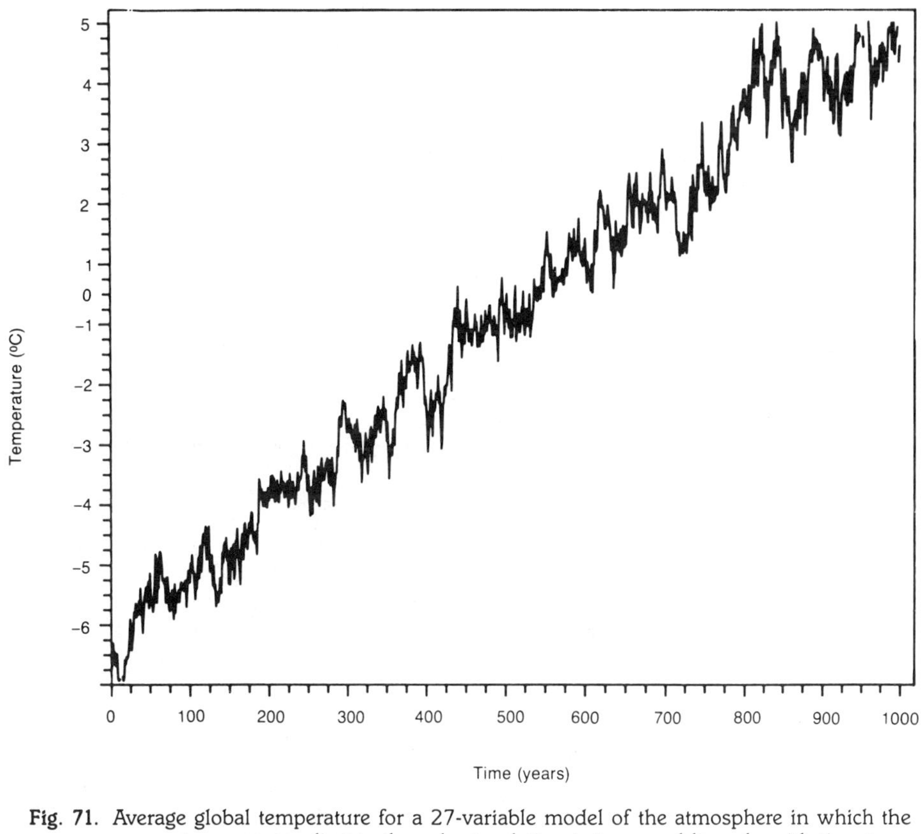

Fig. 71. Average global temperature for a 27-variable model of the atmosphere in which the parameter corresponding to the solar insolation is increased linearly with time to provide a ten-degree increase in temperature over 1,000 years.

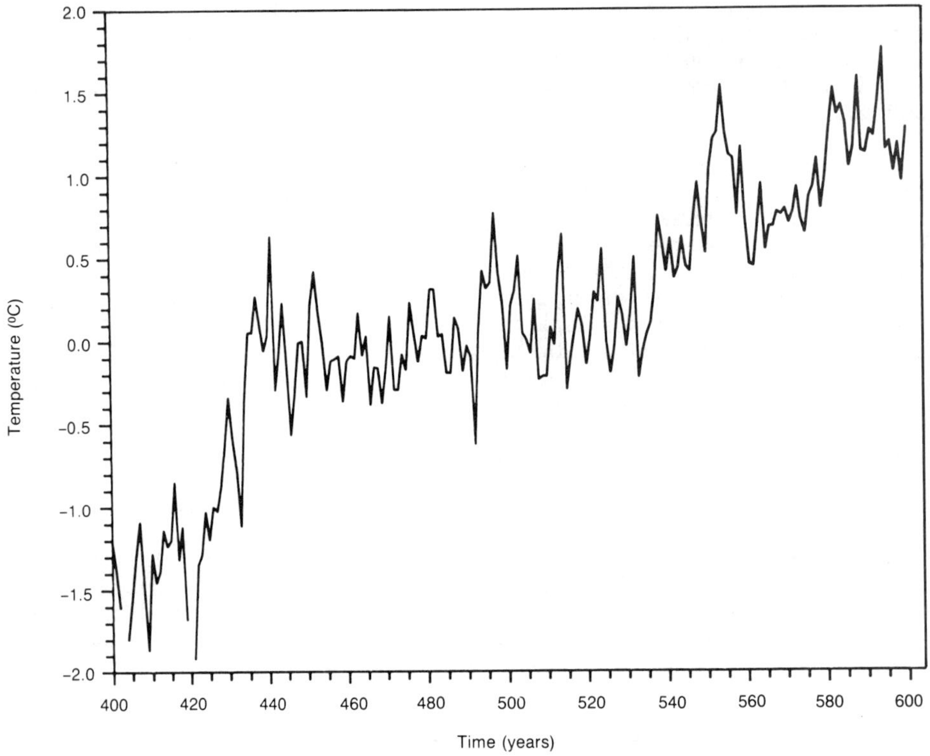

Fig. 72. Global mean temperature for the segment 400 to 600 years shown in Fig. 71.

The attribution problem can be dealt with in part by sharpening the prediction of the models. At some stage, it may be possible to compare the predictions of the models with observations in a statistical way. At present, however, this does not appear realistic. Further, we should attempt to determine if observations on other elements of the atmosphere-ocean system could provide better estimates of any trend in global climate than do measurements of surface air temperature.

Ocean Acoustic Thermometry

Munk and Forbes (1989) have suggested an alternative means of detecting warming trends using very long acoustic paths through the ocean. The travel times of acoustic signals over very long paths (15,000 km) are about 10,000 seconds. For a given source, the observed travel time can be determined to within 1 ms. At the axis of a typical sound-wave guide in the ocean, the sound speed increases by 4 ms per 1°C temperature rise. A mean warming of 0.005°C/yr leads to a reduction in travel time by 140 ms/year for a 1,500-km transmission path. The effect of changes in ocean composition (principally salinity) on the travel time is less by an order of magnitude. The question that remains is whether or not the variation resulting from the chaotic, internal variability of the oceans can be distinguished from an imposed, longer-term trend.

Munk and Forbes (1989) suggest placing a sound source at Heard Island in the Southern Indian Ocean; a source at South Georgia Island in the Atlantic would also achieve the desired long-range paths. Using the observed sound-speed variation as a function of geographic location and ocean depth, one obtains the minimum travel time paths from Heard Island to various oceanographic stations. These minimum travel-time paths are shown in Fig. 73. An acoustic source in the vicinity of Heard Island has direct oceanic paths through all five ocean basins:

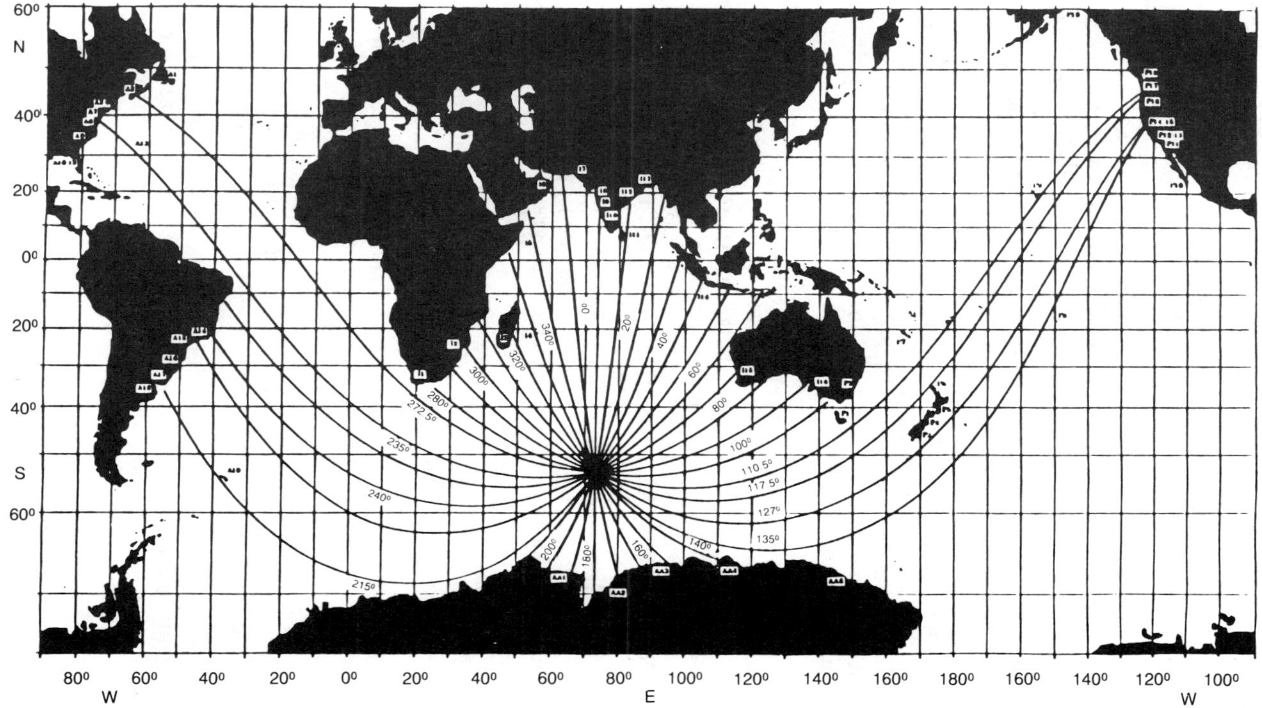

Fig. 73. Refracted acoustic rays traveling a path from a source at Heard Island in the Southern Indian Ocean (source: Munk and Forbes, 1980). The important point is that there are unimpeded paths, both to northern-latitude stations in the Atlantic (Bermuda) and stations in the eastern Pacific.

westward to South Georgia, Brazil, and Bermuda; eastward to Tasmania, New Zealand, Tahiti, Hawaii, San Francisco, and Oregon; northward to Indonesia; and southward to the Antarctic.

Both the acoustic velocity and the depth of the sound channel vary geographically. The sound velocity is low (1,440 ms^{-1}) in the Antarctic Ocean and reaches high values (1,500 ms^{-1}) in the waters coming from the Mediterranean off Gibraltar. The depth of the axis of the sound channel rises from 1 km near Perth, Australia, to about 200 m in the Antarctic waters and again descends to 1 km at mid latitudes. Using data from 8,000 hydrographic stations, Munk et al. (1988) have mapped the position and velocity of the axis of the sound channel, which permits the calculation of the geodesic paths shown in Fig. 73. Measurements of the time taken to travel the long path provide an estimate of the average sound speed and thus of the temperature along the path. Because global warming would warm the ocean to the kilometer depths of the sound channel axis, coupled ocean-atmosphere GCMs can be used to obtain estimates of the effect of global warming on the upper ocean.

Coupled ocean-atmosphere GCMs, developed by Schlesinger and Xingjian (1988), indicate a strong latitudinal dependence for ocean warming. Surface warming increases from the tropics toward the mid latitudes of both hemispheres, with greater depth of penetration in the mid and high latitudes; in the tropics, the penetration is minimized by the upwelling of cold water.

A coupled ocean-atmosphere GCM simulation by Bryan et al. (1988) emphasizes the asymmetry between the two hemispheres; in the Southern Hemisphere, deep warming is greater than shallow warming, but the total amount of heat being added to the oceans is about the same in the two hemispheres. Using a 0.8% per year rate of CO_2 buildup and the GCM described by Manabe and Stouffer (1988) to account for gases other than CO_2, Bryan et al. (1988) obtain an average rate of warming along the Heard Island-Bermuda path (16,000 km) of 4.4 m °C/yr; the Heard Island-Coos Bay, Oregon, path (18,000 km) of 14.6 m °C/yr; and the Heard Island-San Francisco path (17,000 km) of 16.7 m °C/yr (Munk and Forbes, 1989). Because tomographic measurements can provide travel times that have an accuracy of 1 ms at 10^4-km ranges, there should be no problem in detecting the calculated warming. As in all geophysical problems, the difficulty is in distinguishing the longer-term trend from the noise, which will have irregularities at low frequencies.

The internal irregularity of the ocean can be analyzed using computer modeling. Semtner and Chervin (1988) have developed a multi-level, primitive-equation model that exhibits global mesoscale variability. In the southern oceans, the region south of Cape Hope shows a high level of irregularity. As a result, there is much greater fluctuation in the Heard Island-Bermuda path (see Fig. 74) than in the eastward paths. The autocorrelation function for the variation in travel time between Heard Island and Bermuda is shown in Fig. 75; the power spectrum is given in Fig. 76. The record shows significant correlation or autocorrelation at all lags, while the power spectrum shows the characteristic increase toward low frequencies. For comparison, the travel time variation between Heard Island and Coos Bay, Oregon, and its power spectrum are shown in Figs. 77 and 78.

The effect of background noise on estimates of a trend is illustrated in Table 16. The trends have been estimated using the technique of minimizing the absolute deviation. The estimates obtained for San Francisco and Coos Bay are satisfactory, but the background noise leads to errors of a factor of two in the trend for the Heard Island-Bermuda path.

Given the estimates of the background variance provided by the Semtner and Chervin (1988) model, we can calculate the length of time required to

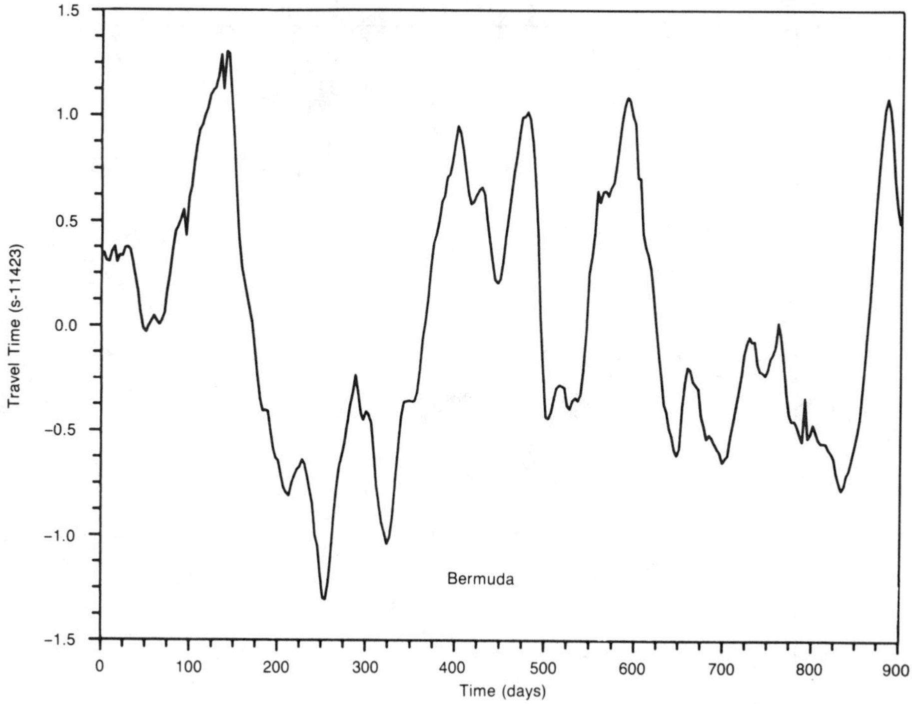

Fig. 74. Variation of calculated travel time for an acoustic path from Heard Island to Bermuda due to ocean irregularities modeled by Semtner and Chervin (1988). A trend of 4 m°C/yr warming due to greenhouse warming has been added to the trend.

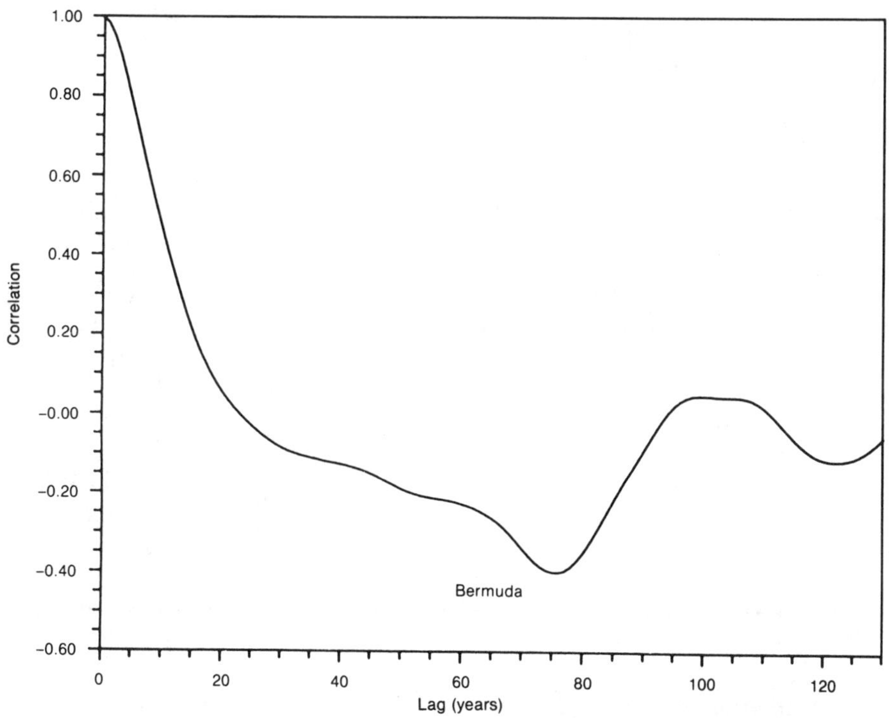

Fig. 75. Autocorrelation of the travel-time curve between Heard Island and Bermuda shown in Fig. 74.

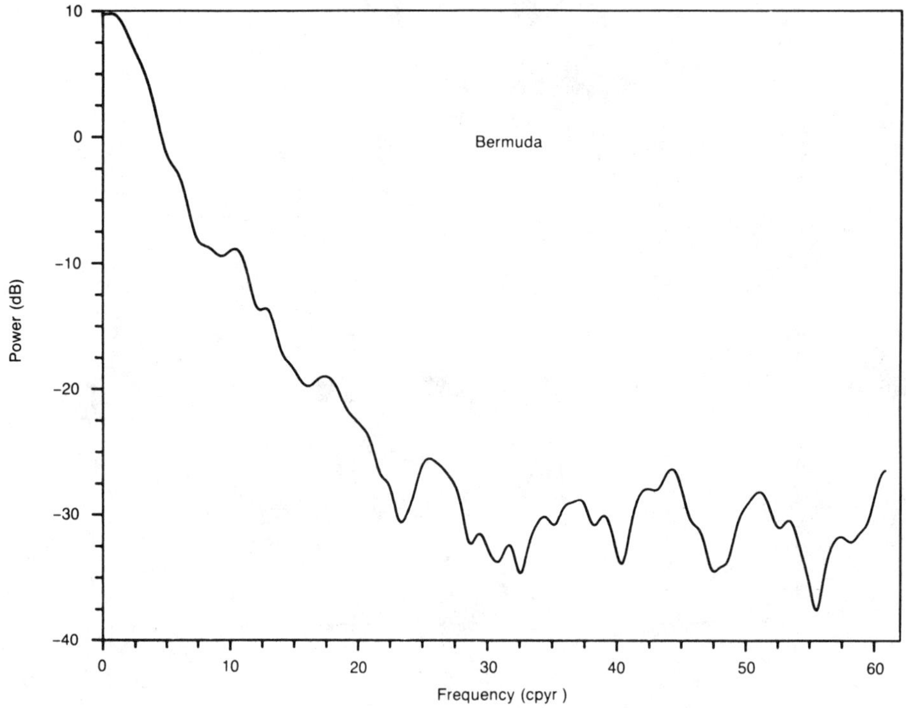

Fig. 76. Power spectrum of acoustic travel-time variation from Heard Island to Bermuda shown in Fig. 74.

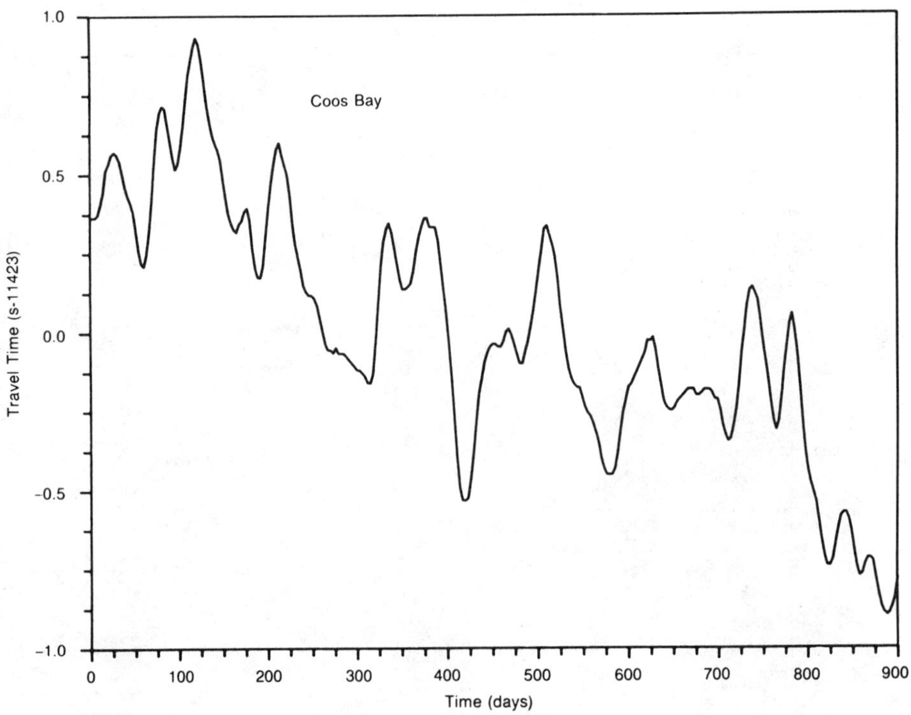

Fig. 77. Travel-time variation for an acoustic path between Heard Island and Coos Bay, Oregon. A trend corresponding to a greenhouse warming of 14.6 m°C/yr has been added to the record.

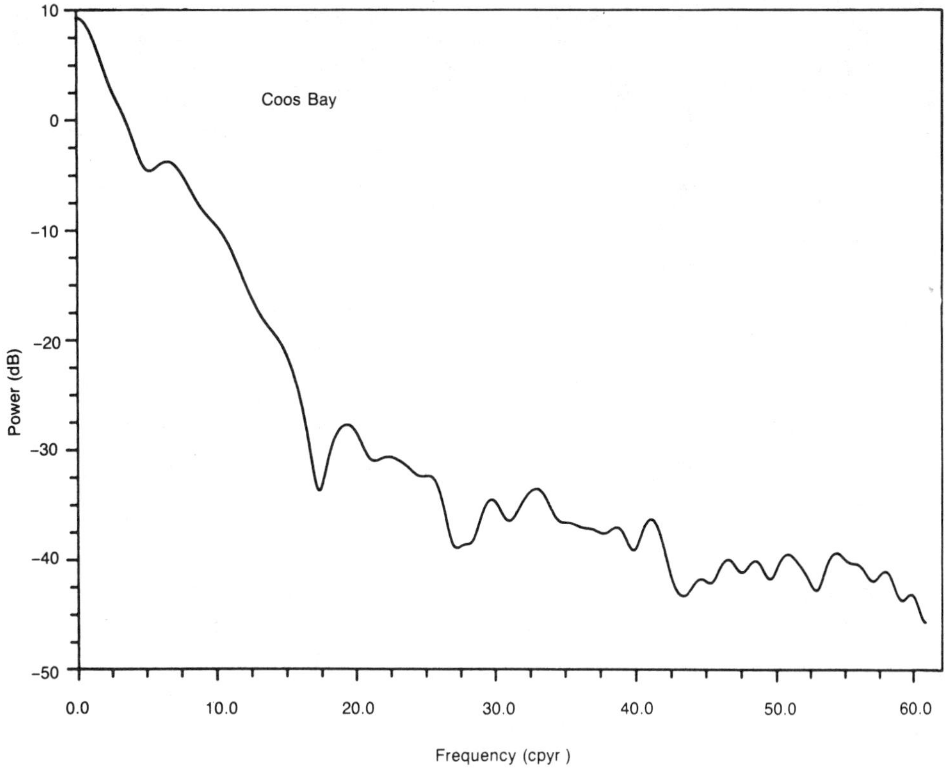

Fig. 78. Power spectrum of acoustic travel-time variation on path between Heard Island and Coos Bay, Oregon, shown in Fig. 77.

establish a trend at the 95% confidence level. Table 17 shows that with records covering about a decade, warming could be detected. Taking into account the high standard deviation of the estimates, it would be possible to detect a warming trend due to a 0.8% per year buildup of greenhouse gases from changes in the paths going from Heard Island to either Coos Bay or San Francisco. An observing interval of 25 years would be required if only the Heard Island-Bermuda path were used. These times are short compared with the times required to determine a trend from the noisy record of global-mean, surface-air temperatures.

Table 16. Trend analysis of ocean acoustic travel times using minimization of absolute deviation.

Receiving Site	Calculated Trend (s/yr)	Actual Trend (s/yr)
Coos Bay	-0.457	-0.467
San Francisco	-0.489	-0.505
Bermuda	-0.277	-0.125

Table 17. Length of record required to detect a trend in acoustic travel time variations due to a 0.8% per year buildup of greenhouse gases from Heard Island to various sites.

Receiving Site	Trend (s/yr)	Time Required (years)
Bermuda	0.125	25.5 (± 3.0)
Coos Bay	0.467	3.6 (± 2.4)
San Francisco	0.505	4.0 (± 2.4)

The ability to measure acoustic travel times very precisely opens another avenue to detection of global warming. The acoustic method has the advantage of averaging over long path lengths. This capability aids in attribution, because comparison with model calculations involves large-scale averaging with a lesser dependence on the details of ocean-atmosphere interactions and point variability. The possibility of monitoring several ocean basins further increases the statistical robustness of the method.

Summary Comments on the Detection of Greenhouse Warming

The most important question concerning greenhouse warming is not whether the infrared gases will produce a recognizable global warming, but when they will do so. The underlying physics that connects changes in atmospheric composition to changes in radiative flux trapped within the atmosphere is well understood. Much less understood are the great variety of feedback processes involving the hydrologic cycle and the biosphere. As climate models improve, the statistical techniques outlined in this section can be used to test the hypothesis that changing atmospheric composition is indeed warming the atmosphere and oceans. It appears that there is no threshold at which the concentration of greenhouses gases becomes critical and causes a sudden shift in climate. Instead, we can anticipate gradual, but possibly accelerating, changes in the future. Only continuing observation, coupled with further model development, will tell us whether the greenhouse warming has truly begun.

REFERENCES

Angell, J., 1988, Variations and trends in tropospheric and stratospheric global temperatures 1958-1987, *J. Climate*, **1**:1,296-1,313.

Arnold, V., 1978, *Mathematical Methods of Classical Mechanics*, Springer-Verlag, New York.

Atkinson, T., Briffa, K., Coope, G., Joachim, M., and Perry, D., 1985, Climate calibration of coleopteran data, *in*: "Handbook of Holocene Paleontology and Paleohydrology," B. Berglund, ed., J. Wiley, New York.

Bacastow, R., Keeling, C., and Whorf, T., 1985, Seasonal amplitude increase in atmospheric CO_2 concentration at Mauna Loa, Hawaii, 1959-1982, *J. Geophys. Res.*, **90**:10,529-10,540.

Barnola, J., Raynaud, D., Korotkevich, Y., and Lorius, C., 1987, Vostok ice core provides 160,000-year record of atmospheric CO_2, *Nature*, **329**:408-414.

Berger, A., 1978a, Long-term variations of caloric insolation resulting from earth's orbital elements, *Quat. Res.*, **9**:139-167.

Berger, A., 1978b, Long-term variations of daily insolation and Quaternary climatic changes, *J. Atmos. Sci.*, **35**:2,362-2,367.

Birkhoff, G., 1927, *Dynamical Systems*, American Mathematical Society, Providence, RI.

Berner, R., Lasaga, A., and Garrels, R., 1983, The carbonate-silicate geochemical cycle and its effect on atmospheric carbon dioxide over the past 100 million years, *Amer. J. Sci.*, **283**:641-683.

Blake, D., and Rowland, F., 1988, Continuing worldwide increases in tropospheric methane, 1978-1987, *Science*, **239**:1,129-1,131.

Bolin, B., 1986, How much CO_2 will remain in the atmosphere? *in*: "The Greenhouse Effect: Climate Change and Ecosystems," B. Bolin, B. Döös, J. Jäger, and R. Warrick, eds., J. Wiley, New York.

Broecker, W., 1981, Glacial to interglacial changes in ocean and atmosphere chemistry, *in*: "Climate Variations and Variability: Facts and Theories," A. Berger, ed., Reidel, Boston.

Broecker, W., 1982, Glacial to interglacial changes in ocean chemistry, *Progress in Oceanog.*, **11**:151-197.

Broecker, W., and Peng, T., 1986, Carbon cycle: 1985 glacial to interglacial changes in the operation of the global carbon cycle, *Radiocarb.*, **28**:309-327.

Bryan, K., Manabe, S., and Spelman, M., 1988, Interhemispheric asymmetry in the transient response of a coupled ocean-atmosphere model to a CO_2 forcing, *J. Phys. Oceanogr.* **18**:851-867.

Budyko, M., Ronov, A., and Yanshin, A., 1987, *History of the Earth's Atmosphere*, Springer-Verlag, New York.

Carslaw, H., and Jaeger, J., 1959, *Conduction of Heat in Solids*, Oxford, London.

Chung, Y., 1988, The variations of atmospheric carbon dioxide at Alert and Sable Island, Canada, *Atmos. Environ.*, **22**:383-394.

Cicerone, R., and Oremland, R., 1988, Biogeochemical aspects of atmospheric methane, *Global Biogeochemical Cycles*, **2**:299-327.

Claypool, G., and Kvenvolden, K., 1983, Methane and other hydrocarbon gases in marine sediment, *Ann. Rev. Earth Planet. Sci.*, **11**:299-327.

Crowley, T., 1983, The geologic record of climate change, *Rev. Geophys. Space Physics*, **21**:853.

Craig, H., and Chou, C., 1982, Methane: The record in polar ice cores, *Geophys. Res. Lett.*, **9**:1,221-1,224.

Cramér, H., 1946, *Mathematical Methods of Statistics*, Princeton Univ. Press, Princeton, NJ.

Detwiler, R., and Hall, C., 1988, Tropical forests and the global carbon cycle, *Science*, **239**:42-47.

Feister, U., and Warmbt, W., 1987, Long-term measurements of surface ozone in the German Democratic Republic, *J. Atmos. Chem.*, **5**:1-21.

Francis, N., Naiman, R., and Melillo, J., 1985, Nitrogen fixation in subarctic streams influenced by the beaver *(Castor canadensis)*, *Hydrobiologia*, **121**:193-202.

Fraser, P., Khalil, M., Rasmussen, R., and Steele, L., 1984, Tropospheric methane in the mid-latitudes of the southern hemisphere, *J. Atmos. Chem.*, **1**:125-135.

Fraser, P., Pearman, G., and Hyson, P., 1983, The global distribution of atmospheric carbon dioxide, 2: A review of provisional background observations, 1978-1980, *J. Geophys. Res.*, **88**:3,591-3,598.

Friedli, H., Lötscher, H., Oeschger, H., Siegenthaler, U., and Stauffer, B., 1986, Ice core record of the $^{13}C/^{12}C$ ratio of atmospheric CO_2 in the past two centuries, *Nature*, **324**:237-238.

Friedli, H., and Siegenthaler, U., 1988, Influence of N_2O on isotope analyses in CO_2 and mass-spectrometric determination of N_2O in air samples, *Tellus*, **40B**:129-133.

Gornitz, V., Lebedeff, S., and Hansen, J., 1982, Global sea level trend in the past century, *Science*, **215**:1,611-1,614.

Halter, B., and Peterson, J., 1981, On the variability of atmospheric carbon dioxide concentration at Barrow, Alaska, during summer, *Atmos. Environ.*, **15**:1,391-1,399.

Hansen, J., Fung, I., Lacis, A., Rind, D., Lebedeff, S., Ruedy, R., Russell, G., and Stone, P., 1988, Global climate changes as forecast by Goddard Institute for Space Studies: Three-dimensional model, *J. Geophys. Res.*, **93**:9,341-9,364.

Hansen, J., Lacis, A., Rind, D., Russell, G., Stone, P., Fung, I., Ruedy, R., and Lerner, J., 1984, Climate sensitivity: Analysis of feedback mechanisms, *in*: "Climate Processes and Climate Sensitivity," J. Hansen and T. Takahashi, eds., American Geophysical Union, Washington, D.C.

Hansen, J., and Lebedeff, S., 1987, Global trends of measured surface air temperature, *J. Geophys. Res.*, **92**:13,345-13,372.

Hansen, J., and Lebedeff, S., 1988, Global surface air temperatures: Update through 1987, *Geophys. Res. Lett.*, **15**:323-326.

Hao, W., Wofsy, S., McElroy, M., Beer, J., and Toqan, M., 1987, Sources of atmospheric nitrous oxide from combustion, *J. Geophys. Res.*, **92**:3,098-3,104.

Hays, J., Imbrie, J., and Shackleton, N., 1976, Variations in the earth's orbit: Pacemaker of the ice ages, *Science*, **194**:1,121-1,132.

Hoefs, J., 1987, *Stable Isotope Geochemistry* (3rd Edition), Springer-Verlag, New York.

Houghton, R., Boone, R., Fruci, J., Hobbie, J., Melillo, J., Palm, C., Peterson, B., Shaver, G., Woodwell, G., Moore, B., Skole, D., and Myers, N., 1987, The flux of carbon from terrestrial ecosystems to the atmosphere in 1980 due to changes in land use: Geographic distribution of the global flux, *Tellus*, **39B**:122-139.

Houghton, R., Hobbie, J., Melillo, J., Moore, B., Peterson, B., Shaver, G., and Woodwell, G., 1983, Changes in the carbon content of terrestrial biota and soils between 1860 and 1980: A net release of CO_2 to the atmosphere, *Ecological Monographs*, **53**:235-262.

Houghton, R., and Woodwell, G., 1989, Global climatic change, *Scientific American*, **260**:4, 36-44.

Hunt, J., 1979, *Petroleum Geochemistry and Geology*, W. H. Freeman, San Francisco.

Imbrie, J., Hays, J., Martinson, D., McIntyre, A., Mix, A., Morley, J., Pisias, N., Prell, W., and Shackleton, N., 1984, The orbital theory of Pleistocene climate: Support from a revised chronology of the marine $\delta^{18}O$ record, *in*: "Milankovitch and Climate, Part 1," A. Berger et al., eds., Reidel, Dordrecht.

Jones, P., Bradley, R., Diaz, H., Kelly, P., and Wigley, T., 1986a, Northern Hemisphere surface air temperature variations: 1851-1984, *J. Climate Appl. Meteor.*, **25**:161-179.

Jones, P., Wigley, T., and Wright, P., 1986b, Global temperature variations between 1861 and 1984, *Nature*, **322**:430-434.

Jones, P., Wigley, T., Folland, C., Parker, D., Angell, J., Lebedeff, S., and Hansen, J., 1988, Evidence for global warming in the past decade, *Nature*, **332**:790.

Karl, T., and Jones, P., 1989, Urban bias in area-averaged surface air temperature trends, *Bull. Amer. Meteor. Soc.*, **70**:265-270.

Karl, T., J. Tarpleg, R. Quayle, H. Diaz, D. Robinson, and R. Bradley, 1989, The recent climate record: What it can and cannot tell us, *Rev. Geophys.*, **27**:405-430.

Keeling, C., 1973, Industrial production of carbon dioxide from fossil fuels and limestone, *Tellus*, **25**:174-198.

Keeling, C., Bacastow, R., Carter, A., Piper, S., Whorf, T., Heimann, M., Mook, W., and Roeloffzen, H., 1989, A three-dimensional model of atmospheric CO_2 transport based on observed winds: 1. Analysis of observational data, *in*: "Aspects of Climate Variability in the Pacific and the Western Americas," H. Peterson, ed., *Geophysical Monograph* **55**, American Geophysical Union, Washington, D.C., 165-236.

Khalil, M., and Rasmussen, R., 1983, Increase and seasonal cycles of nitrous oxide in the earth's atmosphere, *Tellus*, **35B**:161-169.

Khalil, M., and Rasmussen, R., 1988, Nitrous oxide: Trends and global mass balance over the last 3000 years, *Annals. Glaciol.*, **10**:73-79.

Komhyr, W., Gammon, R., Harris, T., Waterman, L., Conway, T., Taylor, W., and Thoning, K., 1985, Global atmospheric CO_2 distribution and variations from 1968-1982 NOAA/GMCC CO_2 flask sample data, *J. Geophys. Res.*, **90**:5,567-5,596.

Kutzbach, J., and Street-Perrott, F., 1985, Milankovitch forcing of fluctuations in the level of tropical lakes from 18 to 0 kyr BP, *Nature*, **317**:130-134.

Kvenvolden, K., 1988, Methane hydrate—A major reservoir of carbon in the shallow geosphere? *Chem. Geol.*, **71**:41-51.

Lachenbruch, A., and Marshall, B., 1986, Changing climate: Geothermal evidence from permafrost in the Alaskan Arctic, *Science*, **234**:689-696.

Lamb, H., 1977, *Climate: Present, Past and Future*, **2**, Methuen, London.

Lloyd, C., 1984, Pre-Pleistocene paleoclimates: The geological and paleontological evidence; modeling strategies, boundary conditions, and some preliminary results, *in*: "Advances in Geophysics," **26**, B. Saltzman, ed., Academic Press, New York.

Lorenz, E., 1963, Deterministic nonperiodic flow, *J. Atmos. Sci.*, **20**:130-141.

Lorenz, E., 1984a, Irregularity: A fundamental property of the atmosphere, *Tellus*, **36A**:98-110.

Lorenz, E., 1984b, Formulation of a low-order model of a moist general circulation, *J. Atmos. Sci.*, **41**:1,933-1,945.

Lowe, D., Brenninkmeijer, C., Manning, M., Sparks, R., and Wallace, G., 1988, Radiocarbon determination of atmospheric methane at Baring Head, New Zealand, *Nature*, **332**:522-525.

MacDonald, G., 1982, *The Long-Term Impacts of Increasing Atmospheric Carbon Dioxide Levels*, Ballinger, Cambridge, MA.

MacDonald, G., 1988, Variation in atmospheric carbon dioxide and ice age climate, *in*: "Preparing for Climate Change," Proc. First North American Conference on Preparing for Climate Change: A Cooperative Approach, Government Institutes, Rockville, MD.

MacDonald, G., 1989, Spectral analysis of time series generated by nonlinear processes, *Rev. Geophys.* **27**:449-469.

MacDonald, G., 1990, Role of methane clathrates in past and future climates, *Climatic Change* (in press).

Manabe, S., and Stouffer, R., 1988, Two stable equilibria of a coupled ocean-atmosphere model, *J. Climate*, **1**:841-866.

Manning, M., Lowe, D., Melhulsh, W., Sparks, R., Wallace, G., and Brenninkmeijer, C., 1989, The use of radiocarbon measurements in atmospheric studies, *Radiocarbon*, **31** (in press).

Marland, G., and Boden, T., 1989, Carbon dioxide releases from fossil-fuel burning, Statement before Senate Committee on Energy and Natural Resources, July 26, 1989, Washington, D.C.

Marland, G., and Rotty, R., 1984, Carbon dioxide emissions from fossil fuels: A procedure for estimation and results for 1950-1982, *Tellus*, **36B**:232-261.

Mercer, J., 1978, West Antarctic ice sheet and CO_2 greenhouse effect: A threat of disaster, *Nature*, **271**:321-325.
Milankovitch, M., 1920, *Théorie Mathématique des Phénomènes Thermiques Produits par la Radiation Solaire*, Gauthier-Villars, Paris.
Munk, W., and Forbes, A., 1989, Global ocean warming: An acoustic measure? *J. Phys. Oceanography* (in press).
Munk, W., O'Reilly, W., and Reid, J., 1988, Australia-Bermuda sound transmission experiment (1960) revisited, *J. Phys. Oceanogr.*, **18**:1,876-1,898.
Naiman, R., Melillo, J., and Hobbie, J., 1986, Ecosystem alteration of boreal forest streams by beaver *(Castor canadensis)*, *Ecology*, **67**:1,254-1,269.
Neftel, A., Moor, E., Oeschger, H., and Stauffer, B., 1985, Evidence from polar ice cores for the increase in atmospheric CO_2 in the past two centuries, *Nature*, **315**:45-47.
Nisbet, E., 1989, Some northern sources of atmospheric methane: Production, history, and future implications, *Can. J. Earth Sci.* (in press).
Oltmans, S., and Komhyr, W., 1986, Surface ozone distributions and variations from 1973-1984. Measurements at the NOAA Geophysical Monitoring for Climate Change Baseline Observatories, *J. Geophys. Res.*, **91**:5,229-5,236.
Oremland, R., 1988, Biogeochemistry of methanogenic bacteria, *in*: "Biology of Anaerobic Microorganisms," A. Zehnder, ed., J. Wiley, New York.
Owen, T., Cess, R., and Ramanathan, V., 1979, Enhanced CO_2 greenhouse to compensate for reduced solar luminosity on early earth, *Nature*, **277**:640-642.
Pearman, G., and Hyson, P., 1981, The annual variation of atmospheric CO_2 concentration observed in the northern hemisphere, *J. Geophys. Res.*, **86**:9,839-9,843.
Peltier, W., and Tushingham, A., 1989, Global sea level rise and the greenhouse effect: Might they be connected? *Science*, **244**:806-810.
Peterson, B., and Melillo, J., 1985, The potential storage of carbon caused by eutrophication of the biosphere, *Tellus*, **37B**:117-127.
Poincaré, H., 1892, *Les Méthodes Nouvelles de la Méchanique Céleste*, Gauthier-Villars, Paris; reprinted Dover, New York, 1957, 3 vols.
Prinn, R., Cunnold, D., Rasmussen, R., Simmonds, P., Alyea, F., Crawford, A., Fraser, P., and Rosen, R., 1987, Atmospheric trends in methylchloroform and the global average for the hydroxyl radical, *Science*, **238**:945-950.
Rasmussen, R., and Khalil, M., 1981, Atmospheric methane (CH_4) : Trends and seasonal cycles, *J. Geophys. Res.*, **86**:9,826-9,832.
Rasmussen, R., and Khalil, M., 1984, Atmospheric methane in the recent and ancient atmospheres: Concentrations, trends, and interhemispheric gradient, *J. Geophys. Res.*, **89**:11,599-11,605.
Raynaud, D., Chappellaz, J., Barnola, J., Korotkevich, Y., and Lorius, C., 1988, Climatic and CH_4 cycle implications of glacial-interglacial CH_4 change in the Vostok ice core, *Nature*, **333**:655-657.
Rinsland, C., Levine, J., and Miles, T., 1985, Concentration of methane in the troposphere deduced from 1951 infrared solar spectra, *Nature*, **318**:245-249.
Rotty, R., 1987, A look at 1983 CO_2 emissions from fossil fuels (with preliminary data for 1984), *Tellus*, **39B**:203-208.
Roxburgh, I., 1980, Long-term variation in the solar constant, *in* "Sun and Climate," CNES/SNRS/DGRST Conference Proceedings, 261-268.
Ruddiman, W., and Wright, H., 1987, *North America and Adjacent Oceans During the Last Deglaciation*, Geological Society of America, Boulder, CO.
Schlesinger, M., and Xingjian, J., 1988, The transport of CO_2-induced warming into the ocean: An analysis of simulations by the OSU coupled ocean-atmosphere General Circulation Model, *Climate Dynamics*, **3**:1-17.
Semtner, A., and Chervin, R., 1988, A simulation of the global ocean circulation with resolved eddies, *J. Geophys. Res.*, **93**:15,502-15,522.
Shackleton, N., 1977, Carbon-13 in Uvigerina: Tropical rainforest history and the equatorial Pacific carbonate dissolution cycles, *in*: "The Fate of Fossil Fuel CO_2 in the Oceans", Marine Science Series, **6**, N. Andersen and A. Malahoff, eds., Plenum Press, New York.

Shackleton, N., Backman, J., Zimmerman, H., Kent, D., Hall, M., Roberts, D., Schnitker, D., Baldauf, J., Desprairies, A., Homrighausen, R., Huddlestun, P., Keene, J., Kaltenback, A., Krumsiek, K., Morton, A., Murray, J., and Westberg-Smith, J., 1984, Oxygen isotope calibration of the onset of ice-rafting and history of glaciation in the North Atlantic region, *Nature*, **307**:620-623.

Shackleton, N., Imbrie, J., and Hall, M., 1983, Oxygen and carbon isotope record of East Pacific core V19-30; Implications for the formation of deep water in the late Pleistocene North Atlantic, *Earth Planet. Sci. Lett.*, **65**:233-244.

Shackleton, N., and Pisias, N., 1985, Atmospheric carbon dioxide, orbital forcing, and climate, *in*: "The Carbon Cycle and Atmospheric CO_2 : Natural Variations Archean to Present," E. Sundquist and W. Broecker, eds., *Geophysical Monograph*, **32**, American Geophysical Union, Washington, D.C., 303-317.

Siegenthaler, U., 1983, Uptake of excess CO_2 by an outcrop-diffusion model of the ocean, *J. Geophys. Res.*, **88**:3,599-3,608.

SORG, 1988, Stratospheric ozone 1988, Second Report of the UK Stratospheric Review Group, HMSO, London.

Stauffer, B., Fischer, G., Neftel, A., and Oeschger, H., 1985, Increase of atmospheric methane recorded in Antarctic ice core, *Science*, **229**:1,386-1,388.

Stauffer, B., Lochbronner, E., Oeschger, H., and Schwander, J., 1988, Methane concentration in the glacial atmosphere was only half that of the preindustrial Holocene, *Nature*, **332**, 812-814.

Tyurnin, B., 1984, Factors determining numbers of river beaver *(Castor fiber)* in the European north, *Sov. J. Ecology*, **14**:43-50.

Volz, A., and Kley, D., 1988, Evaluation of the Montsouris series of ozone measurements made in the nineteenth century, *Nature*, **332**:240-242.

Wahlen, M., Tanaka, N., Henry, R., Deck, B., Zeglen, J., Vogel, J., Southon, J., Shemesh, A., Fairbanks, R., and Broecker, W., 1989, Carbon-14 in methane sources and in atmospheric methane: The contribution from fossil carbon, *Science*, **245**:286-290.

Walsh, J., 1984, The role of ocean biota in accelerated ecological cycles: A temporal view, *BioScience*, **34**:499-507.

Wigley, T., and Raper, S., 1987, Thermal expansion of sea water associated with global warming, *Nature*, **330**:127-131.

Zaloumnis, E., 1987, Don't pull the plug, *African Wildlife*, **41**:210-217.

Zeikus, J., and Winfrey, M., 1976, Temperature limitation of methanogenesis in aquatic sediments, *Appl. Env. Microbiol.*, **31**:99-107.

MODELING CLIMATE CHANGE

Dennis L. Hartmann

Department of Atmospheric Sciences
University of Washington
Seattle, Washington 98105, USA

INTRODUCTION

The climate of the earth continually changes. Although environmental conditions on earth have been adequate for life to exist for 3.5 billion years or more, the climate and biology of earth have evolved together, with gradual changes occurring in the climate and in the particular organisms best suited for coexistence. Early in earth's history, primitive photosynthetic organisms such as cyanobacteria developed in the oceans. Photosynthesis in the oceans produced oxygen in the atmosphere, which led to the development of the protective ozone layer in the stratosphere and the eventual dominance by aerobic organisms near the earth's surface. The development and evolution of life were accompanied by profound changes in atmospheric composition and corresponding changes in surface climate.

During the last several centuries, human influence on the climate of the earth has become increasingly important. Through burning of fossil carbon fuels, production of certain industrial chemicals, and modification of land surfaces, humans are now capable of inducing changes in the earth's climate that are more rapid and significant than any that have occurred for more than a million years. These changes are inadvertent, in that they are produced as by-products of efforts to produce goods and services rather than direct attempts to modify climate. Nonetheless, climate changes over the next century are likely to be highly significant for humans and for the other species that occupy the earth.

In order to evaluate how our current activities will shape the future, and to plan for that future, it is essential to have the means to predict the effect of these activities on climate. In addition to mean temperature and precipitation, which are the primary climatic variables, prediction of the variability and reliability of weather elements is necessary. Climatic extremes often determine the design parameters of water and energy supply systems for human settlements. The plant and animal species of marine and terrestrial ecosystems are also strongly affected by weather extremes, and by the mean climate of which they are a part.

This chapter describes the physical principles and processes that must be taken into account when developing a model of the earth's climate. Delineation of model details is kept to a minimum, except where they are necessary to illustrate the level of uncertainty associated with current efforts to model future climate changes. An attempt is made to describe in simple terms the complex interactions between the many components of the climate system.

The chapter consists of five sections, the first of which provides an introduction to the physics of earth's climate and the role of the greenhouse effect

Global Climate and Ecosystems Change, Edited by G. MacDonald
and L. Sertorio, Plenum Press, New York, 1990

in maintaining warmth near the surface. The concept of climate sensitivity is next introduced and used to identify the key parameters and processes for setting the climate and determining the magnitude of change expected from increasing levels of greenhouse gases. In the third section, the basic elements of models used to predict climate change are described. Also included are the methods that have been used to assess the ability of these models to predict climate changes. The fourth section presents the equilibrium response of climate models to a doubling of carbon dioxide. The final part of this paper describes the transient response of climate models and explores the role of the ocean in determining the temporal evolution of the response to increases in greenhouse gas concentrations.

BASIC MAINTENANCE OF THE EARTH'S SURFACE CLIMATE

Global Energy Balance

On earth, the temperature is such that the energy the planet emits is balanced by the solar energy it absorbs. Energy generated in the interior of the earth is too small to have much of an effect on the surface temperature. The simplest climate model solves for the global-mean terrestrial temperature by equating the absorbed solar energy with the energy emitted by the earth. The flux of energy from the sun per unit area, at the mean distance of the earth from the sun, is about 1,367 $\pm$ 2 Wm^{-2}, which we will denote as S. Of this amount, about 30% is reflected by the planet without being absorbed. The fraction reflected is called the planetary albedo and is usually denoted by α. As a working hypothesis, we may assume that the earth emits energy like a blackbody at its emission temperature, and therefore follows the Stefan-Boltzmann Law. The energy balance for the planet can be written as follows:

Absorbed solar energy = emitted terrestrial energy

$$\frac{S}{4}(1-\alpha) = \sigma T_e^4. \tag{1}$$

The factor of 4 in Eq. (1) represents the ratio of the emission area to the shadow area of a spherical planet. The Stefan-Boltzmann constant is $\sigma = 5.67 \times 10^{-8}\ Wm^{-2}K^{-4}$. If we solve Eq. (1) using an albedo of 0.3, we find that the emission temperature is $T_e = 255$ K (−18°C). This is considerably lower than the observed global-mean surface temperature of $T_S = 288$ K (+15°C). The difference between the emission temperature required for global energy balance and the actual surface temperature is explained by the so-called "atmospheric greenhouse effect."

The Atmospheric Greenhouse Effect

To illustrate how the greenhouse effect works, one can elaborate the simple global-mean climate model by adding an atmosphere with the right properties. The key aspect that makes earth's atmosphere like a greenhouse is its relative transparency to solar radiation. In a constructed greenhouse, the building walls hold in solar energy by preventing air currents from carrying it away. In the atmospheric greenhouse effect, energy is prevented from escaping easily from the surface by the atmosphere's near-opacity to surface-emitted terrestrial energy. Let us suppose then that we cover the surface of our model planet with an atmosphere that is transparent to solar radiation but opaque to terrestrial radiation. The atmosphere, which we will represent by a thin layer at a single temperature, absorbs all of the radiation emitted by the surface and emits like a blackbody at its temperature. The flux terms are represented by arrows in Fig. 1.

The energy balance equation for the top of the atmosphere is identical to that for the planet without an atmosphere and implies that the atmospheric temperature must equal the emission temperature:

$$\frac{S}{4}(1-\alpha) = \sigma T_A^4 = \sigma T_e^4. \tag{2}$$

The energy balance for either the atmosphere or the surface can be used to determine the surface temperature in equilibrium:

$$\text{Absorbed energy} = \text{emitted energy}$$

$$\text{atmosphere} \quad \sigma T_S^4 = 2\sigma T_A^4 \tag{3}$$

$$\text{surface} \quad \frac{S}{4}(1-\alpha) + \sigma T_A^4 = \sigma T_S^4. \tag{4}$$

The balances expressed by Eqs. (2) through (4) require that

$$T_S^4 = 2T_e^4, \tag{5}$$

which for $T_e = 255$ K gives $T_S = 303$ K. The surface temperature in equilibrium is considerably warmer than the emission temperature because of the added downward flux of terrestrial radiation from the atmosphere (see Eq. (4)). It is necessary for the atmosphere to be transparent to solar radiation for the greenhouse effect to warm the surface. One can see this by assuming that all of the solar radiation is absorbed in the atmosphere, and then by solving for the

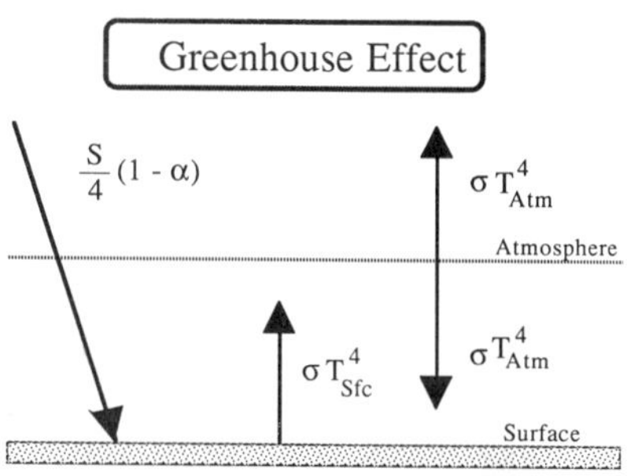

Fig. 1. Schematic diagram of the global-average radiative energy fluxes when the atmosphere is modeled as a layer that is transparent to solar radiation, but opaque to terrestrial radiation.

equilibrium temperatures in the simple model above. The reader can verify that, in this case, the surface and the atmosphere are both at the effective temperature of 255 K. If the atmosphere becomes absorbent rather than transparent, the surface temperature in radiative equilibrium will decrease, even though the atmosphere remains at the same temperature.

Through observation or accurate calculation, one can verify that the essential elements of the atmospheric greenhouse effect, illustrated by the simple model above, actually operate in the earth's atmosphere. An energy-flow diagram for the atmosphere is shown in Fig. 2. The numerical values given for the energy fluxes are percentages of the global average insolation at the top of the atmosphere (342 Wm^{-2}). Of the solar energy available at the top of the atmosphere, approximately 50% reaches the surface and is absorbed there, 30% is reflected, and 20% is absorbed in the atmosphere. Despite the efficient transmission of solar energy through the atmosphere, nearly twice as much energy reaches the surface in the form of emission from the atmosphere as from the sun. Less than 10% of the radiative energy emission from the surface escapes to space without being absorbed by the atmosphere. The atmosphere is an equally effective emitter of energy, and emits 50% more energy in the form of terrestrial radiation than it absorbs. The atmosphere loses about 30 units of energy (about 103 Wm^{-2}) continually through radiative fluxes, while the surface gains an equivalent amount. Radiative imbalances are offset by non-radiative fluxes of energy from the surface to the atmosphere. Of the 30 units of non-radiative flux from the surface, 25 are accounted for by latent heat transport. When water that is evaporated from the surface by solar energy condenses and rains out, the latent heat of vaporization is left behind in the atmosphere.

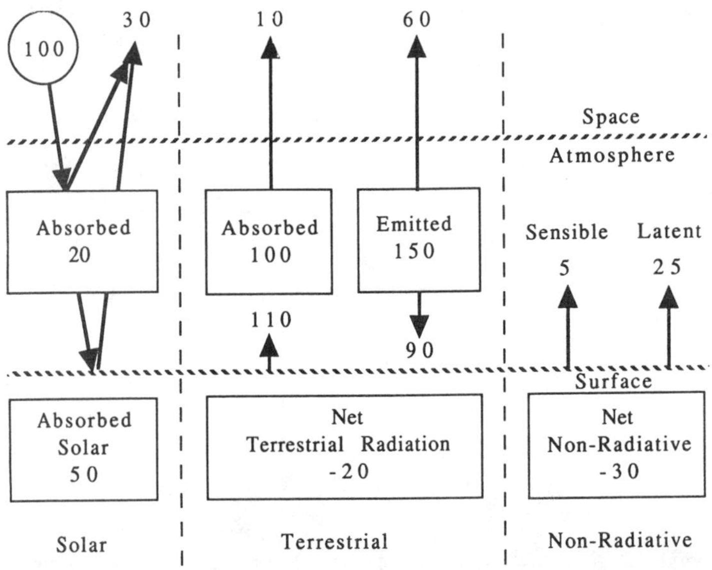

Fig. 2. Energy-flux box diagram showing radiative and non-radiative exchanges between the surface, the atmosphere, and space. Units are percentages of the global-average insolation (100% = 342 Wm^{-2}).

Substantial vertical heat flux by convection in the atmosphere is a concomitant of the greenhouse effect. Because a large fraction of the solar energy retained by the planet is absorbed at the surface, the heating continuously creates air that is less dense near the surface. The heated air rises and is replaced by sinking air that has been cooled in the upper troposphere and has a lower potential density. This vertical stirring keeps the troposphere well mixed and drives a hydrologic cycle in which water is continually circulated among the oceans, the atmosphere, and the land surface. The circulation of air and water is vital for most of the life-forms that have evolved on earth.

Selective Absorption of Radiation

The atmospheric greenhouse effect depends upon the molecular composition of the atmosphere and the way each molecular species interacts with radiation. Figure 3 is a diagram that shows the emission spectra of a body at 6,000 K, which is the effective temperature of the sun, and a body at 250 K, which is close to the emission temperature of the earth. Because of the great difference in the temperature of these two bodies, the emission spectra are almost completely distinct. The sun emits energy primarily at visible and near-infrared wavelengths, while the earth emits in the thermal infrared. Thus, very little overlap exists between the frequencies at which the sun and earth emit energy.

Molecules absorb and emit energy by moving from one energy level to another according to rules described by quantum mechanics. The energy contained in a photon of radiation varies linearly with the frequency of the radiation. Thus, high-frequency, short-wavelength solar radiation consists of photons with higher energy than the photons associated with terrestrial radiation. The primary gaseous constituents of the atmosphere, molecular nitrogen (N_2, 78%) and molecular oxygen (O_2, 21%), are symmetric molecules with no vibrational and rotational transitions. These molecules interact strongly only with very energetic photons corresponding to ultraviolet and shorter wavelengths and

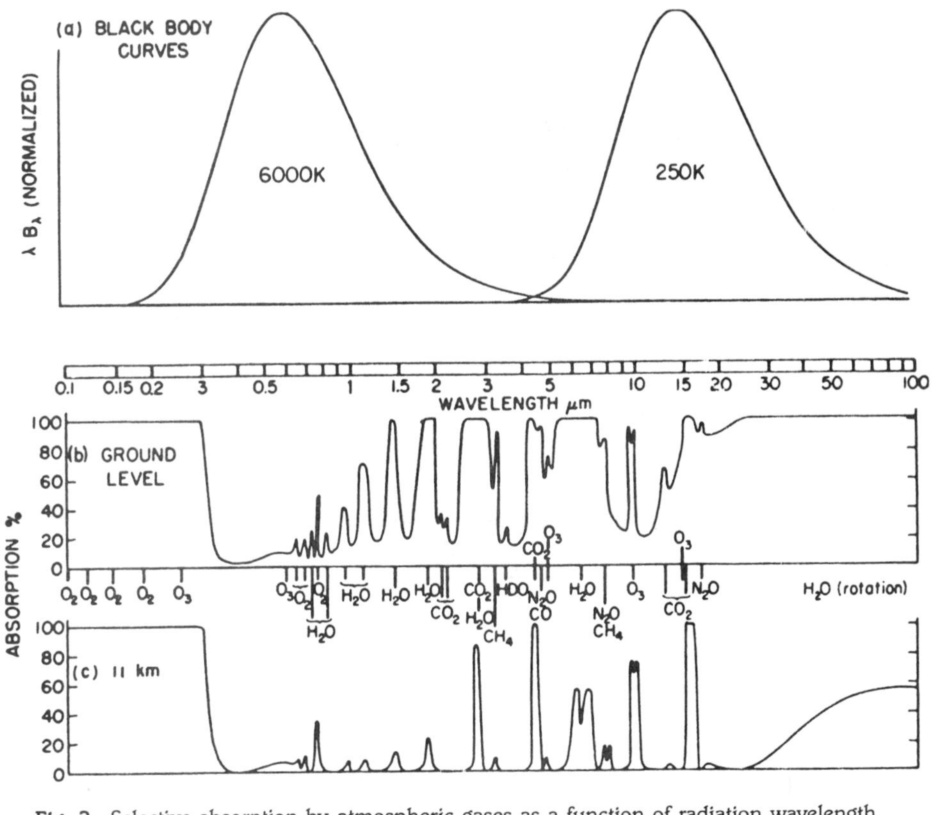

Fig. 3. Selective absorption by atmospheric gases as a function of radiation wavelength. Top: Normalized blackbody emission curves for solar (~ 6000 K) and terrestrial (~ 250 K) radiation. Middle: Percent absorption between the ground and the top of the atmosphere. Bottom: Percent absorption between 11 km and the top of the atmosphere. Conditions are typical of mid-latitude conditions with a solar zenith angle of 50° (source: Goody and Yung, 1989).

are therefore transparent to the bulk of the sun's energy. More complex molecules made up of three or more atoms may have vibration/rotation or pure rotation bands. The differences between the vibrational and rotational energy levels of these molecules correspond to the frequencies of terrestrial radiation, and so they efficiently absorb and emit terrestrial radiation.

The most important absorber of terrestrial and solar energy in the atmosphere is water vapor. Water vapor is a bent triatomic molecule with a permanent dipole moment. In addition to strong vibration/rotation bands centered near 6.3 μm, water vapor has a broad band of densely packed, pure-rotation absorption lines at wavelengths longer than about 12 μm. Water also has a number of absorption bands in the near-infrared wavelengths of solar radiation. After water vapor, the most energetically important absorber of terrestrial radiation is carbon dioxide, which has a strong vibration/rotation band near 15 μm. The third primary radiative gas is ozone, which absorbs strongly at 9.6 μm.

As a consequence of the frequency-specific interaction of radiation with the gaseous constituents of the atmosphere, relatively little gaseous absorption of visible radiation occurs; the atmosphere is nearly transparent to light, which constitutes a large fraction of the energy coming from the sun. In the frequency range corresponding to visible radiation (0.4 to 0.7 μm), the absorption by the atmosphere is less than 10% (see Fig. 3). At near-infrared wavelengths, especially those longer than 1 μm, absorption by water vapor begins to become important. For wavelengths longer than about 4 μm, corresponding to the emission of energy by the earth, the atmosphere is almost completely opaque. The exception is the region between about 8 and 12 μm, where the absorption by atmospheric gases is weak, with the notable exception of ozone. This region is called the "atmospheric window," or often the "water vapor window." It is in this region where some small fraction of the radiation emitted by the surface is able to escape directly to space. Increased concentrations of trace gases that have strong absorption bands at wavelengths in the range of 8 to 12 μm have a particularly strong potential for enhancing the greenhouse effect of the atmosphere, because the atmosphere is currently nearly transparent at these wavelengths. Examples of such gases are the chlorofluorocarbons (CFCs), whose concentrations are increasing because of human activities (Ramanathan et al., 1985).

Clouds of water and ice in the atmosphere, when sufficiently thick, are virtually opaque to terrestrial radiation, so they also play an important role in the atmospheric greenhouse. Because clouds are approximately blackbodies for terrestrial radiation, they absorb terrestrial radiation at all frequencies. Table 1 gives the approximate contributions of the principal terrestrial radiation absorbers to the atmospheric greenhouse effect for the preindustrial atmosphere. Note that the gaseous, liquid, and solid forms of water contribute nearly 80% of the current or natural greenhouse, with smaller contributions from carbon dioxide and ozone.

Clouds and the Net Radiation Balance

Clouds also have a very important effect on absorbed solar energy through

Table 1. Approximate contributions to the atmospheric greenhouse effect in the current atmosphere.

Water Vapor	~ 50%
Clouds of Water and Ice	~ 28%
Carbon Dioxide	~ 18%
Ozone	~ 4%

their relatively high albedos. The effects of clouds on the energy balance at the top of the atmosphere can be seen by looking at the average and cloud-free radiation budget quantities as measured by the Earth Radiation Budget Experiment (ERBE) (Barkstrom and Smith, 1986; ERBE, 1986; Hartmann et al., 1986). The cloud-free radiation budget parameters were determined by time-space averaging those pixels from the ERBE scanning instrument that were identified as being "clear" (Wielicki and Green, 1989; Ramanathan et al., 1989; Harrison et al., 1989; Hartmann and Doelling, 1990).

Contour plots of outgoing long-wave radiation (OLR), which is the energy emitted by the earth and transmitted to space, are shown in Fig. 4. The upper plot shows the distribution of average OLR with the effects of clouds included, and the lower plot shows the ERBE estimate of the fluxes in cloud-free regions; both are for the month of January 1986. Regions with values less than 220 Wm^{-2} are heavily shaded. Lighter shading indicates OLR greater than 280 Wm^{-2}. Along the equator, over South America, Africa, and Indonesia, are three regions with very low OLR. These are regions of intense convection and high clouds, whose cold tops radiate relatively small amounts of energy to space. In the cloud-free climatology, the low OLR values in the convective regions are largely absent.

Figure 5 shows the albedo values for the month of January 1986. High albedos occur over convective regions in the tropics, over the Sahara and Arabian Deserts, and in high latitudes. When the effects of clouds are removed, only the high albedos associated with bright desert surfaces and snow and ice cover remain. The cloud-free albedos of land areas are generally above 15% and exceed those of adjacent oceans.

If average and clear-sky radiation budget components are averaged over the globe and the annual cycle, the net effect of clouds on the earth's energy balance can be calculated as the difference between them. The effect of clouds on the energy balance can be called the cloud radiative forcing of the climate, and is defined as follows:

$$\text{Cloud radiative forcing} = \text{average} - \text{clear-sky} \tag{6}$$

Estimates of the annual-average energy budget components and cloud radiative forcing are given in Table 2. These estimates are based on ERBE data for the months of April, July, and October 1985, and January 1986. We would expect the global annual-average net radiation to be very small. The relatively large value of 5 Wm^{-2} in Table 1 represents a sampling or calibration error and gives the approximate uncertainty that must be attached to these numbers. These values do not correspond exactly to those of Harrison et al. (1989) because of adjustments made over snow-covered areas, but they are very similar.

It can be seen in Table 2 that clouds reduce the terrestrial emission by about

Table 2. ERBE estimates of globally and annually averaged radiation budget components and cloud radiative forcing (Wm^{-2}).

	Average	Clear-Sky	Cloud Forcing
Outgoing Long-Wave	235	267	−33
Albedo (percent)	30%	15%	15%
Absorbed Solar	240	293	−53
Net Radiation	5	26	−20

OLR
JANUARY 1986

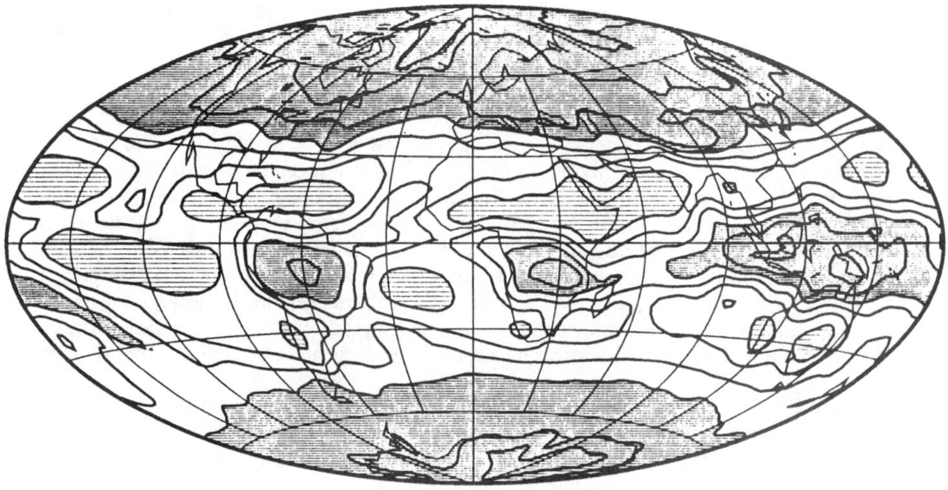

W/m**2

OLR CLEAR SKY
JANUARY 1986

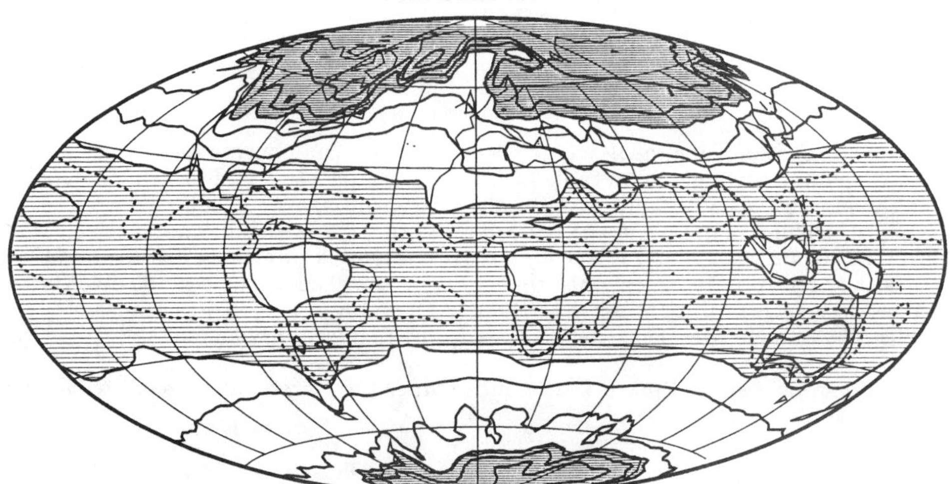

W/m**2

Fig. 4. Contour plot of outgoing terrestrial radiation for the month of January 1986. Top: Average for all scene types. Bottom: Average for cloud-free scenes only. The contour interval is 20 Wm^{-2}. Areas with values greater than 280 Wm^{-2} are lightly shaded, and areas with values less than 220 Wm^{-2} are heavily shaded. The dashed contour in the lower figure is 290 Wm^{-2}. The earth is shown in Hammer equal-area projection.

ALBEDO
JANUARY 1986

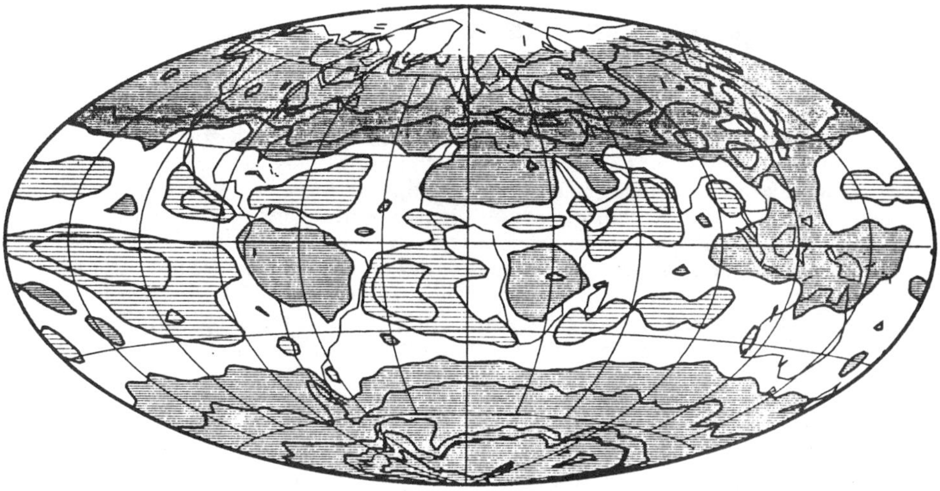

PERCENT

ALBEDO CLEAR SKY
JANUARY 1986

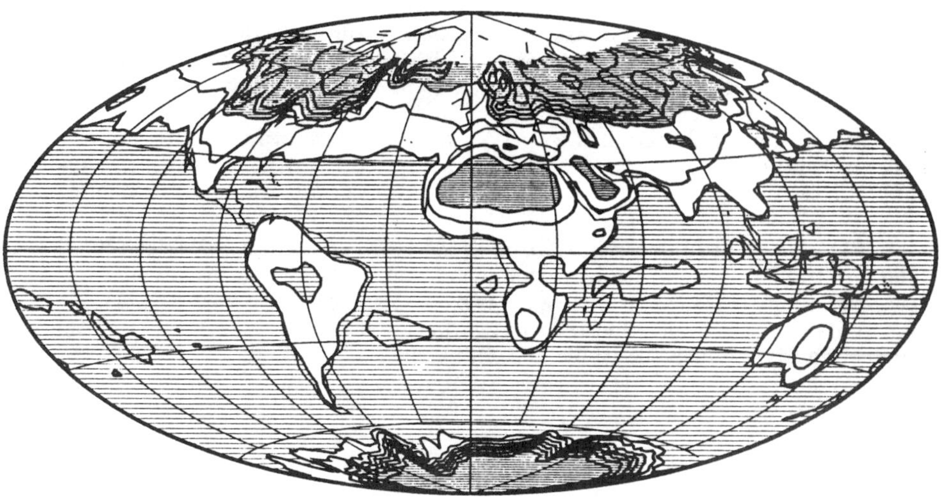

PERCENT

Fig. 5. Contour plot of planetary albedo for the month of January 1986. Top: Average of all scene types. Bottom: Average of cloud-free scenes. The contour interval is 5% from zero to 20%, and 10% from 20 to 100%. Values greater than 30% are heavily shaded. Light shading is used for values less than 20% in the top plot and 15% in the lower plot.

30 Wm^{-2}, which represents the magnitude of their greenhouse warming effect. Clouds double the earth's albedo from 15% to 30%, corresponding to a net loss of absorbed solar radiation of about 50 Wm^{-2}. The albedo effect of clouds thus outweighs their greenhouse effect and results in a net loss of energy of about 20 Wm^{-2} under present conditions.

One-Dimensional Radiative-Convective Climate Models

Because so much of the physics that controls earth's surface temperature is involved in the vertical transport of energy in the atmosphere, a globally averaged model with altitude as the only spatial coordinate furnishes a good starting point. If the concentrations and characteristics of the radiatively active gases and particles are known, one can calculate the radiative fluxes explicitly. The primary radiative gases are water vapor, carbon dioxide, and ozone, as mentioned previously, although a large number of other trace gases are gaining importance as their atmospheric concentrations increase. These other gases include methane, chlorofluorocarbons, nitrous oxide, and a host of other minor constituents. Obviously, the specification of clouds and their structure and optical properties plays a key role in determining the thermal equilibrium of the atmosphere. Also important are aerosols, surface albedo, and insolation.

Among the problems that one must face in formulating a one-dimensional model of the atmospheric temperature profile is the importance of non-radiative fluxes in the vertical redistribution of energy. The importance of such fluxes is clearly shown in Fig. 2. Of the solar radiation absorbed by the surface, 60% is returned to the atmosphere by latent and sensible heat carried away from the surface by the fluid motions of the atmosphere. The physics of these advective processes cannot be represented explicitly in a globally averaged model; they must be represented in an *ad hoc* fashion.

Manabe and Strickler (1964) performed a classic study of the vertical profile of temperature in thermal equilibrium. To account for the vertical transport of energy by atmospheric motions, they proposed a convective adjustment. In their calculations, the decrease of temperature with altitude (lapse rate) is not allowed to exceed a critical value. If radiation makes the lapse rate greater than the critical value, the temperature profile is adjusted in such a way that energy is conserved. This process implicitly requires a vertical redistribution of heat, which mimics the effect of convection. In practice, convective adjustment within one-dimensional models fixes the tropospheric lapse rate at the prescribed maximum value.

Figure 6 shows three profiles calculated by Manabe and Strickler (1964). The pure radiative-equilibrium profile, where no convective adjustment is made, has both a lapse rate, for which small vertical displacements are unstable, and a temperature discontinuity between the surface and the air in contact with the surface, and is therefore unstable and unrealizable. The thermal equilibrium calculation with a lapse rate adjusted to be no more than 6.5 K km^{-1} is much more realistic.

Radiative heating rates for the thermal equilibrium calculation are shown in Fig. 7. The net heating rate in the troposphere is approximately 1.5 K day^{-1} throughout the troposphere and is zero in the stratosphere. The troposphere is continually cooled by radiative processes. Energy lost radiatively is balanced by an upward flux of energy from the surface, which has a positive net radiation balance. Non-radiative energy exchange is modeled by the convective adjustment. The amount of energy moved from the surface to the atmosphere by the convective adjustment is about 105 Wm^{-2}, the same quantity implied in Fig. 2 for non-radiative energy fluxes at the surface. The net cooling in the troposphere is nearly equal to the long-wave cooling by water vapor alone. Infrared cooling by carbon dioxide is approximately balanced by heating produced by the absorption of solar radiation by water vapor. The stratosphere is in radiative equilibrium

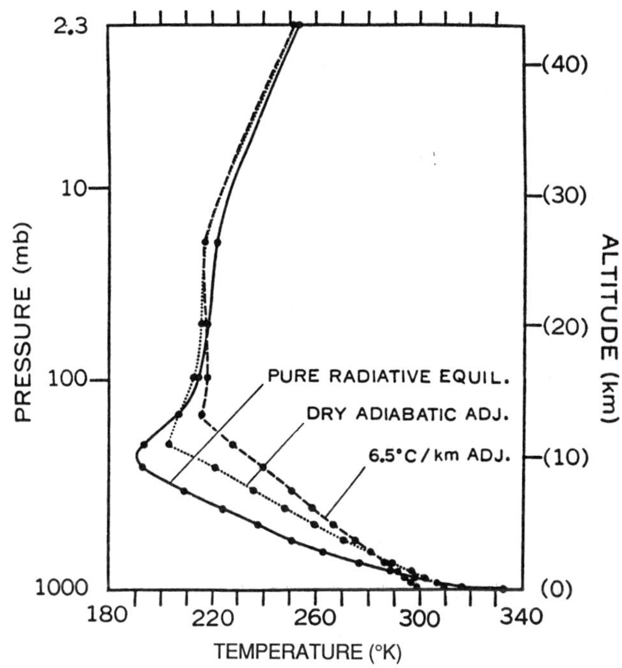

Fig. 6. Radiative-convective equilibrium temperature profiles for critical lapse rates of 6.5 K/km and 10 K/km, together with the pure, radiative-equilibrium calculation without convective adjustment (source: Manabe and Strickler, 1964).

because convective adjustment is not required. In the stratosphere, the radiative balance is primarily between long-wave cooling by carbon dioxide and solar absorption by ozone.

The relative roles of the primary gaseous absorbers can be seen in Fig. 8, which shows the thermal equilibrium profiles obtained by considering water vapor alone, water vapor plus carbon dioxide, and these two plus ozone. Clearly, most of the greenhouse effect is provided by water vapor, because adding 300 ppm of carbon dioxide increases the surface temperature by only about 10 K. Adding ozone in these calculations has only a tiny effect on surface temperature, but a profound effect on the stratosphere. The temperature inversion at the tropopause exists only because of the absorption of solar radiation by ozone in the stratosphere.

Manabe and Strickler (1964) found it necessary to introduce clouds into their calculation to produce the observed global-mean temperature profile. They specified cloudiness very crudely in terms of clouds at three levels with fixed albedos and solar absorptivities and assumed that they were blackbodies for terrestrial radiation. Figure 9 shows the thermal equilibrium profiles for their clear-sky calculation, the profile from their calculation with clouds, and the observed global-mean temperature profile, as represented by the U.S. Standard Atmosphere. Adding clouds reduced the suface temperature from nearly 300 K in

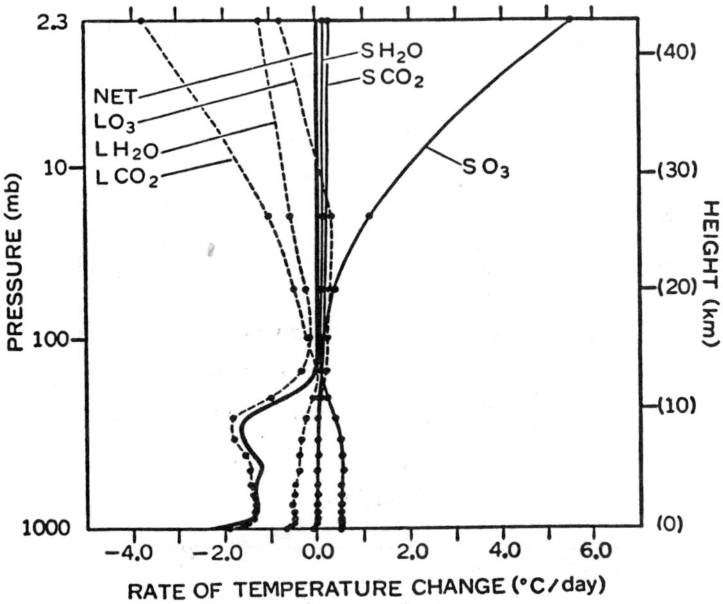

Fig. 7. Profiles of the heating rates by various gases in radiative-convective equilibrium for an atmosphere without clouds and a critical lapse rate of 6.5 K/km. The heavy solid line represents the net heating by all constituents. Prefix S indicates solar radiation and prefix L indicates long-wave or terrestrial radiation (source: Manabe and Strickler, 1964).

the clear-sky calculation to the observed value of 288 K. By using a simple relationship between forcing and temperature change, to be described in the next section, we can infer their cloud forcing. If we adopt a linear relationship between forcing and temperature change with a coefficient of 0.25 $K/(Wm^{-2})$, we find their net cloud forcing must have been about -40 Wm^{-2} to have produced a temperature drop of 10 K. This amount is double the cloud forcing inferred from ERBE data for the real atmosphere. Part of their need for such a large negative cloud radiative forcing can be explained by the relatively large solar constant they used, which was 2 Langleys per minute, or 1,395 Wm^{-2}. This is about 30 Wm^{-2} higher than the currently accepted value, and has the effect of reducing their required cloud forcing by about 8 Wm^{-2} and their clear-sky equilibrium temperature by about 2 K.

CLIMATE SENSITIVITY AND FEEDBACKS

The climate system is very complex, and over long time scales, many elements of the climate system can change in response to an initial perturbation.

A fundamental question is one of climate sensitivity: how much will the global mean climate change in response to a specified forcing? In analyzing the sensitivity of the global climate, it is helpful to determine which climate-system elements have the greatest potential for affecting the magnitude of the change. Such a determination will identify those aspects of the climate change problem that should be addressed first and those elements of the climate system that require accurate treatment in climate models.

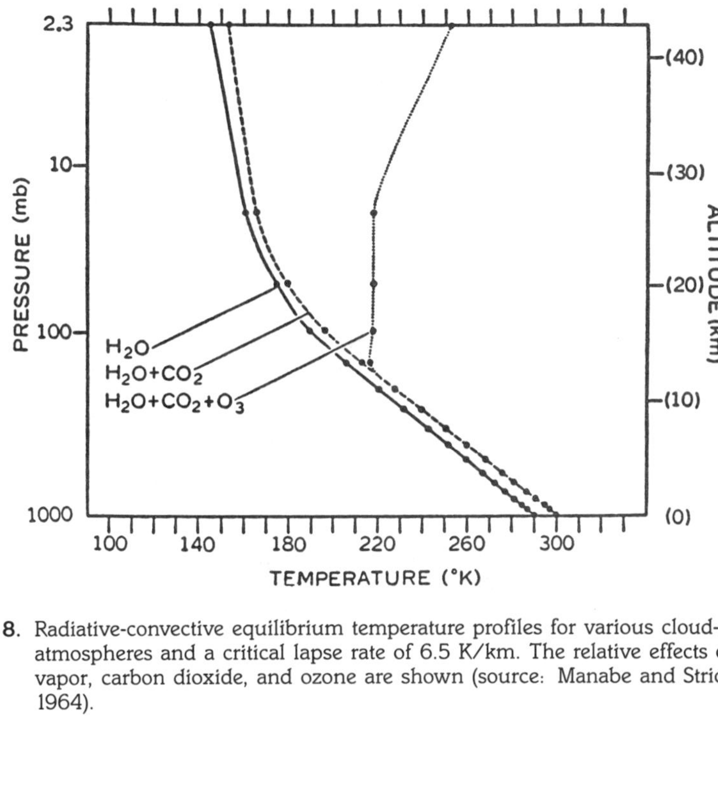

Fig. 8. Radiative-convective equilibrium temperature profiles for various cloud-free atmospheres and a critical lapse rate of 6.5 K/km. The relative effects of water vapor, carbon dioxide, and ozone are shown (source: Manabe and Strickler, 1964).

The problems of sensitivity and feedbacks can be expressed most succinctly with mathematical equations. Suppose we consider the classic problem in climate sensitivity analysis of determining how much the global-mean surface temperature will change if we alter the solar constant by a specified amount. This change is expressed as a total derivative of the global-mean surface temperature with solar constant:

$$\frac{dT_S}{dS} = ? \tag{7}$$

Many aspects of the climate respond when the solar constant is varied, and many of these will feed back on the magnitude of the temperature change resulting from the solar constant variation. Examples are relative humidity, clouds, ice cover, vegetation, ocean circulation, and many more. We represent these other variables as x_i. In this case, we can calculate the total derivative using the chain rule,

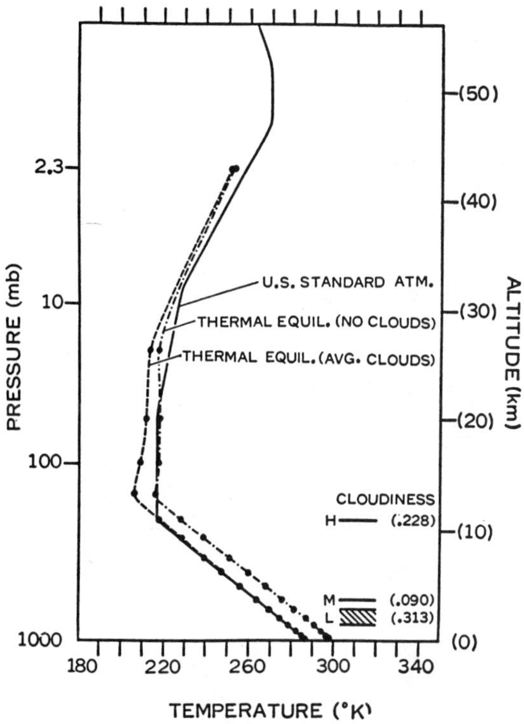

Fig. 9. Radiative-convective equilibrium temperature profiles with and without clouds compared with the U.S. Standard Atmosphere (source: Manabe and Strickler, 1964).

$$\frac{dT_S}{dS} = \frac{\delta T_S}{\delta S} + \sum_{i=1}^{N} \frac{\delta T_S}{\delta x_i} \frac{dx_i}{dS}. \tag{8}$$

The partial derivative standing alone on the right in Eq. (8) represents the direct effect of the solar constant change on the surface temperature, and the summation represents the accumulation of all the feedbacks associated with each of the climatic variables x_i. The total number of possible feedback variables N is not generally small, but it is hypothesized that, by including only those variables with the strongest feedbacks, the number of variables can be kept manageably small without seriously affecting the quality of the prediction. In any case, it is necessary to evaluate the importance of each potential feedback process in order to establish its priority for inclusion in a climate model.

Sensitivity of the Global-Mean, Energy-Balance Model

As a starting point for evaluation, one can consider the sensitivity of the global-mean, energy-balance model. The balance is as expressed in Eq. (1), except that we allow the emitted energy to be an unknown function of the surface temperature $F(T_S)$. The albedo is also assumed to be parametrically related to the global-mean surface temperature. R is the net radiation, which is required to be zero in equilibrium:

$$R = \frac{S}{4}(1 - \alpha(T_S)) - F(T_S) = 0. \tag{9}$$

Suppose that the equilibrium climate is forced to change by adding a perturbation Q to the energy balance. Q may be the radiation imbalance introduced either by a sudden doubling of the carbon dioxide concentration, or by a change in solar constant. We can again use the chain rule to express the change in net radiation to the perturbation forcing:

$$\frac{dR}{dQ} = \frac{\delta R}{\delta Q} + \frac{\delta R}{\delta T_S} \frac{dT_S}{dQ} = 0. \tag{10}$$

Because the total derivative of net radiation must be zero in going from one equilibrium state to another, we can rearrange Eq. (10) to obtain an expression for the total change in temperature for a particular forcing Q:

$$\frac{dT_S}{dQ} = \frac{-\frac{\delta R}{\delta Q}}{\frac{\delta R}{\delta T_S}} = -\left(\frac{\delta R}{\delta T_S}\right)^{-1} = \left(S\frac{\delta \alpha}{\delta T_S} + \frac{\delta F}{\delta T_S}\right)^{-1} = \lambda. \tag{11}$$

We can use this expression to relate the magnitude of the global-mean temperature response to a specified external forcing:

$$\Delta T_S = \lambda \cdot \Delta Q. \tag{12}$$

In the global-mean model, where OLR and albedo are parametrically related to the surface temperature, the feedback factor λ depends on the sensitivities of the albedo and OLR to surface temperature. As a baseline case, we can consider a climate where the albedo is fixed and the OLR varies according to the Stefan-Boltzmann Law, as in Eq. (1). If we further assume that the emission temperature is smaller than the surface temperature by a fixed amount, $T_e = T_S - constant$, then we obtain a baseline value for the OLR feedback parameter,

$$\frac{\delta F}{\delta T_S} = \frac{\delta}{\delta T_S}(\sigma T_e^4) = 4\sigma T_e^3 = 3.8 \text{ Wm}^{-2}\text{ K}^{-1}. \tag{13}$$

The feedback parameter associated with the blackbody function is thus

$$\lambda_{BB} = \left(\frac{\delta F_{BB}}{\delta T_S}\right)^{-1} = 0.26 \text{ K } (\text{Wm}^{-2})^{-1}. \tag{14}$$

This is the feedback parameter for a climate model in which only the negative feedback of the blackbody emission is included. For planetary temperatures, the increase of energy emission with temperature is the primary stabilizing mechanism. The basic radiative-convective equilibrium model, with absolute humidity, clouds, and albedo all held constant, as considered by Manabe and Strickler (1964) has a sensitivity as described approximately in Eq. (14). We have used this relationship to derive the approximate cloud forcing in their model from the effect that clouds have on the thermal-equilibrium surface temperature.

The instantaneous insertion of a doubled carbon dioxide concentration introduces about a 4 Wm^{-2} reduction in the OLR and, thereby, an equivalent increase in the net radiation. According to Eqs. (12) and (14), the direct effect of this forcing on a climate in which only the blackbody emission feedback operates would be a warming of about 1 K. The climate is not as simple as that, however. Operating in the climate system are a large number of important feedbacks that, cumulatively, can have a substantial effect on the equilibrium climate response to perturbed forcing. Three of the most important feedbacks that can be discussed within the context of a one-dimensional model are relative humidity, ice-albedo, and clouds.

Relative Humidity Feedback

One of the most basic and most powerful feedbacks that operates within the earth's climate system is the relative humidity feedback. From Table 1 we recall that water vapor accounts for about half of the current greenhouse effect. The saturation pressure of water vapor increases rapidly with temperature according to the Clapeyron-Clausius relationship,

$$\frac{de_S}{dT} = \frac{L}{T(a_v - a_l)}. \tag{15}$$

In Eq. (15), e_S is the saturation vapor pressure above a liquid surface, L is the latent heat of vaporization, T is the temperature, and α represents the specific volume of the vapor and liquid forms of water. The Clapeyron-Clausius relation (Eq. (15)) can be manipulated to express the fractional change of saturation vapor pressure, and thereby the specific humidity at saturation q_S, to the fractional change of temperature:

$$\frac{\Delta q_S}{q_S} = \frac{\Delta e_S}{e_S} = \left(\frac{L}{R_v T}\right)\frac{\Delta T}{T} = r\frac{\Delta T}{T}. \tag{16}$$

For terrestrial conditions, $T \sim 260$ K and the factor r is approximately 20. This means that a 1% change in temperature of about 3 K will result in a 20% change in the saturation vapor pressure. When the relative humidity (the ratio of the actual humidity to the saturation humidity) remains fixed, the actual water vapor in the atmosphere will increase by 20% for every 3 K temperature increase. This rapid logarithmic increase of saturation pressure with temperature can be seen more explicitly if we consider the approximate solution to Eq. (15) to be valid in the vicinity of STP:

$$e_S \cong 6.11 \cdot \exp\left\{\frac{L}{R_v}\left(\frac{1}{273} - \frac{1}{T}\right)\right\}. \tag{17}$$

The units of pressure in Eq. (17) are millibars. The exponential dependence of the saturation vapor pressure on temperature expressed by Eq. (17) is shown in Fig. 10.

Manabe and Wetherald (1967) noted that the relative humidity distribution in the atmosphere remains nearly fixed through a seasonal cycle in which temperature changes at high latitudes are rather large. On this basis, they argued that a good first approximation for what would happen during a climate change is to assume that the relative humidity remains constant as the temperature changes. The assumption of fixed relative humidity (FRH), which is certainly better than an assumption of fixed absolute humidity (FAH), gives rise to a very powerful positive feedback, as sketched in Fig. 11. If an external forcing that changes the climate is applied, the change in water vapor (the primary greenhouse gas) will tend to amplify the temperature change. The water vapor feedback leads to an approximate doubling of the sensitivity of earth's climate, as expressed in terms of the feedback factor λ:

$$\lambda_{FRH} = \left(\frac{\delta F_{FRH}}{\delta T_S}\right)^{-1} = 0.50 \text{ K } (\text{Wm}^{-2})^{-1}. \tag{18}$$

The effect of the FRH assumption on the dependence of OLR on surface temperature in a one-dimensional model is shown in Fig. 12. Note that the exponential form of the vapor pressure curve given in Eq. (17) offsets the fourth-power dependence of the blackbody emission, so that the OLR depends approximately linearly on surface temperature over a large range of temperatures. Furthermore, the rate of increase of OLR with temperature is much less than that of a blackbody, giving rise to the increase in sensitivity.

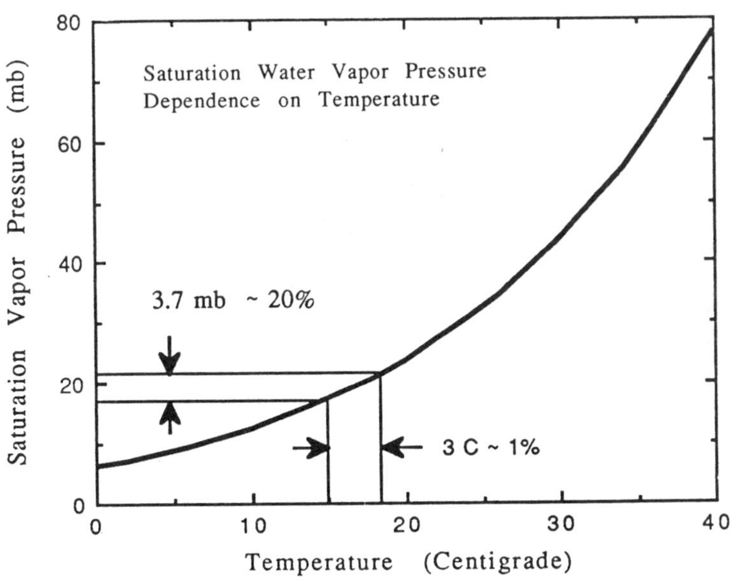

Fig. 10. Plot of the saturation water vapor pressure at standard ambient pressure as a function of temperature calculated from Eq. (17). Relative percentage changes of temperature and water vapor are shown.

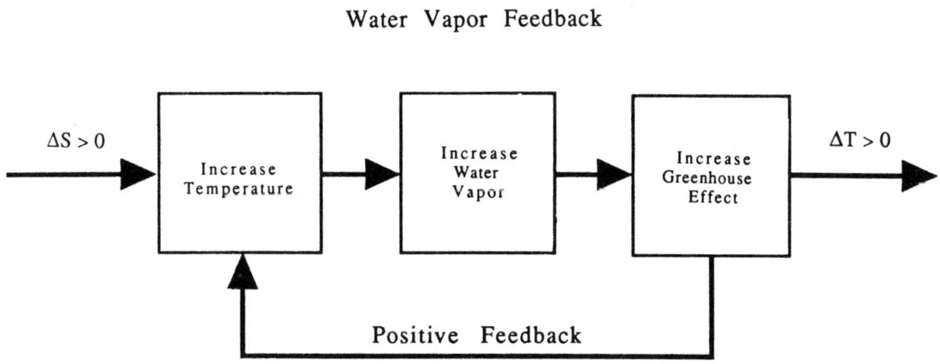

Fig. 11. Schematic diagram of the positive feedback of relative humidity on temperature variations. An initial climate forcing, such as a solar constant increase ΔS, causes a directly forced temperature increase. This temperature increase leads to an increase in the infrared opacity of the atmosphere, because warmer air can hold more water vapor. The greenhouse effect of this added water vapor leads to a further increase in temperature, so that the net temperature change ΔT is larger than the direct effect of the solar constant change and contains a portion contributed by the feedback process.

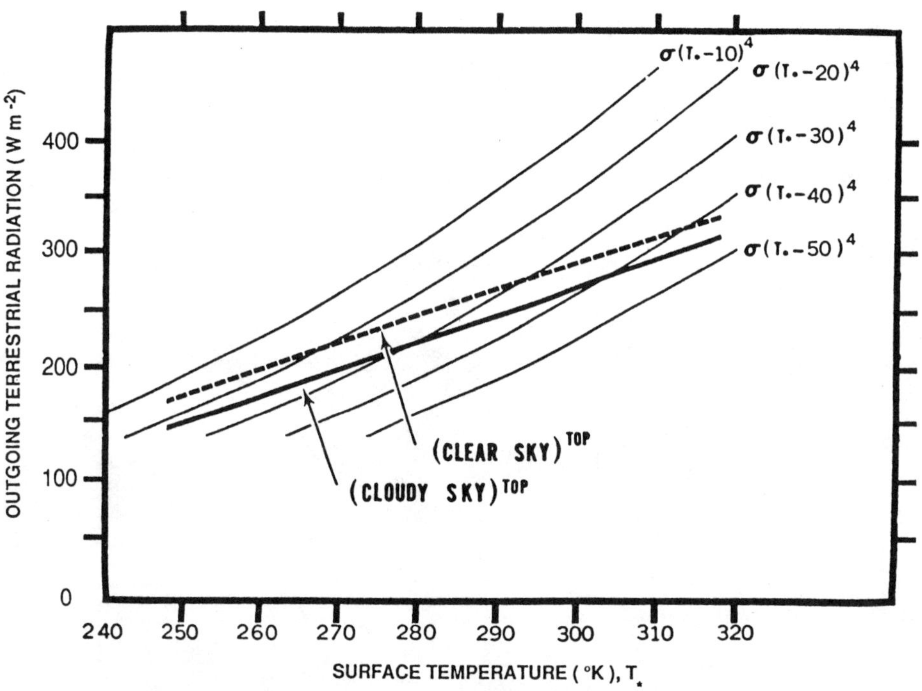

Fig. 12. Outgoing terrestrial radiation at the top of the atmosphere calculated for temperature profiles of the same shape with varying surface temperature and fixed relative humidity. Thin lines show the blackbody emission curves for bodies at temperatures equal to the surface temperature minus the amounts shown (adapted from Manabe and Wetherald, 1967).

Table 3. Surface Albedos of some selected surfaces. Units are percentage of incident solar radiation reflected (source: Kukla and Robinson, 1980).

Surface Type	Range	Typical Value
Ice-Free Surfaces		
Water (0°-60° latitude)	5-20%	7%
Water (60°-90° latitude)	6-44%	10%
Tropical Rainforest	7-15%	10%
Farmland, Common Soils	7-25%	17%
Semi-Deserts	20-33%	22%
Deserts with Light Sand	28-44%	35%
Snow-Covered Surfaces		
Snow over Land	25-70%	45%
Snow over Ice Sheet	75-90%	80%
Sea Ice	25-80%	60%

Ice-Albedo Feedback

The fraction of incident solar radiation reflected by earth's surface varies widely, depending on the nature of the surface. Examples of some typical surface albedo values are given in Table 3. The surface with the lowest albedo is that of oceans at relatively low latitudes. Albedos of land surfaces are generally higher, with moist, densely vegetated surfaces having lower albedos than dry grasslands or deserts. The highest surface albedos are those of snow- or ice-covered surfaces. Each classification in Table 3 has a range of values, depending on the detailed characteristics of the particular surface in question. The albedo of water surfaces depends on the solar zenith angle, the wind speed and related waves and foam, and the suspended matter in the water. Land-surface albedos depend on the nature of the vegetation and the underlying soil and rock. The albedo of snow-covered surfaces depends on the depth, age, and temperature history of the snow cover and the color and texture of the underlying surface. The highest land albedos are those of cold snow surfaces underlain by a layer of permanent ice cover, such as those that occur over the major ice sheets of Greenland and Antarctica.

The ability of snow cover to transform the surface from a relatively low albedo surface to a very high albedo surface forms the basis of ice-albedo feedback. As the temperature becomes colder, the fraction of surface area covered by snow and ice is expected to increase. An increase in snow cover raises the albedo of the planet, which reduces the radiation balance and leads to further cooling. The ice-albedo feedback was employed by Budyko (1969) and Sellers (1969) in simple climate models. These models carry latitude as the only spatial variable and use the energy balance of the earth as the governing equation. Planetary albedo, terrestrial emission, and equator-to-pole energy transport are all related parametrically to the mean surface temperature at each latitude (e.g., North et al., 1981). The models showed a surprising sensitivity, such that a decrease in the solar constant of less than 1% could produce an ice age. The reasons for this sensitivity can be analyzed by considering the feedback parameter λ in Eq. (11).

From the perspective of evaluating the importance of albedo feedback in climate models, the key parameter is the dependence of the planetary albedo on the global-mean surface temperature. Budyko (1969), for example, assumed that the albedo in the absence of snow cover would be 0.32, and that when the annual average temperature fell below −10°C at a latitude, the albedo would increase to 0.62, the albedo assumed for a snow-covered surface. Thus, the albedo contrast associated with snow cover was a uniform 30%.

Lian and Cess (1977) pointed out that Budyko overestimated ice-albedo feedback for three reasons. First, the surface and planetary albedos in the absence of snow cover are already high at the polar latitudes, where perennial snow forms because of solar zenith angle effects. Second, the sun never rises very far above the horizon at polar latitudes, and scattering of solar radiation by the atmosphere and natural surfaces becomes more efficient at high solar zenith angles. Finally, and more importantly, much of the surface albedo contrast associated with snow cover is screened out by overlying clouds. Lian and Cess estimate that only 26% of the surface albedo contrast associated with snow cover is felt at the top of the atmosphere. In addition, the albedo of clouds increases rapidly with solar zenith angle, making the screening of surface albedo changes by clouds particularly effective at high latitudes.

In Table 4, the operative sensitivity of OLR and albedo to surface temperature in Budyko's model is compared with that in models considered more appropriate by the author. Also shown is a matrix of feedback parameters for each combination of sensitivities. The dependence of OLR on surface temperature is that of an atmosphere with fixed relative humidity and does not include potential effects from cloud changes. It can be seen that Budyko used a rather small negative OLR feedback (1.42 $Wm^{-2}K^{-1}$) and a rather large positive ice-albedo feedback ($-2.2 \times 10^{-3}K^{-1}$) compared to current estimates (2.0 $Wm^{-2}K^{-1}$ and $-0.9 \times 10^{-3}K^{-1}$, respectively). These two differences, taken together, produce a much larger feedback parameter ($\lambda = 1.42$ $K/(Wm^{-2})$) than current best estimates would indicate ($\lambda = 0.59$ $K/(Wm^{-2})$). In particular, the enhancement of the feedback parameter by ice-albedo feedback is weaker than anticipated by Budyko (1969) and Sellers (1969) because of zenith angle effects and cloud screening of surface albedo changes.

Table 4. OLR and Albedo dependences on global-mean surface temperature from Budyko (1969) and from current estimates, and a matrix of the feedback parameters λ calculated from Eq. (11). Units of δ are $K/(Wm^{-2})$.

		$\frac{\delta\alpha}{\delta T_S}$	
		Budyko (1969) $-2.2\times10^{-3}K^{-1}$	Best Estimate (1989) $-0.9\times10^{-3}K^{-1}$
$\frac{\delta F}{\delta T_S}$	Budyko (1969) 1.45 $Wm^{-2}K^{-1}$	$\lambda = 1.42$	$\lambda = 0.87$
	Best Estimate (1989) 2.0 $Wm^{-2}K^{-1}$	$\lambda = 0.80$	$\lambda = 0.59$

Cloud Feedback

As discussed earlier in this paper, clouds have a significant effect on the earth's energy balance, and the individual contributions of the albedo and greenhouse effects of clouds are larger than their net effect. Small percentage changes in either of these effects would have the same order-of-magnitude influence on the energy balance as would either a doubling of the carbon dioxide concentration of the atmosphere, or a 2% change in solar constant. Changes in cloud structure that could influence the sensitivity of climate include changes in cloud texture, top height, liquid water content, particle size, fractional area coverage, and the distribution of clouds between land and ocean or high and low latitudes.

A one-dimensional model with fixed relative humidity can be used to assess the sensitivity of the global climate to a variety of influences, including clouds. To do this, we can introduce a new sensitivity parameter β, which relates global-mean, surface-temperature changes to fractional changes in particular climate parameters denoted by x_i:

$$\beta_i = x_i \frac{\delta T_S}{\delta x_i}. \tag{19}$$

Using results published by Manabe and Wetherald (1967), we can estimate the β values associated with a number of climate parameters, including clouds of a specified type. Manabe and Wetherald define three types of clouds—high, middle, and low—which are assumed to be blackbodies for terrestrial radiation, and whose albedo and height are specified according to the values given in Table 5. The specifications of cloud albedo and cloud-top temperature determine the sensitivity of the surface temperature to cloud fractional-area changes. The values in Table 5 are a crude representation of a much richer spectrum of possible cloud properties that occur in nature. The albedo of the earth's surface is assumed to be 0.102 in Manabe and Weatherald's baseline case.

The values of β estimated from figures published in Manabe and Wetherald (1967) are given in Table 6. All the values except two are for one-dimensional thermal equilibrium calculations with fixed relative humidity. From Table 1, we again infer the importance of relative humidity feedback, because it doubles the sensitivity of the surface temperature to changes in the solar constant. By dividing the β values by 100, we obtain the temperature change that would result from a 1% change in the value of the variable in question. We can see, therefore, that a 100% change in carbon dioxide concentration can be balanced by a 10% change in the amount of low cloud of the type specified here. On the other hand, the effect of a 100% change in carbon dioxide can be doubled by a 20% change in the high-cloud amount.

Table 5. Properties of cloud types used in the analysis of Manabe and Wetherald (1967). Heights are given in kilometers above the surface.

Cloud Type	Cloud-Top Height	Cloud-Bottom Height	Cloud Albedo
High	10.0	10.0	0.20
Middle	4.1	4.1	0.48
Low	2.7	1.7	0.69

Table 6. Sensitivity estimates from the one-dimensional radiative-convective equilibrium model of Manabe and Wetherald (1967). FAH indicates a fixed absolute humidity calculation. FRH indicates a result from a fixed relative humidity calculation in which relative humidity feedback is included.

Parameter (x_i)	β_i
Solar Constant (FAH)	70 K
Solar Constant (FRH)	140 K
Carbon Dioxide Concentration (FRH)	3.5 K
Relative Humidity	15 K
Surface Albedo (FRH)	−35 K
High Cloud (FRH)	20 K
Middle Cloud (FRH)	−20 K
Low Cloud (FRH)	−40 K

The difficulty in evaluating cloud feedback is not in calculating the effect of a specific change in cloud properties, but rather in predicting how cloud properties may vary during a climate change. A recent analysis of 14 atmospheric models (Cess et al., 1989) showed that the feedback factor associated with the models' predicted cloud changes varied by a factor of three. Climate change was simulated in the models by changing the specified sea-surface temperature. By specifying the surface temperature, and hence the ice cover, the ice-albedo feedback was eliminated. The sensitivity parameter λ could then be evaluated as the ratio of the change in the specified surface temperature to the resulting change in net radiation. The effects of clouds were isolated in the models by calculating the clear-sky and average radiation balances separately. When the effects of cloud changes were removed, the feedback factor produced by the models ranged from 0.42 to 0.57 K/Wm^{-2}, with a mean value of 0.47. This number is very close to the value obtained from a one-dimensional thermal equilibrium model with fixed relative humidity. Infrared radiation and relative humidity are the main feedbacks allowed to operate in these experiments, so it seems that the 14 models all infer approximately the same sensitivity for these feedbacks. When cloud changes are included, however, the model sensitivities range from 0.39 to 1.11 $K/(Wm^{-2})$, with a mean value of 0.68 and a standard deviation of 0.24. The sensitivity of the models to cloud changes, and the contribution of cloud feedback to total sensitivity, each vary by a factor of almost three among the models. Although the cloud prediction schemes within these models are primitive, this intercomparison reveals both a great potential for cloud changes to influence the sensitivity of the climate and a lack of consensus about the magnitude of the cloud feedback effect.

The Hydrologic Cycle

Water and the hydrologic cycle are at the center of the most important basic feedbacks in the climate system. Eighty percent of the earth's natural greenhouse effect is provided by water vapor and clouds of liquid water and ice. Clouds also account for half of the reflection of solar radiation by the earth. Fifty percent of the solar energy available at the top of the atmosphere passes through the atmosphere and is absorbed at the surface. Eighty percent of the absorbed solar energy evaporates water. This water vapor condenses in the atmosphere, where its latent energy offsets a continuous radiative energy loss by the atmosphere. The water vapor balance of the atmosphere is determined by the upward flux of water in clouds, precipitation microphysics, and water transport by the large-scale circulation patterns of the atmosphere. Earth's greenhouse effect, planetary albedo, and surface water balance are all coupled closely together and strongly affect the response of the climate to perturbations.

CLIMATE SIMULATION: GENERAL CIRCULATION MODELS

Predicting the effect of projected increases in greenhouse gases on the climate over the next century requires a very sophisticated climate model that must accurately treat all of the feedbacks discussed in the previous section and more, and must fully allow for the possible interactions between them. Such complete models do not currently exist. Nonetheless, existing models can produce projections for the coming century from first principles; these can be validated against past climates and must be taken seriously. In this section, we will describe briefly the elements of current climate models and the physical processes incorporated within them. The validation of these models against current and past climate data will then be discussed.

Structure of a Climate Model

A hierarchy of climate models exists, ranging from the simplest energy balance model described at the beginning of this paper to highly sophisticated models that require the largest and fastest available computers. We will focus here on this latter category of model, in which we place our hope of eventually producing reliable forecasts of future regional and global climate changes. For historical reasons, models of this class are referred to as general circulation models, or GCMs. They evolved out of numerical weather prediction models that were first developed during the 1950s and 1960s (Smagorinsky, 1983; Simmons and Bengtsson, 1984; Washington and Parkinson, 1986). Although modern improvements in these models focus primarily on increasingly sophisticated treatment of the physical processes that control feedback mechanisms, predicting both climate change for specified geographic regions and the changed probabilities of localized extreme events is critical. Prediction of regional changes requires that the evolution of individual weather systems continues to be an essential climate model component, even if the important effects of these systems on the mean climate could be incorporated without explicit calculation of their life histories.

The structure of a typical GCM, designed for climate simulation during the 1970s and 1980s, is shown in Fig. 13. A GCM uses the equations of motion to predict the temporal evolution of wind fields, which transport heat and moisture in the atmosphere. Models of vertical transport of heat by radiation and small-scale convection are included. The latter is a highly simplified, parametric model, because the motions in these small-scale systems cannot be explicitly calculated in a global model with current or foreseeable computer technology.

A GCM must also predict temperature, moisture content, and snow cover for land surfaces. Vegetation on land plays a key role in mediating the flow of moisture between the atmosphere and the soil, but vegetation is not explicitly carried as a variable in most current climate models (Dickinson, 1983, 1984). Land surface is generally represented as one or more layers with fixed moisture-holding capacities. The actions of water-stressed plants in limiting evapotranspiration are simulated by reducing evaporation efficiency as the water content of the soil becomes small.

For the calculations described in this section, the ocean is modeled as a mixed layer with a specified heat capacity but no currents. Heat cannot be transferred horizontally in such an ocean model, and communication with the deeper layers of the ocean is also suppressed. Sea ice plays a very important role in climate sensitivity and must also be carried as a variable. Sea ice is generally predicted using a thermodynamic model in which ice thickness is determined by the requirement to keep the ice/water interface at the freezing point of water. The insulating effect of sea ice can be related to the thickness of the ice and the overlying snow cover.

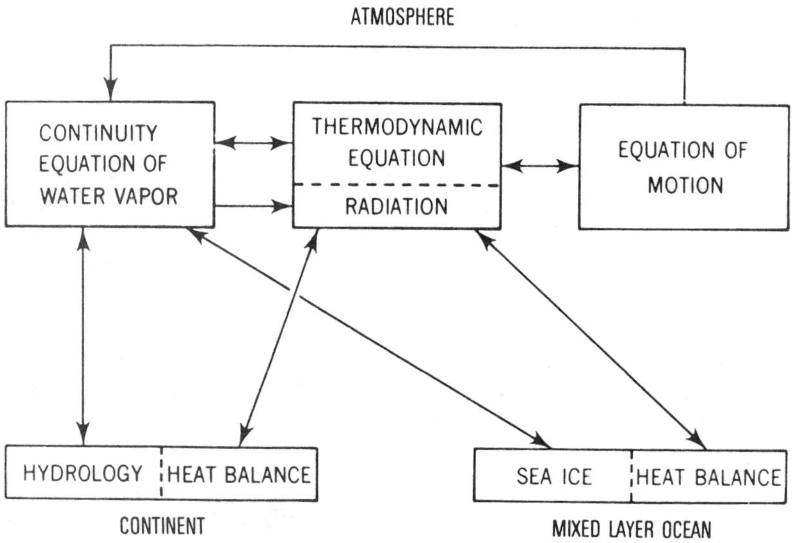

Fig. 13. Box diagram showing the structure of a typical climate model of the 1970 to 1990 period (source: Manabe and Stouffer, 1988).

Validation of Climate Models

After calibration and adjustment, a GCM can be made to reproduce the current climate of earth. GCMs can simulate not only the annual mean climate, but also the response to the annual cycle in insolation. The diurnal cycle places an additional constraint on these models and is particularly important for testing the land-surface portion of the model. Over land, the diurnal variation of temperature depends sensitively on the buffering of temperature change by evaporation. Although the ability of climate models to simulate the current climate is, of course, essential, it tells us relatively little about how well these models might predict climate change. Because of the highly simplified way in which many small-scale physical processes must be treated, the number of adjustable parameters in a GCM is large. It is therefore uncertain whether or not these simplifications are adequate to treat cases where the climate changes significantly. Even if simplifications are adequate to this task in principle, it is possible to adjust the parameters in such a way that the current climate is well simulated, but the sensitivity of the model is not realistic.

Ice Age Simulations

For the reasons outlined above, it is important to test the realism of the model against an example of significant climate change. From geologic evidence and a variety of natural recording systems, we know that the earth has undergone a series of glacial/interglacial cycles during the last million years or so. We are currently in an interglacial period, in the sense that the volume of continental sheet ice is near its minimum value of the last million years. The last glacial maximum, when the ice volume was near its maximum value, occurred about 18,000 years ago (18 kyr BP). A significant amount of information on the state of the climate system during the years between the last glacial maximum and the present is available from paleoclimatic data. This information includes ice sheet positions and approximate depths, sea level, minimum and maximum sea-ice extent, sea-surface temperatures, land vegetation, lake levels, and atmospheric composition. In addition, the parameters of earth's orbit, and thereby the distribution of insolation with season and latitude, are known with precision. During the last glacial maximum, sea level was lower than at present by about 100 meters because of the large volume of ice suspended above sea level in continental ice sheets. In addition, snow-free land albedos were higher because of generally drier conditions over most land masses, especially in the subtropics. The carbon dioxide concentration was about 200 ppm, compared to the current value of about 350 ppm and the preindustrial value of about 280 ppm.

Paleoclimatic data can be used to test climate models. For example, Broccoli and Manabe (1987) considered the importance of four factors to the earth's climate during the last glacial maximum. These factors were the changes that occurred in continental ice sheets, sea level, atmospheric carbon dioxide concentration, and snow-free land albedo. They conducted four experiments, the first of which was a control simulation of the current climate. The second simulation included changes in continental ice and sea level to make them characteristic of the last glacial maximum. The third experiment added the effect of the snow-free land albedo changes, and the final experiment added the carbon dioxide decrease.

This sequence of experiments can answer two questions: By comparing experiments, can we estimate the relative importance of ice sheet formation, vegetation changes, and atmospheric carbon dioxide concentration as feedbacks that determine the amplitude of the temperature change between current and glacial conditions? And is the global-mean, sea-surface temperature decrease, resulting from the changes in ice, vegetation, and CO_2 concentration, close to the decrease inferred from ocean core data? The response of the model climate on regional scales to the imposed changes can be compared to paleoclimatic data. This comparison allows a measure of validation for a model's ability to simulate climate change.

The changes in global-mean, sea-surface temperature associated with the four climate experiments performed by Broccoli and Manabe (1987) are shown in Table 7. The averages were taken over only those grid boxes that were designated as ocean in all four experiments. These calculations show that the temperature changes produced by ice sheet growth and CO_2 decrease are about the same size. The inference is that the positive feedbacks associated with ice sheet and atmospheric changes are of about the same order of magnitude. The temperature change produced by vegetation and soil changes is estimated to be smaller—only about 10% of the total change in sea-surface temperature. Similar conclusions are reached if the surface air temperature from the entire globe is used. Table 7 also shows the sea-surface temperature changes inferred from deep-sea cores and published by CLIMAP (1976). The changes produced by the model are slightly larger than those inferred by CLIMAP but are within the uncertainty of these estimates. The differences between the hemispheres are predicted by the model and clearly relate to the fact that the ice sheets are concentrated in the Northern

Table 7. Sea-surface temperature changes produced by the model of Broccoli and Manabe (1987) in response to the gradual introduction of climate forcings that are characteristic of the last glacial maximum.

	Global	N.H.	S.H.
Glacial Ice Sheets and Sea Level	−0.8	−1.6	−0.2
Glacial Snow-Free Land Albedo	−0.2	−0.3	−0.2
Glacial Carbon Dioxide	−1.0	−0.7	−1.1
All Glacial Changes Combined	−1.9	−2.6	−1.5
CLIMAP (18 kyr BP - present)	−1.6	−1.9	−1.3

Hemisphere. The ice sheets in the Northern Hemisphere have a relatively small effect on Southern Hemisphere temperatures. The simultaneity of ice ages in the two hemispheres appears to be enforced by the change in carbon dioxide concentration and the associated changes in the global greenhouse effect.

The latitudinal distribution of sea-surface temperature changes from the four experiments of Broccoli and Manabe (1987) are shown in Fig. 14. Also shown are the differences in sea-surface temperature between 18 kyr BP and the present values inferred by CLIMAP. Comparison of these values with the total ice age simulation reveals that the model agrees very well with observations, except in the subtropics and in high northern latitudes. The discrepancies may be related to the fact that the model has no ocean circulation. The polar amplification of the temperature change is well simulated and corresponds to the movement of the sea-ice margin in the model. Comparison of the temperature changes for individual experiments reveals that the introduction of ice sheets has a strong effect in the Northern Hemisphere, as expected, but that ice sheets have very little effect on the temperature of the Southern Hemisphere. This suggests that ice-albedo feedback, by itself, cannot explain the observed global simultaneity of ice ages. In order to produce the global nature of ice ages, the changes in atmospheric carbon dioxide concentration are necessary. Figure 14 shows that most of the Southern Hemisphere cooling in the model ice age is produced by the CO_2 reduction.

Orbital Parameters and Climate History

The knowledge that the earth has undergone periods of remarkable glacial advance and retreat began to spread in the 1840s. Almost immediately, it was proposed that variations in earth's orbit could drive glacial waxing and waning (Imbrie and Imbrie, 1979). Three key parameters control the distribution of insolation with latitude and season: obliquity, eccentricity, and longitude of

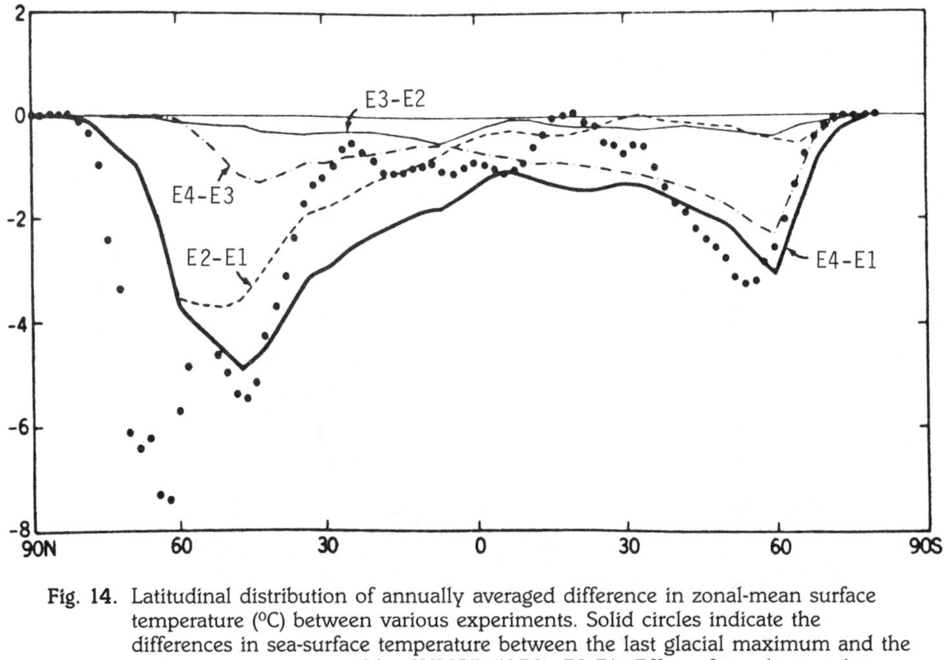

Fig. 14. Latitudinal distribution of annually averaged difference in zonal-mean surface temperature (°C) between various experiments. Solid circles indicate the differences in sea-surface temperature between the last glacial maximum and the present, as reconstructed by CLIMAP (1976). E2-E1: Effect of ice sheet and sea level changes. E3-E2: Effect of snow-free land albedo changes. E4-E3: Effect of carbon dioxide concentration decrease from 300 to 200 ppm. E4-E1: Effect of all changes combined (source: Broccoli and Manabe, 1987).

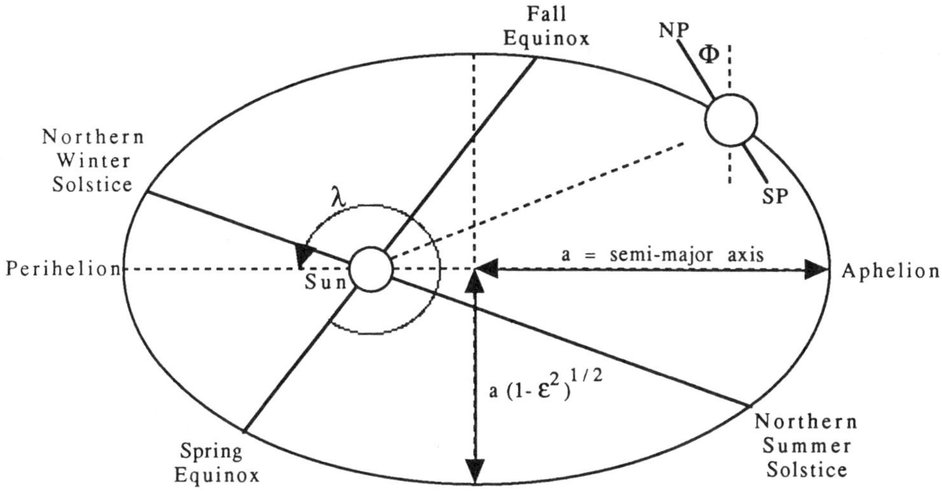

Fig. 15. Schematic diagram of the orbital parameters that determine the seasonal and latitudinal distribution of insolation. The parameters are eccentricity (ϵ), obliquity (ϕ), and longitude of perihelion (λ).

perihelion (see Fig. 15). The obliquity ϕ is the angle between the axis of rotation of the earth and the normal to the plane of the earth's orbit about the sun. The current value of the obliquity is 23.45°, and it varies between about 22.5° and 24.5°, with a period of about 41,000 years. As the obliquity increases, it increases the insolation near the poles during the summer season. This, in turn, increases the amplitude of the annual cycle and the annual average insolation near the poles.

The deviation of earth's orbit from a circular path is measured by the eccentricity ε. The eccentricity of earth's orbit varies from about 0.06 to zero, with dominant time scales of 100,000 and 400,000 years. Currently, the eccentricity value is 0.016, which makes the insolation vary by about 7% between perihelion, the earth's closest approach to the sun, and aphelion, when earth is farthest from the sun. The final parameter, the longitude of perihelion λ, measures the phase of the seasons relative to the closest approach to the sun. Longitude of perihelion cycles on a time scale of about 20,000 years. Eccentricity and longitude of perihelion together determine the distribution of seasonal insolation between the hemispheres. The effect can be measured by the combined parameter $\varepsilon \sin \lambda$. When the eccentricity is large and earth passes closest to the sun during the Northern Hemisphere summer solstice, the Northern Hemisphere receives an anomalously large amount of insolation. The effect is enhanced in high latitudes if the obliquity is also relatively large. Because of geometric factors, the obliquity cycle of 40,000 years is more important in high latitudes, while the longitude of the perihelion cycle is more important in the subtropics.

Milankovitch (1920; 1941) proposed that the orbital parameters controlled the succession of ice ages by driving the high-latitude ice sheets in the Northern Hemisphere, which contains most of the land area. He argued that when the orbital parameters are arranged such that Northern Hemisphere summertime insolation at high latitudes is above average, then summer melting of ice is enhanced and ice sheets retreat, leading to an interglacial. The opposite conditions would lead to an ice age. When the summer insolation is low, the winter insolation is high. Because the temperature during winter in high latitudes is well below the freezing point of water, wintertime warming would enhance the snowfall rate through the temperature dependence of vapor pressure, as discussed earlier. Thus, the orbital parameter changes could simultaneously induce decreased summer snowmelt and increased winter snowfall, which favor the growth of ice sheets.

By comparing insolation variations with time series of global ice volume inferred from oxygen isotope variations in deep-sea cores, Hays et al. (1976) have proved that the orbital parameters play a key role in the timing of glacial age successions. Imbrie and Imbrie (1980) showed that the relationship between the orbital parameters and the global ice volume was approximately as predicted by the mechanism proposed by Milankovitch.

Widespread evidence indicates that the climate of the Sahara Desert region was moist in the relatively recent geologic past. Street-Perrott and Harrison (1984) suggest that lake levels were greatly elevated in the Saharan region around 6,000 years ago. The presence of much more moisture in the Sahara at about this time is corroborated by fossil evidence of hippopotamus and crocodile, as well as plant species that are indicative of a very different climate. Climate model simulations suggest that these changes were forced in the Sahara at least in part by the much different orbital parameters in effect at that time (COHMAP, 1988). Ten thousand years ago, the longitude of perihelion caused the Northern Hemisphere summer solstice to occur at perihelion, which is just the opposite of current conditions in which the Southern Hemisphere receives more solar radiation.

The annual variation of precipitation over Africa is such that the latitude of heaviest precipitation follows the sun. Extension of the near-equatorial maximum in precipitation into the Saharan region occurs at northern summer solstice. Solar heating of the surface drives convection, which produces rainfall. The increased insolation of the Northern Hemisphere at summer solstice provides additional heating of the surface that can push the summertime convection deeper into the Saharan region. On the time scale of centuries to millennia, modest variations in precipitation can lead to changes in vegetation and soil that reinforce and solidify the climate change; these lead further to a substantial contraction in the area of the desert. Because the albedos of moist, vegetated surfaces are less than those of dry land or sand, the vegetation increase is accompanied by a further increase of surface solar heating, which feeds back positively on the initial forcing provided by the orbital parameter change.

A comparison between observed lake levels and model-simulated moisture budgets over the last 18,000 years for the northern subtropics is shown in Fig. 16. The model correctly predicts a moist stage between 9 and 6 kyr BP, which is driven in the model by the enhanced insolation of the Northern Hemisphere during this period. Comparisons between other observed and modeled climate changes over the last glacial cycle by COHMAP (1988) also reveal good agreement. One concludes that, in general, current climate models can be used to simulate and understand past climate changes. The good simulation of past climates allows a measure of confidence that current climate models can furnish useful predictions of future climate changes that may result from increases in CO_2 and other greenhouse gases.

EQUILIBRIUM RESPONSE TO GREENHOUSE GAS INCREASE

As a baseline test of the sensitivity of the climate to increasing abundances of greenhouse gases, one can consider the equilibrium response to a doubling of the carbon dioxide concentration of the atmosphere from 300 to 600 ppm. In reality, the greenhouse warming will be forced by changes in the concentrations of many species in addition to carbon dioxide (Ramanathan et al., 1985; Hansen et al., 1988). The principal gases that will contribute to the greenhouse warming are carbon dioxide ($\sim$ 50%), chlorofluorocarbons ($\sim$ 30%), methane ($\sim$ 15%), and nitrous oxide ($\sim$ 3%). The contribution to greenhouse warming by ozone changes is uncertain, but could also be significant. Based upon current trends, one can estimate that the total greenhouse warming would be equivalent to a doubling of carbon dioxide in about 50 years. Such projections are uncertain, however, because of the difficulty in projecting the global rates at which carbon fuels will be used and CFCs released. Carbon dioxide doubling in models is a surrogate climate

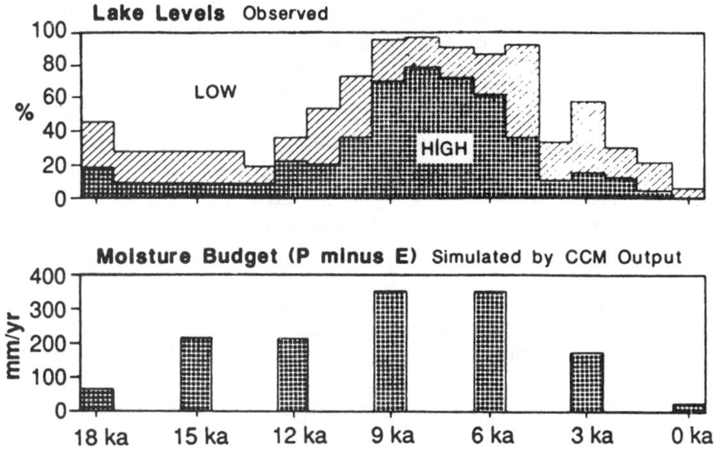

Fig. 16. Observed status of lake levels (top) and model-simulated moisture budget of precipitation (P) minus evaporation (E). Model data are calculated at 3,000-year intervals between 18 kyrbp (ka) and the present (source: COHMAP, 1988).

change experiment that approximates the more complex mix of gases that will contribute to the actual greenhouse gas warming.

The models used to perform this experiment have typically been atmospheric GCMs with a mixed-layer ocean model (Schlesinger and Mitchell, 1987). The models predict surface and air temperature and include feedbacks associated with atmospheric water vapor content and sea ice; they may or may not include changes in cloud amount and/or cloud optical properties. Predictions of cloud optical properties within climate models are still in their infancy (Charlock and Ramanathan, 1985; Slingo, 1989; Mitchell et al., 1989). The equilibrium response is the climate that results after the transients have died away and a new steady climate state has been established. In realistic projections of climate change, one must deal with the transient nature of the greenhouse gas increase and the climate's response to it. Some of these issues will be addressed in the next section.

Climate models predict equilibrium, global-mean surface temperature responses to CO_2 doubling that range from approximately 2 to 5 K, with the consensus forecast being about 4 K at the time this paper was written. Of this 4 K change, only a little over 1 K is directly associated with the forcing provided by the carbon dioxide increase. About two-thirds of the surface temperature response is related to feedbacks involving water vapor, surface ice, and clouds. The least certain of these is cloud feedback. Current models generally predict changes in cloud amount and cloud-top altitude, but not in the optical properties of the clouds. Most of the models predict a decrease in the fractional area of clouds and an increase in the average cloud-top altitude, both of which are positive feedbacks. However, the liquid water content of clouds and cloud particle characteristics could also change and would have important effects on the radiation balance (e.g., Somerville and Remer, 1984; Charlson et al., 1987). For example, if the liquid water content of a cloud increases as the cloud tops rise during global warming, then the increased cloud albedo associated with the higher liquid water content could offset the greater greenhouse effect associated with the higher cloud top. Much model development and validation is required before the optical property changes of clouds can be confidently modeled. This work is necessary, however, because such changes could play a major role in determining the size of the climatic response to CO_2 doubling.

Polar Amplification

Climate models are in agreement on several key aspects of the response to increases in greenhouse gas concentrations. One of these is the polar amplification of the warming. Figure 17 shows a plot of the surface temperature response to CO_2 doubling versus latitude and season that was produced by the model of Hansen et al. (1984), which has an equilibrium, global mean response to CO_2 doubling of 4.2 K. The response in high latitudes during the winter season is about three times the global mean warming and causes the annual mean response to increase with latitude. Changes in the thickness of sea ice have much to do with this aspect of the response to greenhouse gas increases.

Manabe and Stouffer (1980), who obtained a response similar to that pictured in Fig. 17, concluded that the seasonal and latitudinal structure was primarily related to sea ice changes. Untersteiner (1984) described the characteristics of water ice at the surface: in a warmed climate, sea ice forms later in the fall, melts later in the spring, and leads and melt ponds are more prevalent in the summer. When more open water is exposed in the summer season, the albedo is lowered and much more solar radiation is absorbed. However, greater absorption does not lead to a significant increase in summer temperatures, because of the large heat-storage capacity of the seas. The large amount of heat stored in the open water during the summer season does delay the formation of sea ice in the fall and causes the average thickness of the ice to decrease. Because sea ice and snow that may fall on it insulate surface air from the underlying water,

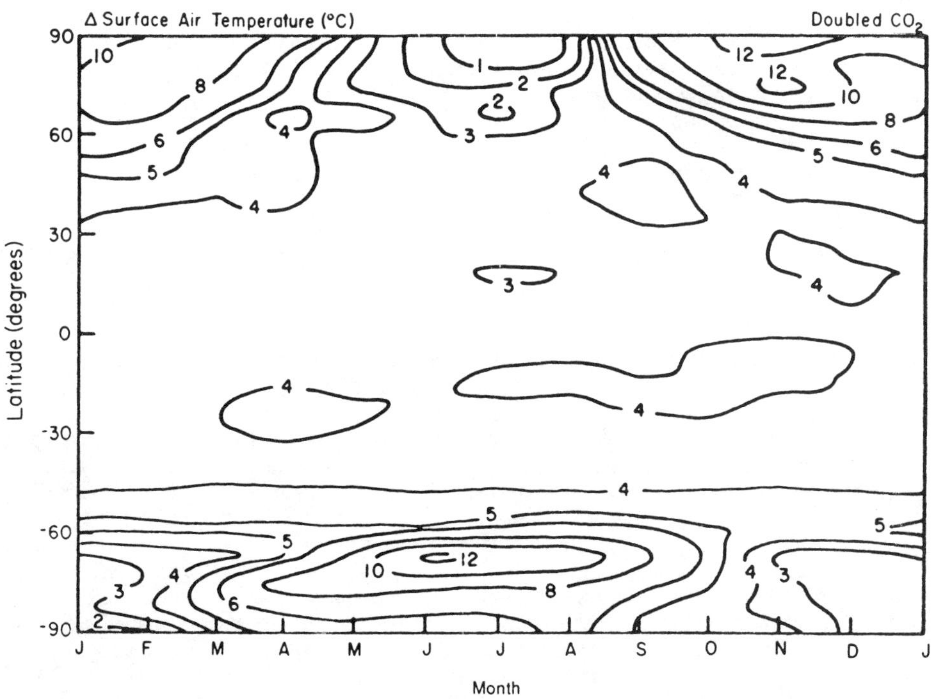

Fig. 17. Distribution of longitudinally averaged equilibrium response of surface air temperature to carbon dioxide doubling as a function of latitude and season. The model has a global-mean equilibrium sensitivity to doubled carbon dioxide of 4.2°C (source: Hansen et al., 1984).

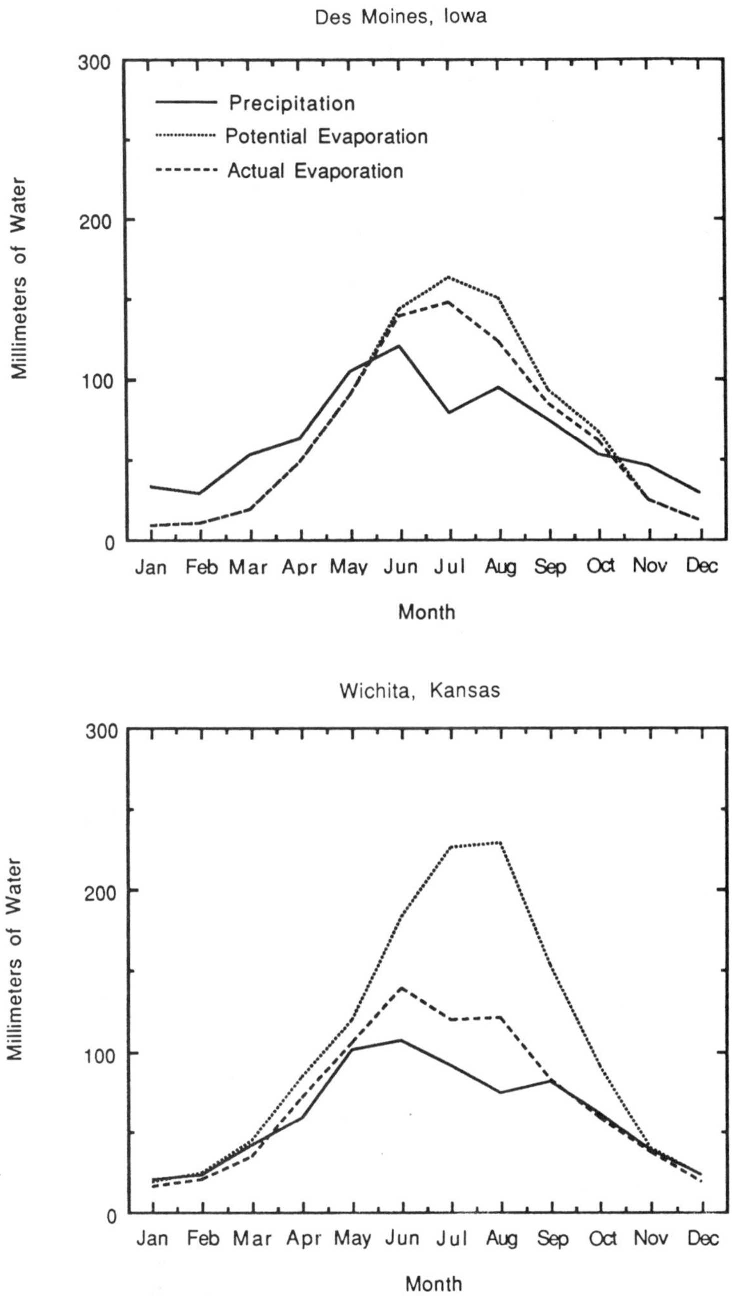

Fig. 18. Annual hydrologic cycle components at Des Moines, Iowa, and Wichita, Kansas. Precipitation in excess of evaporation indicates runoff or water storage in the soil. Potential evaporation, the rate at which water could be evaporated from a moist surface in excess of actual evaporation, indicates dry conditions (data from Eagleman, 1976).

which must be near the freezing point, the air above sea ice can fall very much below freezing during the winter months when a layer of ice is present. When this layer of sea ice is thin or absent, the flux of heat from the polar ocean can keep wintertime surface air relatively warm.

Thus, the high albedo and high insulation value of sea ice, and the high heat-storage capacity of open ocean can explain the key aspects of the seasonal and latitudinal structure of the greenhouse warming. The structure is different in the two hemispheres because the North Pole is covered by ocean, whereas the South Pole is covered by land. The largest response in the Southern Hemisphere occurs in the winter season at those latitudes where seasonal sea ice forms. The proper simulation of sea ice in climate models is therefore critical to the accurate simulation of climate change. The large climate changes at high latitudes predicted by the models, and the corresponding changes in sea-ice extent, have significant implications for the ecology of polar regions.

Soil Moisture Changes

Several climate models predict increased dryness in mid-latitude land areas as a result of global warming caused by greenhouse gases (e.g., Manabe and Wetherald, 1987; Hansen et al., 1988). As described earlier in this paper, the water-carrying capacity of air increases rapidly with temperature. This means that the capacity of air both to provide water to the surface through precipitation and remove it through evaporation is increased. In areas where conditions normally produce intense precipitation, we expect the precipitation intensity to increase further. In areas with marginal soil moisture or seasonal drought, we expect the probability and intensity of drought to increase. In mid-latitude continental areas, where much of the world's population and agricultural productivity are located, the soil generally gains moisture during the winter season. Soil moisture increases in winter because evaporation is not equivalent to rain and snowfall; there is insufficient energy to provide the latent heat of vaporization. In the springtime, temperatures warm, snow melts, and increased insolation provides the energy necessary to evaporate water from the surface. This annual cycle is illustrated in Fig. 18, which shows precipitation, potential evaporation, and actual evaporation for two locations in the American Midwest. In the midwestern United States, precipitation peaks in the summer season when the temperature and insolation are also favorable for plant growth. This concurrence is part of what makes the Midwestern states such an efficient food-producing region. Des Moines, Iowa, is in a very favorable climatic regime that supports corn and livestock farming. Its hydrologic diagram (Fig. 18) indicates only a small excess of potential evaporation over actual evaporation during the core of the summer season, which indicates that the soil stays relatively moist all through the growing season. Des Moines can be contrasted with Wichita, Kansas, where the precipitation is similar, but potential evaporation greatly exceeds actual evaporation, limiting agriculture to wheat and small-grain farming. At Wichita there is no wintertime storage of soil moisture; evaporation equals precipitation for most of the year, and the summer is very dry. Though the precipitation is highly similar at the two locations, subtle differences in insolation and altitude cause the ground hydrology, and the agriculture it can support, to differ greatly. In view of the hydrologic cycle, experience with major climate shifts during the 1930s and 1950s in these regions suggests that the ground hydrology in the American Midwest may be sensitive to modest climate anomalies.

Climate models generally use very simple models of ground hydrology that do not accurately take account of the active role of vegetation in the soil moisture balance or the variable water-holding capacity of land surfaces. Nonetheless, these models do simulate the basic annual variation of ground hydrology indicated in Fig. 18. In response to global warming, snow melts earlier and summer depletion of the soil moisture reservoir begins earlier. Consequently, by the core of the summer season, when plants such as corn should be coming to maturity, these models show much drier soil in doubled-CO_2 experiments than in control runs.

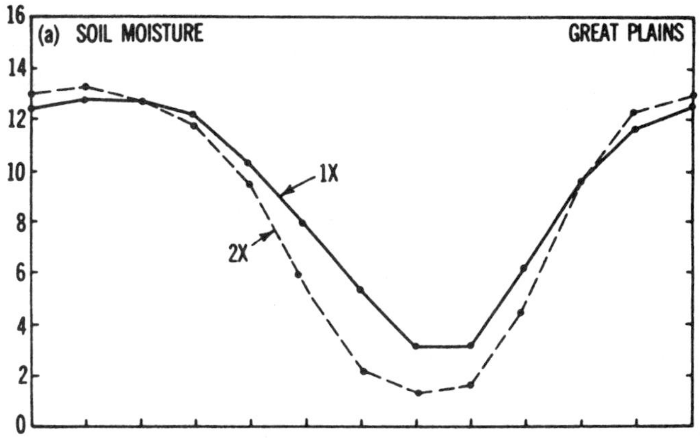

Fig. 19. Simulation of the annual cycle of soil moisture content (cm) for the Great Plains region for present conditions (1X) and for doubled carbon dioxide (2X) (source: Manabe and Wetherald, 1987).

Figure 19 shows an example from Manabe and Wetherald (1987) of the annual variation of soil moisture simulated for the Great Plains of the United States, both for current conditions and for doubled CO_2. The doubled-CO_2 experiment shows a more rapid decline of soil moisture during May and June, which results in a halving of the soil moisture content during the months of July, August, and September. In Manabe and Wetherald's model, reduced soil moisture in summer leads to local reductions of cloud cover. These reductions in cloudiness increase the insolation incident on an already dry surface and feed back positively on the summer dryness. Summer soil moisture reductions and increased probability of summer drought of the type indicated by these model projections would have serious implications for agriculture and water resource allocation in the Great Plains regions.

In the examples shown in Fig. 18, it is notable that soil moisture in winter is about the same for both experiments, despite an increase in wintertime precipitation in the doubled-CO_2 experiment. Part of the reason for this must be the fixed soil water capacity of 15 cm assumed in these experiments. If the soil water capacity were greater, or increased with the warming, then increased storage of water in the soil during winter might offset the earlier snowmelt and enhanced summertime evaporation, which produce the summertime drying. Much more work is necessary to adequately understand and model soil moisture responses to global warming.

TRANSIENT RESPONSE TO GREENHOUSE GAS INCREASE

Over the next century, the concentrations of greenhouse gases in the atmosphere will gradually increase as a result of fossil fuel burning, industrial chemical production, deforestation, and other natural and anthropogenic physical, biological, or chemical changes. In response to the increasing infrared opacity of the atmosphere, the climate of the earth will warm. The rate and timing of this warming is important for a variety of reasons. If the warming is sufficiently gradual, human and biological systems may be able to adapt without catastrophic consequences. If the warming is rapid, ecosystems may not be able to adapt or migrate quickly enough to survive, and the infrastructure of human civilization may become maladjusted to the climate on a time scale shorter than its normal replacement time. In addition, by understanding and monitoring the rate of warming we may be able to validate our projections of the future climate and adjust our activities accordingly.

Simple Concepts of Transient Response

We can illustrate some of the basic aspects of the climate response to warming by considering a very simple model in which the climate is perturbed from an initial balanced state by a forcing $F(t)$. The system is assumed to have a heat capacity C, and relaxes toward its equilibrium state with a rate given by λ^{-1},

$$C\frac{\delta T}{\delta t} = -\frac{1}{\lambda}T + F(t). \tag{20}$$

T represents the deviation of the temperature that is produced by the time-dependent forcing $F(t)$. The units of $F(t)$ are Wm^{-2}. Note that the meaning of λ in Eq. (20) is exactly that of the sensitivity parameter defined in Eqs. (11) and (12): it is the relationship between a specified forcing and the temperature response. In Eq. (20), λ represents the net effect of all the feedbacks operating in the system. The simplest form of the forcing function is step function forcing, such that $F(t) = A$, for $t > 0$, otherwise $F(t) = 0$. This represents an immediate switch-on of the forcing, as if the carbon dioxide were doubled instantaneously. The solution to Eq. (20) in this case is

$$T(t) = \lambda \cdot A\left[1 - \exp\left(-\frac{t}{C\lambda}\right)\right]. \tag{21}$$

Note that the temperature exponentially approaches an equilibrium response to the forcing of λA on a time scale given by $C\lambda$. Thus, both the amplitude of the response and the time it takes to achieve it are proportional to the sensitivity parameter λ. If more positive feedbacks are added to increase λ, the response will be larger, but it will take longer to achieve it. This fact is at the heart of the difficulty of evaluating the sensitivity of climate from observed temperature changes.

Equally important in determining the response time is the heat capacity of the system, C. The heat-storage capacity of the climate system on the time scale of a century is contained mostly in the upper layers of the ocean. The first 50 to 100 m of the ocean are stirred by the winds, so that thermal communication between this mixed layer and the atmosphere is relatively efficient. If we take a mean mixed-layer depth of 70 m, then the heat capacity is about 3×10^8 Jm^{-1} K^{-1}. The heat-storage capacity of the atmosphere roughly equals that of the first two meters of the ocean and is therefore negligible if an oceanic mixed layer underlies the atmosphere. Because of the low thermal conductivity of land, the thermal capacity of land surfaces is comparable to or less than that of the atmosphere. In consequence, land areas respond much more quickly to seasonal and diurnal forcing than do ocean areas. The slower response time over the oceans in the transient greenhouse warming will likely lead to regional anomalies associated with land-sea differences.

In order to see how uncertainties in climate sensitivity and release rates may affect early detection of global warming, we may consider solutions to Eq. (20) for various conditions. We will consider the case in which a greenhouse gas increase leads to a 4 Wm^{-2} forcing of the climate in 50 years' time using two scenarios of release—one exponential and one linear—each of which reaches the 4 Wm^{-2} level in 50 years. Taking the heat capacity of a 70-m mixed layer, two sensitivities will be considered: one corresponds to the basic OLR plus relative humidity feedbacks ($\lambda = 0.5$ K per Wm^{-2}), while the other assumes substantial additional positive feedbacks from clouds and ice cover ($\lambda = 1.0$ K per Wm^{-2}). The results of these four calculations are shown in Fig. 20. It is interesting that the differences in the response to linear and exponential growth of the forcing are rather small until long times have elapsed. This is largely because we have fixed the forcing at 50 years to be 4 Wm^{-2}. It is significant that the responses for the two sensitivities do not begin to differ markedly until after year 20. If the uncertainty in measuring global mean trends is 0.5°C because of natural variability and measurement errors, then one could not distinguish the sensitivity of the climate from the observed trend until about 25 years into the warming, even though the difference in the temperature response at 50 years differs by a factor of two. From such simple considerations it seems that one cannot hope to determine the sensitivity of the climate by observing the early response to known greenhouse gas increases.

The Ocean in Climate

The ocean serves as the primary reservoir of heat for the climate system, actively participates in the horizontal transport of heat, and provides an abundant source of water. In addition, the chemistry and biology of the ocean are central to atmospheric composition. The physical aspects of the ocean's role in determining climate have begun to be incorporated in climate models. We can simplify our discussion of the physical aspects of the ocean by dividing it into three components: the mixed layer, the wind-driven circulation, and the deep thermohaline circulation. All of these components interact strongly to produce the net flow and transport of heat in the ocean.

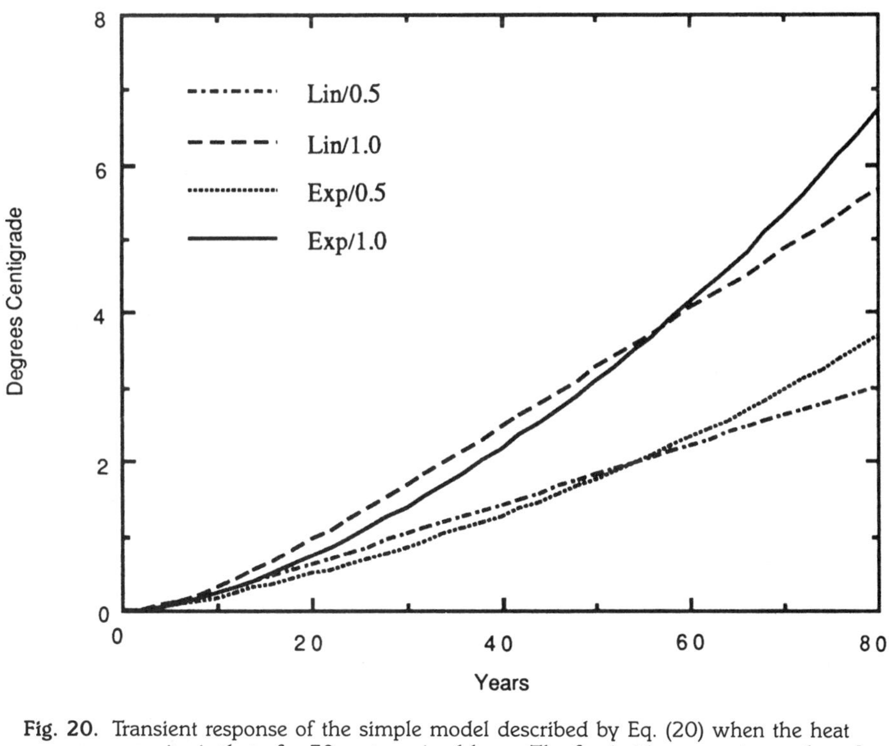

Fig. 20. Transient response of the simple model described by Eq. (20) when the heat capacity is that of a 70-meter mixed layer. The forcing increases to a value of 4 Wm^{-2} at year 50, following either a linear or exponential growth rate. Two model sensitivities are shown: $\lambda = 0.5$ K/(Wm^{-2}), and $\lambda = 1.0$ K/(Wm^{-2}), corresponding to equilibrium responses of 2 K and 4 K, respectively.

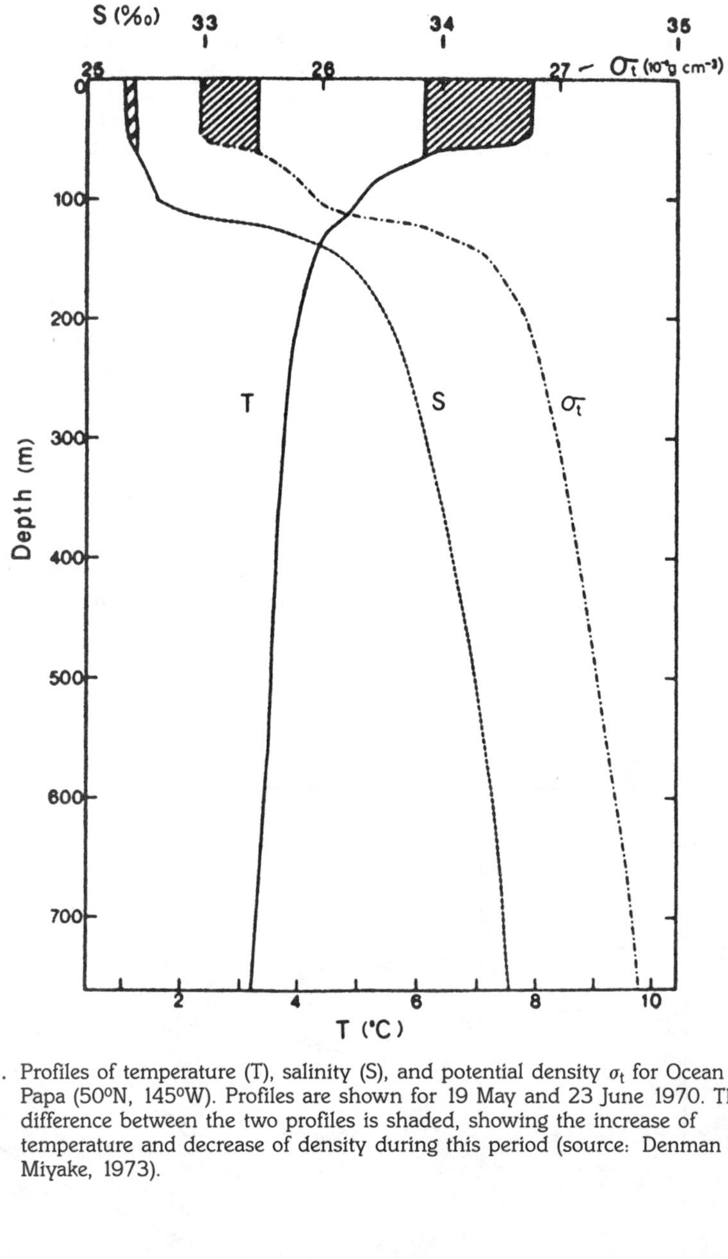

Fig. 21. Profiles of temperature (T), salinity (S), and potential density σ_t for Ocean Station Papa (50°N, 145°W). Profiles are shown for 19 May and 23 June 1970. The difference between the two profiles is shaded, showing the increase of temperature and decrease of density during this period (source: Denman and Miyake, 1973).

The Mixed Layer

The mixed layer is the uppermost layer of the ocean in contact with the atmosphere. Because of stirring by wind stresses, the temperature, salinity, and density are independent of depth in the upper layer of the ocean. The efficiency of stirring in the mixed layer allows it to respond relatively quickly to changes in the heat balance at the surface. Figure 21 shows the mixed-layer structure in the North Pacific Ocean and its response to thermal driving in the springtime. Because the deeper layers of the ocean respond on much longer time scales, it is a reasonable first approximation to use a mixed layer to represent the thermal capacity of the ocean on time scales of less than a century.

Fig. 22. Distribution of surface currents in the Atlantic Ocean during February-March (source: Sverdrup et al., 1942).

Hansen et al. (1988) considered the transient response to particular scenarios of greenhouse gas increase in a climate model with a mixed-layer representation of the oceanic thermal response. They concluded that the global greenhouse warming should reach observable levels in the 1990s. Forecasts of regional effects are more uncertain, but the signal will be most easily detected in those regions with a large equilibrium response, such as sea-ice margins or continental interiors, or regions where natural variability is low, such as subtropical oceans. Uncertainties in these forecasts include those associated with the basic sensitivity of the model, the rate of greenhouse gas increase, and the change in heat transports in the oceans. The potentially important role of heat transport within the ocean will be considered in the remainder of this section.

Wind-Driven Circulations of the Oceans

Wind stresses drive currents in the ocean. In mid latitudes, westerly winds that increase with latitude drive a slow, equatorward drift of surface waters. The Atlantic and Pacific Oceans return water to high latitudes, while conserving vorticity, by generating western boundary currents. Both the Gulf Stream and the Kuroshio Currents transport large amounts of warm, saline water, making substantial contributions to the poleward flux of heat by the oceans (Vonder Haar and Oort, 1973). The current systems for the Atlantic Ocean are characterized in Fig. 22; these currents are strongest near the surface and weaken rapidly in the top 1,000 meters. In the Southern Hemisphere, ocean circulations are fundamentally different, because the distribution of continents allows a circumpolar ocean current. Wind stresses are responsible for the relatively cold ocean temperatures along the equator and the eastern margins of subtropical oceans. On the west coasts of continents, offshore or equatorward winds drive offshore drifts of the surface waters, resulting in upwellings of cold, nutrient-rich water.

Thermohaline Circulation of the Oceans

The thermohaline circulation is driven by density variations and extends to the very bottom of the ocean. Tracer studies show that the circulation of the deep ocean is very slow, with recirculation of deep ocean waters requiring 500 years or more, depending on location. The complexity of the deep circulation is illustrated in Fig. 23. Because currents are very weak at depth, the flows indicated by the arrows have been inferred from water temperature, salinity, and oxygen content, rather than direct observation. The basic picture these parameters give is consistent with that derived from other chemical and radioactive tracers in the ocean. Tropical and mid-latitude oceans are topped off by a warm layer of water that is maintained by insolation. Water of the intermediate and deep layers of the ocean is formed at high latitudes, where it can be cooled and the salinity increased by evaporation, or by the exclusion of salt from sea ice.

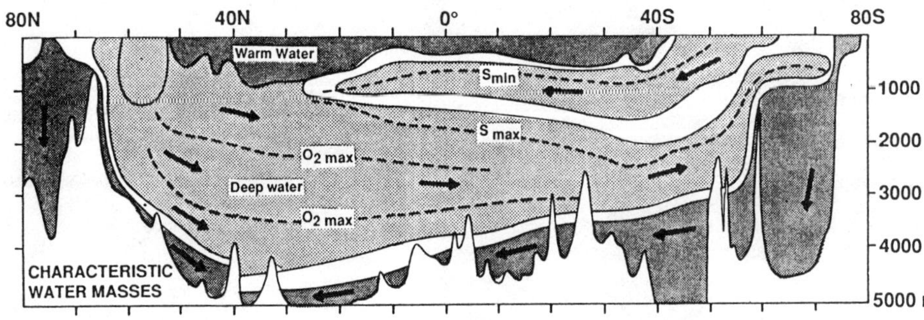

Fig. 23. Schematic cross-section of the deep thermohaline circulation of the Atlantic Ocean indicating water mass types and inferred flow directions (adapted from Dietrich et al., 1980).

Much of the volume of the intermediate and deep waters of the Atlantic is occupied by water that sinks near 60°N. Here, warm, saline water transported poleward by the Gulf Stream encounters cold, dry air from North America during winter. The resulting heat exchange and evaporation produce water that is cold and saline enough to sink to great depths. At temperatures near 0°C, the density of seawater is controlled primarily by salinity (Gill, 1982). Water that reaches the bottom of the ocean is formed around Antarctica, where the density of seawater is increased during the seasonal formation of sea ice. A tongue of cold, fresh water that flows northward under the warm surface water appears to be generated in the mid latitudes of the Southern Hemisphere.

Note that the deep water formed in the Arctic flows southward and then resurfaces near 60°S. This upwelling of deep water around Antarctica may play an important role in the transient response to greenhouse warming. Cold water flows upward from depth in this region, providing communication with the large heat capacity of the deep ocean. Because of the continual upwelling of cold water from depth, the surface water in the Antarctic cannot warm as rapidly as in locations where the deep ocean is insulated from surface conditions by a layer of warm, low-density surface water.

Climate model simulations by Bryan et al. (1988) that include deep ocean currents have shown that the transient response to greenhouse warming is greatly delayed by the upwelling of deep water in the southern oceans. Their model has realistic ocean depths and continent distributions and explicitly calculates ocean currents and transport. In their experiment, Bryan et al. introduced doubled CO_2 instantaneously at year zero, and allowed their coupled atmosphere-ocean model to respond in time. In the Northern Hemisphere, the temperature responded more or less as expected from our simple model of the transient response to instantaneous forcing given in Eq. (21), with a rapid warming to about 60% of the equilibrium response in less than ten years. This was followed by a much more gradual linear increase with time, which is not predicted by Eq. (21) and is related to the deep-ocean response. A latitude-versus-time plot of the normalized response of the zonally averaged sea-surface temperature in their model is shown in Fig. 24. Near 60°S, the temperature actually cools slightly with time in response to the CO_2 doubling. Because the subtropics warm faster than the polar

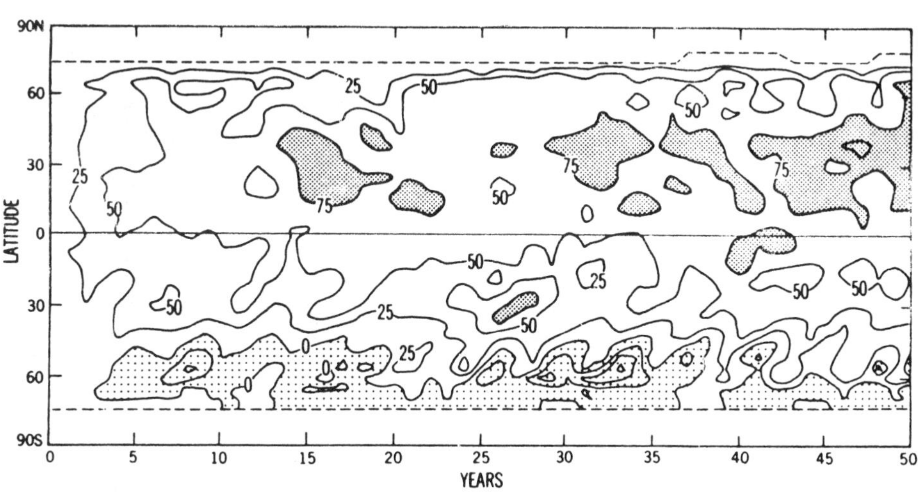

Fig. 24. Longitudinally averaged, normalized response of sea-surface temperature to instantaneous doubling of carbon dioxide concentration at year zero. The units are percentages of the long-time equilibrium response. Negative values are lightly stippled; dashed lines indicate the boundary of polar sea ice (source: Bryan et al., 1988).

regions, the meridional temperature gradient and associated zonal component of geostrophic wind increases. This results in a transient increase in the zonal wind stress in mid latitudes, which helps to drive the upwelling near 60°S and the associated cooling. The active ocean currents produce a transient climate response to warming that is asymmetric between the hemispheres. Because of the complexity of atmosphere-ocean coupling, a forcing that has a global warming effect in equilibrium may produce a transient cooling in particular regions. This makes the observational identification of the greenhouse warming and the prediction of regional effects doubly difficult.

The model employed by Bryan et al. (1988) did not accurately simulate the formation of deep water in the North Atlantic Ocean, so that the hemispherical asymmetry of the climate response to instantaneous doubling of CO_2 may have been exaggerated. Paleoclimatic evidence strongly suggests that the rate of deep water formation in the North Atlantic, and the concomitant poleward flux of warm, saline water to 60°N, have varied dramatically in the past. During the last glacial maximum, the Gulf Stream did not penetrate as far poleward, and the sea-ice margin advanced far south of its current climatological position. At the time of the glacial maximum, the modern processes that we believe lead to the formation of deep water in the North Atlantic could not have operated in the same way.

Oxygen isotope data from Greenland ice cores and other data show an oscillation in the climate following the last ice age. The climate first warmed rapidly to create the Alleröd warm interval at about 13 kyr BP. At this time, North Atlantic sea ice retreated to near-interglacial conditions. This was followed rapidly by a return to cold conditions from 11 to 10.2 kyr BP—known as the Younger Dryas cold interval—when the North Atlantic and Europe returned to near-glacial conditons. Broecker et al. (1985) show evidence to suggest that the Alleröd-Younger Dryas oscillation may be related to transitions in the thermohaline circulation of the North Atlantic, particularly since the oscillation appears to be confined to the North Atlantic region. The magnitude and rapidity of the changes suggest that the concept of two stable equilibria for the North Atlantic may be useful. One has a warm North Atlantic with an active thermohaline circulation, and the other a cold North Atlantic region in which the poleward flow of warm surface water in the Gulf Stream does not extend beyond 50°N.

The two states of the North Atlantic suggested by Broecker et al.'s interpretation of paleoclimatic data were simulated by Manabe and Stouffer (1988). Manabe and Stouffer found that their coupled atmosphere-ocean climate model would sustain deep water formation in the North Atlantic if the initial conditions favored it; likewise, the model would sustain a cold North Atlantic without a thermohaline circulation if the thermohaline circulation was not present initially. The difference between the two stable states included a climate response that was restricted to the North Atlantic region, much as suggested by the paleoclimatic data. The surface air temperature difference between the two states of the Manabe and Stouffer model is shown in Fig. 25. Warming is most intense in the North Atlantic but spreads to warm the entire north polar region. A small cooling may be noted in the vicinity of the Antarctic Circumpolar Current.

The results of coupled atmosphere-ocean models suggest that when ocean dynamics are taken into account, rather important regional variations in the rate of temperature rise due to greenhouse gas increases may be expected. This expectation is supported by paleoclimatic data, which suggest that the regional anomalies in climate that occurred during the decline of the last ice age were most likely associated with changes in ocean circulation. The existence of multiple equilibrium solutions of coupled atmosphere-ocean climate models suggests that the coupled system may be sensitive to perturbations large enough to cause a transition from one climate state to another. The warming after the last glacial maximum was apparently rapid enough to cause several oscillations of the North Atlantic system. The warming to be induced by rapid greenhouse gas release

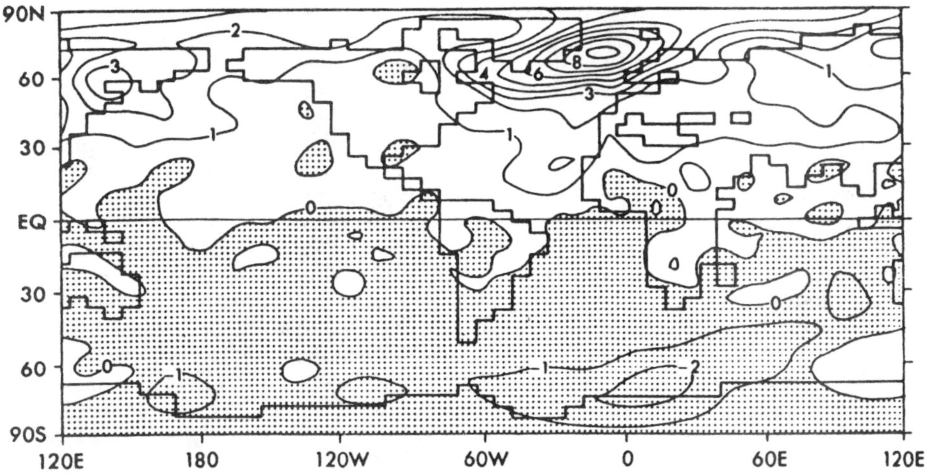

Fig. 25. Difference in surface air temperature (°C) between experiments with active and inactive North Atlantic thermohaline circulation (source: Manabe and Stouffer, 1988).

would be of comparable magnitude to, and more rapid than, that of the glacial-interglacial transition.

SUMMARY AND CONCLUSIONS

The climate of the earth can be modeled mathematically using explicit equations of physics for processes that occur on the scale of the model grid resolution (~ 100 km) and simplified parameterizations for processes that occur on unresolved scales. Such models can faithfully simulate the observed annual cycle of surface temperature. Comparison with paleoclimatic data shows that, to some degree, these models are capable of simulating past climate changes, which suggests that they may also be useful for forecasting the climate response to greenhouse gas increases that are expected over the next century. The equilibrium response of global-mean surface temperature to greenhouse gases expected within the next hundred years is predicted to be on the order of 5°C, which is equal to the change between the last glacial maximum 18,000 years ago, when thick ice sheets covered much of North America and Europe, and the present. The warming in high latitudes would be larger than the global average. The consequences of a warming of this magnitude would be very significant for natural ecosystems and for human energy utilization, agriculture, and health.

Uncertainties in forecasts of the climate response to greenhouse gas increases arise from several sources. The basic sensitivity of the climate system remains uncertain within a factor of two. At present, the most important contributor to this uncertainty is the extent and magnitude of cloud feedback. Climate models that predict cloud optical properties must be developed and validated against observations.

One of the most important consequences of global greenhouse warming predicted by climate models is the dessication of mid-latitude, agricultural land areas. Models currently used to predict ground moisture and its interaction with climate are highly simplified and in need of improvement. The active role of vegetation in land hydrology must be incorporated into climate models; changes in vegetation must be predicted and allowed to affect the model climate.

The transient response to greenhouse gas increases depends on the rate at which climate forcing by these gases is increased and on the transient response of the ocean, which carries most of the thermal capacity of the climate system and about half of the horizontal transport. Coupled atmosphere-ocean climate models need further refinement and testing. The connection between oceanic heat transport and sea-ice formations is particularly important.

Climate models need many improvements if we are to increase the confidence we place in their predictions of climate changes resulting from greenhouse gas increases. Much of the temperature change predicted by current climate models stems from feedback processes internal to the climate system, rather than from a direct response to the greenhouse gas forcing. An enhanced research effort on the critical feedback processes, and the incorporation of these feedbacks into global climate models, are needed. Although at the present time climate modelers seem to be focusing on possible negative feedbacks that would mitigate the warming effect of greenhouse gas increases, many positive feedback mechanisms also remain to be investigated thoroughly. Because current models predict climate changes that would be highly disruptive if the current rate of increase of greenhouse gases continues, it would seem prudent to take those actions to reduce greenhouse gas emissions that are not too costly and may have other benefits. Examples are rapid reduction in the release of long-lived chlorofluorocarbons into the atmosphere, conservation of fossil fuels, and development of alternative, environmentally benign energy sources.

REFERENCES

Barkstrom, B. R., and Smith, G. L., 1986, The Earth Radiation Budget Experiment: Science and implementation, *Rev. Geophys.*, **24**:379-390.

Broccoli, A. J., and Manabe, S., 1987, The influence of continental ice, atmospheric CO_2 and land albedo on the climate of the last glacial maximum, *Clim. Dyn.*, **1**:87-99.

Broecker, W. S., Peteet, D. M., and Rind, D., 1985, Does the ocean-atmosphere system have more than one stable mode of operation? *Nature*, **315**:21-26.

Bryan, K., Manabe S., and Spelman, M. J., 1988, Interhemispheric asymmetry in the transient response of a coupled ocean-atmosphere model to a CO_2 forcing, *J. Phys. Ocean.*, **18**:851-867.

Budyko, M. I., 1969, The effect of solar radiation variation on the climate of the earth, *Tellus*, **21**:611-619.

Cess, R. D., Potter, G. L., Blanchet, J. P., Boer, G. J., Ghan, S. J., Kiehl, J. T., Le Treut, H., Li, Z.-X., Liang, Z.-X., Mitchell, J. F. B., Morcrette, J.-J., Randall, D. A., Riches, M. R., Roeckner, E., Schlese, U., Slingo, A., Taylor, K. E., Washington, W. M., Wetherald, R. T., and Yagai, I., 1989, Interpretation of cloud-climate feedback as produced by 14 atmospheric general circulation models, *Science*, **245**:513-516.

Charlock, T. P., and Ramanathan, V., 1985, The albedo field and cloud radiative forcing produced by a general circulation model with internally generated cloud optics, *J. Atmos. Sci.*, **42**:1,408-1,429.

Charlson, R. J., Lovelock, J. E., Andreae, M. O., and Warren, S. G., 1987, Atmospheric sulphur: Geophysiology and climate, *Nature*, **326**:655-661.

CLIMAP, 1976, The surface of the ice-age earth, *Science*, **191**:1,131-1,137.

COHMAP, 1988, Climate changes of the last 18,000 years, *Science*, **241**:1,043-1,052.

Denman, K. L., and Miyake, M., 1973, Upper layer modification at Ocean Station Papa: Observations and simulation, *J. Phys. Ocean.*, **3**:185-196.

Dickinson, R. E., 1983, Land-surface processes and climate: Surface albedos and energy balance, *Adv. in Geophys.*, **25**:305-353.

Dickinson, R. E., 1984, Modeling evapotranspiration for three-dimensional climate models, *in*: "Climate Processes and Climate Sensitivity," J. E. Hansen and T. Takahashi, eds., American Geophysical Union, Washington, D.C.

Dietrich, G., Kaile, K., Kraus, W., and Siedler G., 1980, "General Oceanography: An Introduction," Wiley Interscience, New York.

Eagleman, J. R., 1976, "The Visualization of Climate," Lexington Books, Lexington, MA.

ERBE Science Team, 1986, First data from the Earth Radiation Budget Experiment (ERBE), *Bull. Amer. Meteor. Soc.*, **67**:818-824.

Gill, A. E., 1982, "Atmosphere-Ocean Dynamics," Academic Press, Orlando, FL.

Goody, R. M., and Yung, Y. L., 1989, "Atmospheric Radiation," Oxford University Press, Oxford.

Hansen, J., Lacis, A., Rind, D., Russell, G., Stone, P., Fung, I., Lerner, J., and Ruedy, R., 1984, Climate sensitivity: Analysis of feedback mechanisms, *in*: "Climate Processes and Climate Sensitivity," J. E. Hansen and T. Takahashi, eds., American Geophysical Union, Washington, D.C.

Hansen, J. E., Fung, I., Lacis, A., Rind, D., Lebedeff, S., Ruedy, R., and Russell, G., 1988, Global climate changes as forecast by Goddard Institute of Space Studies three-dimensional model, *J. Geophys. Res.*, **93**:9,341-9,364.

Harrison, E. F., Minnis, P., Barkstrom, B. R., Ramanathan, V., Cess, R. D., and Gibson, G. G., 1989, Seasonal variation of cloud radiative forcing derived from the Earth Radiation Budget Experiment, *J. Geophys. Res.*, submitted.

Hartmann, D. L., and Doelling, D., 1990, On the net radiative effectiveness of clouds, *J. Geophys. Res.*, submitted.

Hartmann, D. L., Ramanathan, V., Berroir, A., and Hunt, G., 1986, Earth radiation budget data and climate research, *Rev. Geophys.*, **24**:439-468.

Hays, J. D., Imbrie, J., and Shackleton, N. J., 1976, Variations in the earth's orbit: Pacemaker of the ice ages, *Science*, **194**:1,121-1,132.

Imbrie, J., and Imbrie, J. Z., 1980, Modeling the climatic response to orbital variations, *Science*, **207**:943-953.

Imbrie, J., and Imbrie, K. P., 1979, "Ice Ages: Solving the Mystery," Enslow Publishers, Short Hills, NJ.

Kukla, G., and D. Robinson, 1980, Annual cycle of surface albedo, *Mon. Wea. Rev.*, **108**:56-68.

Lian, M. S., and Cess, R. D., 1977, Energy-balance climate models: A reappraisal of ice-albedo feedback, *J. Atmos. Sci.*, **34**:1,058-1,062.

Manabe, S., and Stouffer, R. J., 1980, Sensitivity of a global climate model to an increase in the CO_2 concentration in the atmosphere, *J. Geophys. Res.*, **85**:5,529-5,554.

Manabe, S., and Stouffer, R. J., 1988, Two stable equilibria of a coupled ocean-atmosphere model, *J. Climate*, **1**:841-866.

Manabe, S., and Strickler, R. F., 1964, Thermal equilibrium in the atmosphere with a convective adjustment, *J. Atmos. Sci.*, **21**:361-385.

Manabe, S., and Wetherald, R. T., 1967, Thermal equilibrium of the atmosphere with a given distribution of relative humidity, *J. Atmos. Sci.*, **24**:241-259.

Manabe, S., and Wetherald, R. T., 1987, Large-scale changes in soil wetness induced by an increase in atmospheric carbon dioxide, *J. Atmos. Sci.*, **44**:1,211-1,235.

Milankovitch, M., 1920, "Théorie Mathématique des Phénomènes Thermiques Produits par la Radiation Solaire," Gauthier-Villars.

Milankovitch, M., 1941, "Canon of Insolation and the Ice Age Problem," U.S. Department of Commerce, Israel Program for Scientific Translations (1969).

Mitchell, J. F. B., 1989, The "greenhouse" effect and climate change, *Rev. Geophys.*, **27**:115-139.

Mitchell, J. F. B., Senior, C. A., and Ingram, W. J., 1989, CO_2 and climate: A missing feedback? *Nature*, **341**:132-134.

North, G. R., Cahalan, R. F., and Coakley, J. A., 1981, Energy balance climate models, *Rev. Geophys.*, **19**:91-121.

Ramanathan, V., Cess, R. D., Harrison, E. F., Minnis, P., Barkstrom, B. R., Ahmad, E., and Hartmann, D. L., 1989, Cloud-radiative forcing and climate: Results from the Earth Radiation Budget Experiment, *Science*, **243**:57-63.

Ramanathan, V., Cicerone, R. J., Singh, H. B., and Kiehl, J. T., 1985, Trace gas trends and their potential role in climate change, *J. Geophys. Res.*, **90**:5,547-5,566.
Schlesinger, M. E., and Mitchell, J. F. B., 1987, Climate model simulations of the equilibrium climatic response to increased carbon dioxide, *Rev. Geophys.*, **25**:760-798.
Sellers, W. D., 1969, A climate model based on the energy balance of the earth-atmosphere system, *J. Appl. Meteor.*, **8**:392-400.
Simmons, A. J., and Bengtsson, L., 1984, Atmospheric general circulation models: Their design and use for climate studies, *in*: "The Global Climate," J. T. Houghton, ed., Cambridge University Press, Cambridge.
Slingo, A., 1989, Sensitivity of the earth's radiation budget to changes in low clouds, *Nature*, submitted.
Smagorinsky, J., 1983, The beginnings of numerical weather prediction and general circulation modeling: Early recollections, *in*: "Theory of Climate," B. Saltzman, ed., Academic Press, New York.
Somerville, R., and Remer, L. A., 1984, Cloud optical thickness feedbacks and the CO_2 climate problems, *J. Geophys. Res.*, **89**:9,668-9,672.
Street-Perrott, F. A., and Harrison, S. P., 1984, Temporal variations in lake levels since 30,000 yr BP: An index of the global hydrological cycle, *in*: "Climate Processes and Climate Sensitivity," J. E. Hansen and T. Takahashi, eds., American Geophysical Union, Washington, D.C.
Sverdrup, H. U., Johnson, M. W., and Fleming, R. H., 1942, "The Oceans," Prentice-Hall, Inc., Englewood Cliffs, NJ.
Untersteiner, N., 1984, The cryosphere, *in*: "The Global Climate," J. T. Houghton, Cambridge University Press, Cambridge.
Vonder Haar, T. H., and Oort, A. H., 1973, New estimate of annual poleward energy transport by northern hemisphere oceans, *J. Phys. Oceanogr.*, **2**:169-172.
Washington, W. M., and Parkinson, C. L., 1986, "An Introduction to Three-Dimensional Climate Modeling," University Science Books, Mill Valley, CA.
Wielicki, B. A., and Green, R. N., 1989, Cloud identification for ERBE radiation flux retrieval, *J. Climate*, submitted.

THE RESPONSES OF TERRESTRIAL ECOSYSTEMS TO GLOBAL CLIMATE CHANGE

Michael J. Crawley

Centre for Population Biology
Department of Biology
Imperial College, Silwood Park
Ascot, Berks SL5 7PY, UK

INTRODUCTION

In this chapter the aim is to show how modern ecological theory can contribute towards an understanding of the impact of global climate change on terrestrial ecosystems. The main theme concerns the ways in which changes in precipitation and temperature could alter the structure and function of plant and animal communities, and how changes in vegetative cover could alter albedo, surface roughness, and evapotranspiration to influence subsequent climate change.

I shall say nothing about such controversial environmental issues as sea level rise and tropical forest destruction; my main message is that woefully little is known about the detailed consequences of rapid climate change on ecological processes.

ECOLOGICAL DYNAMICS

Although ecological systems are highly complicated, the scientific ideas that underpin modern ecological theory are rather simple. There are two main issues, each of which is represented by two main schools of thought. The first issue is whether ecological systems (communities, ecosystems, or whole biomes) consist of closely integrated, highly coevolved assemblages of mutually interdependent components, or whether they are loose groupings of preadapted, opportunistic fugitives, which are not highly coevolved, and which affect, and are affected by, only a small subset of the other species present. The second issue relates to the nature of interactions between species (be they competitors, predators, or mutualists), and to whether these interactions are reciprocal or asymmetric.

These issues are important, because if ecosystems are tightly coupled, closely integrated systems that are dominated by symmetrical interactions, the consequences of global change are likely to be extremely severe (small perturbations will tend to have large consequences). If, on the other hand, ecosystems are rather loose assemblages that are dominated by asymmetric interactions, there is likely to be substantially more scope for replacement of species' function, following losses of key species as a result of climate change.

The Conventional View

The conventional view of ecology is extreme on both these issues; it holds

Global Climate and Ecosystems Change, Edited by G. MacDonald
and L. Sertorio, Plenum Press, New York, 1990

that ecological systems are closely interwoven, highly structured, and dominated by interactions between species that are tight, continuous, and symmetric. This view can be caricatured by referring to it as the "Many, Complex and Interacting" school of ecology (MC&I for short; Crawley (1986a, p. 12) describes the origin of the term).

Some Victorian poets were greatly taken by the interconnectedness of things; for example,

> *All things by immortal power*
> *Near or far, hiddenly*
> *To each other linked are*
> *That thou canst not stir a flower*
> *Without troubling of a star*
>
> Francis Thompson, 1859-1907, The Mistress of Vision

Charles Darwin was not beyond making up parlor games to display the degree to which things were interlinked in ecological systems. The idea was to construct causal chains that led, through a series of individually plausible links, to an outcome that was highly implausible (and preferably amusing). A famous example shows how the defense of the British Empire depends upon old maids, based on the following reasoning: Old maids, as is well known, keep cats, and cats eat woodmice, which feed on the combs of humble bees. Humble bees are necessary for pollinating clover, which is a vital ingredient of the clover hay required by the horses that are used by the cavalry for the defense of the British Empire.

The distinguished Canadian ecologist, C.S. Holling (1959), proposed a list of desiderata for ecological models and fell into the MC&I school, perhaps unwittingly, when he insisted that all ecological models should strive simultaneously to achieve:

Generality
Realism
Precision
Holism

Few people would argue with the first three criteria, except to point out that they can not be maximized simultaneously, because clear trade-offs are involved (general models, for example, can not be expected to be precise in their predictions). However, the inclusion of "holism" as an objective on a par with generalism was, by some analysts of less distinction than Holling, taken as a spur to go for the "kitchen sink" approach to modeling. In its crudest form, this means that a biologist will include every natural process thought to be involved in an effort to improve the model. These models rapidly become so complex that their behavior is as difficult to understand as the real systems they are intended to depict.

To aim for holism in models of global climate change is likely to be entirely counterproductive. Our only hope of understanding the causes and consequences of global change is to build models that are simple enough to exhibit comprehensible behavior. If the models are realistic and accurate in their predictions, so much the better. But any attempt to reconstruct the holistic complexity of ecosystems in computer simulation models is doomed to failure and is likely to serve only to cloud such important issues as what we can do about global climate change.

The Modern View of Ecology

Nobody disputes the complexity of ecological systems, with their many interacting processes. The full-blown MC&I attitude is a council of despair, however, because it holds that ecosystems are so complicated that it is impossible to understand them well enough to make predictions about how they will behave.

The question at issue is how best to study ecosystems in order to understand their behavior, and, perhaps, to make predictions about their responses to external stresses. The modern view assumes that most of the links in ecological systems are sufficiently weak or asymmetrical to be safely ignored for most practical purposes. The aim is to pare down the numbers of species and interspecific interactions to a manageable subset with which to work. The next step is to inquire about the nature of the interactions within the reduced set: are they symmetrical or asymmetrical, continuous or discontinuous?

Symmetry in Ecological Interactions

The classical theory of interactions between predators and their prey, or between one species and competitor species, evolved from the work of two mathematicians writing in the mid 1920s: the Italian, Vito Volterra (1926), and the American, Alfred Lotka (1925). Their simple, analytical model of predation was based on the assumption that the only process affecting the abundance of a prey population was predation, and that the only factor regulating the abundance of predators was the number of prey available to them. This perfect symmetry was expressed in a pair of differential equations (Table 1). The behavior of the equations was pathological, however, because it predicted neutral cycles with predator and prey numbers cycling indefinitely on a trajectory determined entirely by the initial densities of predators and prey. When the trajectory was disturbed (as by immigration of prey from outside the system), it stayed disturbed, and took on a new trajectory (see May, 1973).

Nevertheless, the equations had an enormous influence on ecological thinking over the following half-century; they were responsible for ingraining the notions that predator-prey interactions were likely to be symmetrical, and that cycles in predator and prey numbers were the norm (see Elton, 1958; Lack, 1954). The concept of symmetry gained further impetus from later theoretical efforts that

Table 1. The Lotka-Volterra equations of predator-prey dynamics.

These symmetrical relationships assume that predator numbers P are determined by prey numbers N, and that prey numbers are determined by predator numbers. In the absence of predators, the prey increase exponentially at a per capita rate a, while in the absence of prey, the predators are assumed to decline exponentially at a per capita rate e. Encounters between predators and prey are assumed to occur by Brownian motion, so that the number of encounters is proportional to the product of their two densities, NP:

$$\frac{dN}{dt} = aN - bNP$$

$$\frac{dP}{dt} = cNP - eP$$

The dynamics of these equations are pathological in that they predict neutral cycles. The system responds to perturbation by adopting a new neutral cycle. The equations can be made substantially more realistic by including density dependence in the prey rate of increase, the functional and numerical responses for the predators, and density dependence in the predator death rate (see Crawley, 1983, for details).

---

showed how modifying the Lotka-Volterra equations to include more realistic biological assumptions (e.g., density-dependence in predator abundance, or food-limitation in the prey) caused much more plausible behavior. The neutral cycles were replaced by a variety of behaviors, including stable equilibria, damped oscillations, and stable-limit cycles (see Crawley, 1983, for examples).

One of the most important insights of the Lotka-Volterra competition equations was the notion that criteria could be established for predicting whether or not a given pair of species could coexist in a stable-equilibrium community. In short, the models predicted stable coexistence only when competition between members of the same species (intraspecific competition) was more influential in their population dynamics than was competition between different species (interspecific competition). On closer examination, however, this prediction simply says that populations need to be self-regulating for coexistence, or, in other words, that the populations must not interact strongly with one another. The model is not especially helpful, therefore, because coexistence is observed in many real systems where there is strong interspecific competition (e.g., Crawley and May, 1987), and in others where the long-term coexistence of apparent competitive equivalents argues for the applicability of non-equilibrium models (e.g., Crawley, 1986a).

The study of mutualistic (i.e., mutually beneficial) interactions received little attention from theoreticians until recently, partly because the Lotka-Volterra models of mutualism gave such ludicrous predictions. What tended to happen was that species A would increase the growth rate of species B, which, in turn, would increase the growth rate of species A, so that a runaway positive feedback ensued, which Robert May graphically described as "an orgy of mutual benefaction" (May, 1981). It is clear from recent field work that most genuine mutualisms are highly asymmetric (Boucher et al., 1984), and that many processes that were traditionally considered as being mutualistic may be thought of more realistically as being parasitic (e.g., lichens, mycorrhizae, etc.).

In any event, the legacy of the Lotka-Volterra equations was profound, and the notion of symmetry in competitive, predatory, and mutualistic interactions survived virtually unchallenged into the mid 1970s.

Asymmetry in Ecological Interactions

Recent field studies have called into question the generality of symmetry in ecological interactions. Most of the careful studies of interspecific competition have shown that one species has a disproportionately great impact on the other (Lawton and Hassell, 1981). Studies of within-species competition have shown that hierarchies are commonplace, with one individual often gaining a disproportionate share of the resources available, compared to a subordinate individual. (In the jargon of theoretical ecology, asymmetric "contest competition" is observed more commonly than symmetric "scramble competition"; see Hassell, 1978, for details.)

Similarly, detailed studies of predator-prey interactions have shown that asymmetries are more often observed than symmetries (Bergerud, 1988). In plant-herbivore interactions, for example, asymmetries appear to be the rule. Whether the plant has more effect on the herbivore or vice versa varies from case to case. In some cases, the herbivore has more impact on the population dynamics of the plant than the abundance of the plant has on the population of the herbivore. In other cases, the reverse is true. Two examples will make this plain.

The myxoma virus, which was introduced to Britain via France from Brazil in the mid 1950s, reduced the rabbit population to the verge of extinction by the end of the same decade. Rabbit numbers have since rebounded, increasing through the 1980s to regain their former abundance in many areas. The disease is less virulent at present than the introduced strain of myxoma virus (which was so lethal that its transmission was ineffective except in extremely dense rabbit populations), and the rabbit population now has a much higher level of genetic resistance to the disease (myxoma acted as a potent agent of natural selection in favor of those few relatively resistant rabbit genotypes that were present in the original population). During the 1960s, when rabbit numbers were low, a great many former grasslands were invaded by woody plants; these plant communities are now dominated by 30-year-old woodlands.

Current grazing experiments show how the exclusion of rabbits can influence the recruitment of different kinds of trees. In the case of the English oak, *Quercus robur*, both the seeds and the seedlings are highly palatable to rabbits, and at present rabbit densities, tree seedlings only establish successfully within experimental exclosures. In contrast, the sycamore, *Acer pseudoplatanus*, is unpalatable to rabbits and regenerates freely outside the exclosures. Sycamore fails to regenerate successfully inside the exclosures, however, because its relatively small seeds give rise to small, uncompetitive seedlings. Protected from rabbit grazing within the exclosures, grasses and other rapidly growing herbaceous plants out-compete the sycamore; the large-seeded oak can withstand this intense competition (Fig. 1). For both tree species, the interaction with rabbits is asymmetric, because the rabbits have far more influence on the fate of the trees than the trees have on the fate of the rabbits. The reason that rabbits can prevent oak regeneration in open grasslands is that rabbit numbers are held at high levels relative to the abundance of oak seedlings by an abundance of alternative foods (chiefly grasses in this example). Thus, reducing the number of rabbits would lead to a large increase in the recruitment of oak, but reducing the number of acorns would have no appreciable impact on the abundance of rabbits.

Other plant-herbivore interactions are asymmetric, but in the opposite direction: the plant has far more effect on the herbivore than vice versa. A good example is provided by the cinnabar moth, *Tyria jacobaeae*, which feeds on ragwort, *Senecio jacobaea*, a common member of the daisy family (Compositae). The adult moth lays her eggs in batches of 50 or so, and the caterpillars that emerge from these egg batches routinely strip all the leaves and flowers from the plant on which the eggs are laid. The caterpillars then move off in search of plants that escaped egg-laying females, where they can complete their feeding before pupating in the soil. At first sight, it appears as if the insect has a devastating impact on the plant, reducing its reproduction to zero. Once the insects have pupated, however, the plant is able to resume growth. In years when late-summer growing conditions are good, grazed plants can produce substantial flower crops on regrowth shoots. Even in poor years, the plants are often able to replenish their underground food reserves before the onset of winter. Thus, while the number of cinnabar moths is strictly limited by the abundance of ragwort plants (the insects eat no other plant species in most habitats where ragwort is found, and suffer low rates of attack by natural enemies), the plant is affected to a much lesser degree. Because several regrowth shoots can appear from the crown of a single rootstock, grazing by cinnabar moth caterpillars may actually lead to an increase in the plant's population density (Crawley and Gillman, 1989).

This kind of result is sometimes misunderstood and taken to mean that grazing is "good for the plant." From an evolutionary standpoint, such an assumption is untrue, because a plant that is kept free of insects experimentally always leaves more progeny than a grazed plant—no matter how good the conditions for regrowth (Belsky, 1986).

The asymmetry in these two examples can be summarized by noting that it is not possible to predict the abundance of cinnabar moth without knowing the amount of ragwort, but the abundance of acorns need not be known in order to predict the density of rabbits.

As described in the previous section, models of symmetric interaction predict a range of dynamical behaviors, depending upon the parameter values incorporated. It is useful to inquire whether the more realistic, asymmetric models

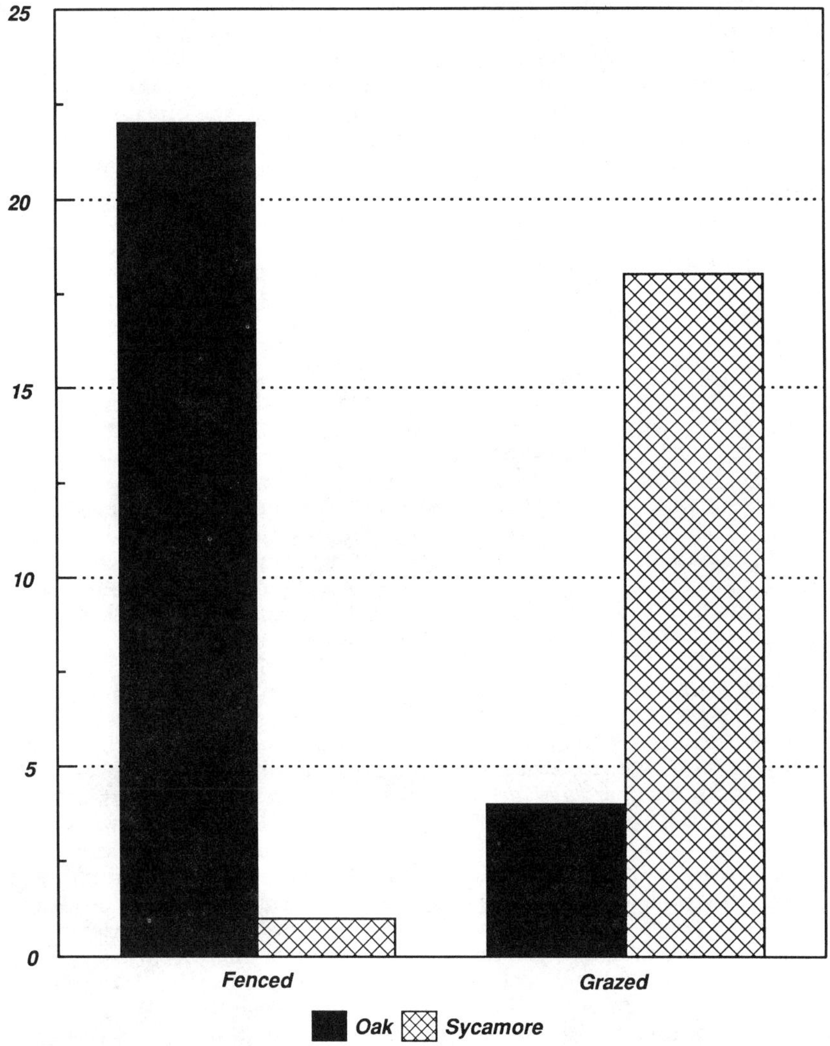

Fig. 1. The relationship between rabbit grazing and tree recruitment. Within fenced rabbit exclosures, the oak seeds produce vigorous seedlings which can compete with the dense, ungrazed vegetation, but the sycamore seeds are lost through interspecific plant competition. Outside the fence, subject to selective herbivory, the oak seeds and seedlings are consumed, whereas the unpalatable sycamore seedlings thrive in the short sward, in conditions of low plant competition.

predict a more restricted range of dynamics. Because asymmetric models tend to be simpler than their symmetric counterparts (i.e., they tend to contain fewer parameters), it might be imagined that their dynamical behavior would be simpler (i.e., would exhibit a more restricted range of dynamical patterns). In fact, the simplest models can display the full range of population dynamics, from stable equilibria through damped oscillations, stable limit cycles, increasing oscillations away from unstable equilibria, and chaos (Crawley, 1983). It is quite clear, therefore, that complex behavior does not require complex models. Quite the contrary: the very simplest, first-order, nonlinear difference equation that predicts the abundance of a population next year on the basis of population this year can exhibit the full range of dynamics for different parameter values (May and Oster, 1976).

SHORT-TERM POPULATION DYNAMICS

One of the most striking aspects of global climate change is its anticipated rate; changes are almost certain to be great enough to influence population dynamics within the next 50 to 100 years. For many of the plants that dominate the world's natural ecosystems, this is less than one generation, yet an understanding of population dynamics is rarely achieved in less than ten generations of intensive study. Thus, it is hopelessly optimistic to expect that the scientific understanding that can be obtained over the next 100 years will enable us to predict the kind and extent of changes in the distribution and abundance of dominant plant species. Nevertheless, population dynamics do need to be studied in order to provide the background understanding that will be vital if we are to have any power to predict the responses of terrestrial organisms, and hence terrestrial ecosystems, to global warming.

Changes in distribution and abundance of plant populations result from the interplay of biological and non-biological factors. Average population density is determined by the relative magnitude of birth and death rates, and by the way in which these rates vary with population density (see Fig. 2). An important question in population dynamics is whether fluctuations in abundance are due

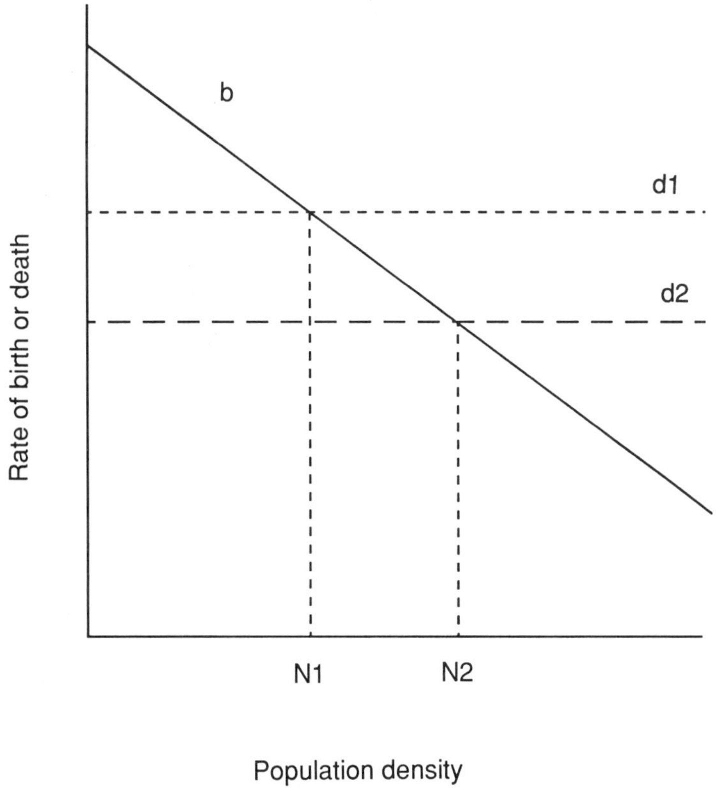

Fig. 2. Population size is determined by the interaction between density-independent and density-dependent processes. In this example, the death rate, d, is density-independent and the birth rate, b, declines linearly with increasing population size (e.g., as a result of reduced food availability). Changes in either the death rate or the birth rate will result in changes in equilibrium population size, but the existence of an equilibrium requires that one or both of the processes be density-dependent.

simply to the vagaries of abiotic conditions (extremes of weather, floods, storms, landslides, and so on), or whether fluctuations are due to interactions between organisms (intrinsic oscillations generated by competition, predation, or mutualism).

Within a given ecosystem, the generation times of the constituent species often vary by several orders of magnitude, from the dominant trees that may live for hundreds of years, down to soil bacteria whose generation times are measured in minutes. The fact that different species can vary in size by many orders of magnitude makes it extremely difficult to support general predictions about how changes in one level (e.g., in bacterial activity) affect changes in another (e.g., the identity of the dominant tree species). Similarly, the spatial scales over which changes in distribution and abundance occur are influenced by the location of physical barriers to dispersal, and by the body sizes and mobilities of the individual organisms. Thus, an individual annual plant may occupy as little as 1 cm^2, while one individual of a clonal grass may cover several hectares. Again, 95% of pollen may fall within 5 m of the parent plant, but 0.001% may travel many kilometers. In considering ecosystem-level changes, therefore, we are dealing with groups of organisms that exhibit at least three orders of magnitude in the temporal scale of their dynamics and eight orders of magnitude in their spatial scales. Attempting to deal simultaneously with dynamics at all these levels is quite impractical. In the context of long-lived organisms, more rapidly responding populations have to be assumed to be in equilibrium; in considering the dynamics of the short-lived species, long-lived organisms must be taken to have constant abundances. So it is with spatial patterns; large-scale dynamics must be based on the assumption that small-scale processes are in equilibrium, while small-scale dynamics are investigated under the assumption that large-scale patterns are constant.

An important question remains, however: to what extent can an understanding of the dynamics of individual species be synthesized into an understanding of the dynamics of ecosystem-level variables such as biomass, diversity, and productivity?

Keystone Species

One way in which simplifications can be made is to focus on so-called "keystone species" (Paine, 1966). These are organisms whose distribution and abundance exert a key influence on ecosystem processes, usually as a result of their feeding activity. Keystone species are most often discovered by removal or exclusion experiments, as when predatory starfish are excluded from sections of a rocky shore, or sea otters are reduced in abundance by human hunters. In these two examples, removal of the important top predator led to burgeoning populations of their prey species (molluscs and sea urchins, respectively) and hence to a drastic reduction in plant populations (smaller seaweeds and kelp, respectively). We do not know how many ecosystems have keystone predators, nor do we know how to detect keystone species other than by direct experimentation.

Rabbits, for example, are keystone herbivores in the acid grasslands discussed earlier, because their presence determines whether or not a grassland will become a woodland. It could be argued that the myxoma virus is the keystone predator in this system, because by controlling the rabbit population, it also controls the regeneration of trees. However, both rabbits (introduced by the Normans) and the myxoma virus (introduced by farmers in the 1950s) are alien species in Great Britain. We do not know whether native herbivores dominated grassland function prior to the introduction of rabbits, or whether native predators, parasites, or diseases kept the native herbivores at densities that were too low to affect plant community structure.

Although the concept of keystone predators is uncomfortably vague, it does allow some broad patterns to be detected. For example, a predator is more likely to be keystone if it: is the only species in its feeding guild (Paine, 1966); exhibits a strong preference for the most influential herbivore in the system (Crawley, 1983); is present in systems with a dominant herbivore, rather than in systems with many more-or-less equally influential herbivores (Lawton and Hassell, 1981); and so on.

Even in cases where "keystone species" have been identified, however, it is not clear what scope there is for replacement if the keystone species should succumb as a result of climate change. We cannot generalize from cases where top-predator species have been identified as keystone species by experimental removal to say that top predators are keystone in all systems where they occur, because it is equally plausible that top predators are food-limited without exerting any regulating influence on the abundance of their prey species (see above).

The simplifications that spring from considering only a limited set of keystone species do not, however, solve the question of synthesis. Even if we could predict how each keystone species would respond to global warming, it is not clear that this would help us to predict the resulting changes at ecosystem level. The bottom-up approach appears to be doomed to failure. Too much information is required, there is too little time to gather it, and there is no guarantee that the information could be integrated sensibly, even if it were to be collected.

LONG-TERM ECOLOGICAL CHANGES

A more fruitful alternative for tackling questions of how ecological systems will respond to global change might be a top-down approach. By asking directly how ecosystem-level attributes such as biomass, diversity, relative abundance, and productivity are affected by changes in the mean, variance, and timing of precipitation, for example, one might be able to predict how herbivore consumption, herbivore biomass, and herbivore productivity (so-called secondary productivity) might be affected (see section on Research Priorities). A major advantage of this approach is that it unifies the time scales involved, and allows one to deal naturally with the kinds of spatial scales at which predictions about climate change might actually be made (i.e., tens or hundreds of square kilometers). To hope for the level of precision in meteorological forecasting that would be required for detailed population-level predictions is unrealistic at present (e.g., patterns of precipitation at a scale of hectares). The time scales of these ecosystem processes are intermediate (decades), and lie between the rapid time scales on which population dynamics occur (months and years) and the slow time scales on which evolutionary and geological changes occur (thousands to millions of years).

There are extreme schools of thought on the nature of change in ecosystems. The MC&I view sees ecosystems as "organisms": tightly coupled, intimately coevolved assemblages that exhibit processes akin to birth, growth, maturity, senescence, and death. An extreme form of this notion is to regard the entire planet earth as a single living organism (the Gaia Hypothesis; Lovelock, 1979). On the other hand, the reductionist view is that ecosystems carry a considerable amount of excess baggage, and that system-level properties can be predicted by concentrating on a small subset of rather loosely coupled species that are preadapted to the prevailing biotic and abiotic conditions and are not intimately coevolved. The debate between these schools of thought has a pedigree stretching back to the 1920s.

Primary Succession

The two contrasting views of community dynamics are encapsulated in the writings of two American plant ecologists working in the 1920s. F.E. Clements (1928) saw plant communities as organic entities akin to individual organisms: "As an organism the (community) arises, grows, matures, and dies. Furthermore, each climax (community) is able to reproduce itself, repeating with essential fidelity the stages of its development. The life-history of a (community) is a complex but definite process, comparable in its chief features with the life-history of an individual plant." In complete contrast, H.A. Gleason (1926) viewed plant communities as random assemblages of preadapted fugitives. To Gleason, the community was "not an organism, but merely a coincidence." Thus, to Clements, the process of succession in which one recognizable plant community is replaced by another represented a strict, necessary, and predictable sequence, leading to the ultimate development of the climax vegetation as dictated by the interaction between climate and substrate. To Gleason, succession was an essentially stochastic process: life forms replaced one another in a reasonably predictable way (e.g., mosses by grasses, grasses by scrub, scrub by trees), but the species composition of any successional stage, and the relative abundance of the plants within any stage, were serendipitous.

Evidence about successional changes comes from two principal sources. The first involves studies of existing gradients within contemporary vegetation. The assumption is made that present-day structure in space reflects historical changes through time. For example, as we walk away from the edge of a lake, we pass through communities of annual plants on the sandy beach, then through a scrub, and then through a range of forest types dominated by progressively longer-lived and more shade-tolerant trees. As we walk inland, the soils become deeper, nitrogen availability and organic-matter content increase, and the average light intensity declines (Tilman, 1988). These spatial zones are interpreted as representing temporal phases in the primary succession from lifeless, sandy beach to species-rich, climax forest. We have no direct evidence for ecological changes of this kind, because they take several thousands of years to occur.

Some contemporary information on primary succession can be obtained by studying the process of soil formation during the colonization of industrial spoil heaps (Marrs et al., 1983). This work demonstrates vividly the importance of soil nitrogen. Before a plant community can produce a typical primary production of 5 t dry matter per hectare, which contains, say, 75 kg of nitrogen per hectare, roughly 1500 kg of nitrogen per hectare must be present in the soil (assuming a turnover rate of 5% per year). The only way to vegetate highly nitrogen-deficient spoil tips is to apply nitrogen fertilizer on a continuing basis, year after year, until such time as the soil capital has built up to the point at which internally recycled nitrogen can supply the needs of primary productivity.

The second important source of information on primary succession comes from plant fossils preserved in peat and lake sediments. These remains are laid down in strata, which can be dated by a variety of techniques, and which allow the approximate structure of former plant communities to be reconstructed. Studies of pollen and other plant remains have enabled us to build a picture of the biological events that are associated with a complete glacial cycle.

A receding glacier leaves behind it a cold, wet, windswept landscape where the major limitations on plant growth are the lack of soil nitrogen and the instability of the substrate. As temperatures rise and the glacier recedes, mosses and lichens colonize the bare ground and begin the twin processes of substrate stabilization and nitrogen enrichment (many of the lichens contain nitrogen-fixing blue-green algae). Small vascular plants then become established: annual herbs and grasses at first, and then perennial grasses. Cold-tolerant shrubs such as willow and juniper dominate this arctic-alpine phase in primary succession, which is known as the cryocratic period.

As warming continues, larger nitrogen-fixing plants such as alder multiply and accelerate the rate of improvement of the soils, which are still of an alkaline reaction. The level of shade increases throughout this period, until a mosaic of woodland and grassland is produced. With temperatures at their interglacial maximum (the mesocratic phase), a "climax forest" develops. Maximum levels of shade are cast, and soil leaching begins to produce a more acidic soil. As temperatures begin to fall, the glaciers start to re-advance from their interglacial minimum. Soil leaching now progresses rapidly, leading to acidification and to the creation of open plant communities, with much development of peat (heathlands dominated by pines and by various ericaceous plants). Eventually, the ice over-runs the degenerating communities, wiping the slate clean, and the cycle begins again.

Because these events occur at a single point in space, the various plant species are assumed to move southwards as the ice advances, then northwards again as the ice retreats. It has become clear from recent research in North America, however, that the conventional wisdom of entire plant communities moving southwards and northwards as intact ecosystems is wrong (Davis, 1981). Pollen data show that each species moves at its own rate, and that plant communities are in a continual state of flux, constantly reshuffling as different species come and go. For example, the communities of trees we see growing together in natural forests today are not necessarily the same as those that grew together at a similar stage during the last interglacial.

In Europe, for instance, the oak (*Quercus spp.*) had returned to all but the extreme northwest of Great Britain by about 7,000 years ago, while lime (*Tilia cordata*) had returned to northern England by 6,000 years ago, but never advanced further northward. As a result, Scottish woodlands, unlike their English counterparts, never experienced codominance by oak and lime during this interglacial. Other plants never made it back to Britain at all. The spruce (*Picea abies*) moved northwards very slowly, only arriving in Norway about 1,000 years ago; it never came sufficiently close to the English Channel to make the crossing to England. Thus, despite the fact that growing conditions are ideal for spruce in much of upland Britain, it never returned under its own devices (the Forestry Commission has gone some way towards rectifying this omission). Some plants made very little northward movement during the present interglacial. The fir (*Abies alba*), for example, remained confined to the Alps and other high mountain ranges in the Pyrenees and the Balkans. Presumably, the warming was so rapid that fir was unable to cross the intervening lowlands before they were clothed with an impenetrable barrier of uninvasible oak/lime forest. Again, a species which could survive at more northerly latitudes was unable to make the journey sufficiently rapidly (i.e., its northward spread was dispersal-limited). This theme is likely to be exhibited frequently under global warming and will be exacerbated by barriers to establishment in man-made agricultural and urban landscapes (see subsection on Data from Fossil Plants).

Secondary Succession

Secondary succession may be a better model for the impact of global warming, because a soil already exists (primary succession can be seen as essentially a soil-making process). There are four main models for the operation of secondary succession: 1) the *inhibition* model assumes that a species is unable to invade the successional community until the resident species that is inhibiting its establishment is debilitated (by pathogens, herbivores, old age, or competition from other plants); 2) the *facilitation* model assumes that a species is unable to invade the successional community until another species has "paved the way for it" by ameliorating conditions to the point at which seedlings of the invading species can recruit and adult plants develop (this model is also know as *relay-floristics*); 3) the model of *initial floristic composition* assumes that all the species are present in the soil seed bank at the outset, and that all we observe during secondary succession is the replacement of the most rapidly growing plants (the annuals) by progressively more slowly growing types, culminating with dominance by the slowest growing plants of all (typically the trees); and 4) the *mixed strategy* model, which suggests that whole successions are not of one type or another, but that the component plant species come and go during the course of community dynamics in species-specific ways (i.e., some species are inhibited by others, some species are facilitated by others, while some species are entirely indifferent, and grow at whatever successional stage they happen to arrive). The first two models emphasize the importance of strict and predictable interactions during successional change, while the last two models emphasize the stochastic and species-specific nature of community dynamics (see Crawley, 1986a, for details).

None of these models distinguishes between different causes for secondary succession, as if any disturbance leading to the destruction of the previous vegetational cover would produce the same sequence of plant species as any other. This is clearly not the case; the kind of disturbance, its intensity, and its history—as reflected by the frequency of disturbances of different intensities—are all likely to be important in influencing which species are found, and their relative abundance. It is not clear how disturbances resulting from global climate change (e.g., a change in the frequency and intensity of rainfall) would differ from other succession-initiating disturbances such as fire, defoliation by herbivores, or cultivation. It is reasonable to assume, however, that climate change will cause successions of a different kind, if only because the species pool from which the organisms are selected will be different.

A number of important questions about climate change and succession are still to be addressed: will global warming wipe the slate entirely clean of the existing species at any one place, and, if not, which species will survive? How rapidly will the slate be wiped clean, and in what order will species disappear? Where will the new species come from, and when will they arrive? How long will individuals of a species persist once conditions have changed to the point where the species can no longer recruit? How will ecosystem-level properties be altered by these changes in species composition and relative abundance?

Gap Regeneration

A third kind of successional process occurs in perennial plant communities, where opportunities for recruitment are normally restricted by the intense shade cast by the mature individuals. Seedlings of many species can only recruit following the opening of a gap in the canopy, caused either by the death of one or more mature plants, or by a physical disturbance such as fire, storm, or landslide (Brokaw, 1985). The size of the gap will influence the response of different plant species: some species regenerate only in the largest gaps, others in much smaller gaps, and a third group recruits seedlings in full shade, where the young plants may remain stunted for many years, only to grow rapidly once a canopy gap opens above them.

Gap regeneration is relevant to questions of global change because of problems of timing: species can only recruit if their propagules are present at the time the gap opens up, or shortly after. Any delay in arrival will allow individuals of other species to occupy the gap, and invasion later on may be impossible due to intense competition, or to predation from natural enemies. Thus, for gap regeneration to lead to the replacement of an existing dominant by a potential dominant that is suitably adapted for the new climatic conditions requires that the propagules of the invading species are readily available. Unless the new species has exceptional powers of dispersal, it is most unlikely that this kind of gradual gap-replacement will occur. It seems more likely that it will be necessary for the existing dominant to disappear (as a result of failure of its own recruitment), creating an early secondary successional stage, and then for the new species to invade this essentially juvenile community in a stepwise fashion from the edge of its existing range.

If, as seems likely, gap regeneration of the replacement dominants does not occur, then dramatic ecosystem-level changes are certain to result from global warming. Forests, for example, will degenerate into scrub, which might then begin succession towards a new kind of mature forest, dominated by species that are capable of recruitment under the prevailing climatic conditions. This replacement is entirely contingent, of course, on the availability of propagules of the new species, and, most importantly, on the availability of space in which to grow. It is a moot point, for example, whether landowners would leave a degenerating woodland alone long enough for it to undergo natural succession to a new kind of woodland; they would probably attempt to turn the resulting scrub to some other, more profitable land use.

At this point, it is worth noting that climate change poses a considerable challenge to nature conservation. Many existing nature reserves are in the wrong place. Estuarine bird reserves will be lost following sea level changes, and the land required to replace them is owned and used for other purposes. Who will buy the land to provide new reserves in future? Inland nature reserves, although safe from rising sea level, also face major problems, because the plant and animal communities they now protect will no longer be appropriate to the prevailing climate. If the existing community is to be preserved, new land will have to be found for it somewhere else (where, again, the land currently belongs to someone and is used for some other economic purpose). If, on the other hand, the existing reserves are kept, will nature reserve managers become interventionists? Will they intentionally introduce species that they anticipate would be appropriate under the new, altered climatic conditions? What subset of species should be introduced? (Will soil microorganisms be imported? What about the diseases and parasites of all the animals—will these be imported, or screened out in quarantine?) These issues have, until very recently, been anathema to conservationists.

DIVERSITY AND THE "BALANCE OF NATURE"

Diversity is almost universally regarded as a good thing. Most conservation programs are explicitly aimed at preserving local diversity of one sort or another (birds, butterflies, corals, rain-forest primates, and so on). It is clear that the motivation for conserving diversity is a moral revulsion at causing species' extinctions, coupled with a deep-seated love of wildlife and wilderness. Economic motives are also expressed from time to time, such as the preservation of the vast, untapped wealth of pharmaceutical and other products that await discovery. But these human motives tell us nothing about the ecological importance of diversity: What is it associated with? Is it functional? Does it affect stability? What are the consequences of reduced diversity? To what extent can ecosystem functions be substituted for by species replacements following local extinction?

The "balance of nature" is another of those attractive, but completely useless concepts that graced ecology in its early days. Some elements of the notion are sound, others are wrong, and others are simply mystical. The notion enshrines the folk wisdom that natural systems are resilient because of a plethora of "checks and balances." It is argued, for example, that plant-eating insects do not strip plants of their leaves, because if the herbivores became too abundant, their own predators would increase to keep them in check. The sound elements of the notion are that it is, indeed, possible for natural enemies to regulate the population densities of herbivorous insects at such low levels that they inflict little perceptible damage on the plants from which they feed. The elements that are wrong relate to the reason for regulation (e.g., that insects are prevented from becoming too abundant "for the good of the species" or, even more extreme, "for the good of the ecosystem"). This is nonsense, of course, and runs counter to the central tenets of Darwinian theory (e.g., it relies on group-selectionist reasoning; Dawkins, 1987). The mystical elements surround the forces that might be responsible for a balance of nature, and hark back to the "stability through diversity" argument. We simply do not know of any such balancing force acting at ecosystem level. On the other hand, we have a plethora of plausible explanations as to why diversity might be maintained in the face of competition (tending to reduce diversity as a result of dominance by the single most competitive species), and despite environmental extremes that might be expected to reduce scarce species to extinction. In fact, it turns out that fluctuating environments are probably a potent force in promoting diversity; extreme events often produce "good years" for certain species, providing them with a buffer against subsequent depletion in "more normal" years (this is called the storage effect; Chesson, 1986). A general class of mechanisms known as frequency dependence tends to promote the maintenance of diversity. These mechanisms require that species do better when they are rare than when they are common; for example, they may not suffer such high rates of attack from their natural enemies when alternative prey species are relatively abundant.

There is no question but that diversity will be lost over the coming decades. Much of the loss will be due to direct habitat destruction by humans. Some will be caused by pollution. More will be caused by hunting and trophy collecting. The loss of species resulting from global climate change is impossible to predict, but is unlikely to be small. Whether or not the loss of many species upsets the balance of nature remains to be seen. Until we know the function of diversity, and the limits to which ecosystem processes can be maintained in the face of diversity losses (by elasticity in the ecological behavior of the surviving species, or by replacement of lost species by invaders from other communities), we shall be in no position to say.

Traditional arguments about the function of diversity are barren and circular. One school of thought held that diversity increased stability. The reasoning was based on nothing more sophisticated than the adage of not putting all your eggs in one basket. If a predator had two prey species, it was less likely to go extinct in the face of prey fluctuations than if it only had one. The opposite school of thought, using a similar depth of reasoning, argued that complexity begat instability. Their insight was based on a celebrated variant of Murphy's Law, that if something can go wrong, it will. Thus, a predator is more likely to fluctuate in abundance if it relies on two prey species than if it relies upon one, because the probability of a fluctuation in either one of the prey is higher than the probability of a fluctuation in any one species.

Evidence was sought by comparing the stability of diverse and simple ecosystems. Tropical systems were said to be more diverse and more stable than temperate systems. Ecologists who took the trouble to measure diversity and stability soon refuted this notion (Wolda, 1978). Not all tropical communities are diverse. Those which are diverse appear to be highly dynamic, and not at all stable in the traditional sense of being "constant in abundance." Many of the most diverse systems appear to have extremely low resilience; they possess little ability to recover following large-scale, severe disturbance. Some very low diversity systems, on the other hand, are highly resilient. The truth is, we do not understand the causes or the consequences of diversity. We know that diversity should be preserved, because species are irreplaceable, and we don't know what they do. That, surely, is reason enough for their preservation. On the other hand, we do not know how to manage the resources available to us for preserving species in the most effective way. We fear that species loss may disrupt the balance of nature, but we don't know what the balance of nature is.

CLIMATE CHANGE AND ECOLOGICAL PROCESSES

Perhaps the most dramatic aspect of global climate change is the rate at which it is likely to happen. Mean temperature changes equivalent to those that separated the warmest and coldest periods of the last ice age are expected to occur in the course of 100 years, instead of tens of thousands of years. If temperature does change this rapidly, then the reassembly of ecological communities will be taking place within the life-span of individual trees, instead of over several generations of trees, as occurred during the advance and retreat of the glaciers.

Data from Plant Fossils

With such rapid global warming, it is by no means clear that the behavior of tree species during the last glacial decline 12,000 years ago will provide any indication about the kinds of changes in distribution and abundance that are likely to occur in the next two centuries.

Horror stories about the consequences of the speed of change abound. My favorite is that in order to keep up with changing mean temperatures, plant species would need to move their boundaries northwards at one meter per hour! The world record for "tree sprinting" after the last ice age is held by spruce in North America, which traveled north at 200 km per century (i.e., about one-fifth of the necessary rate). Most trees moved northwards at substantially less than half this rate (mode about 25 km per century; Davis, 1981). Also, we should note that all the migrations northwards at the end of the last ice age took place through pristine ecosystems. Any migrations that might take place nowadays would have to occur through a landscape dominated by man, and the plants would have to migrate through (or jump over) large tracts of land currently given over to agriculture, forestry, and urbanization. It is quite clear that many species simply will not be able to move from their current refuges across these man-made barriers into positions where the altered climate would permit their continued recruitment (even in the unlikely event that undisturbed habitats were available for colonization in these new locations). Human intervention with ecosystem planning will be necessary on a hitherto unimaginable scale, raising all manner of practical, logistical, and ethical questions. For example, how does one assemble the minimal functional terrestrial ecosystem; would such an ecosystem be self-replacing; what role should genetic engineering play; and so on?

The first requirement is better climatic forecasting. Many ecosystem processes are controlled by the rate of primary production (the annual amount of carbon fixed by green plants per unit area of land surface). Recent work has shown good correlations between log primary production and log herbivore biomass, and between log herbivore consumption and log secondary productivity (McNaughton et al., 1989). Thus, a good deal of information about the ecosystem-level properties of higher trophic levels can be gained from a knowledge of primary productivity alone. Productivity is closely related to actual evapotranspiration, which is an increasing function of air temperature and precipitation (log primary productivity increases linearly with the log of total evapotranspiration). This correlation is embodied in an empirical law known as Rosenzweig's Rule (Fig. 3). Unfortunately, it is not at all clear how changes in global mean temperature will affect the local-scale pattern of precipitation experienced in different parts of the world. Changes in cloud cover, relative humidity, patterns of wind movement, and so on, are extremely difficult (perhaps even impossible) to predict in detail. Thus, we may never be able to predict accurately the spatial distribution of rainfall or the likely frequency and intensity of storms at a spatial resolution that would allow the detailed prediction of terrestrial ecological processes. Until we can predict how mean and variance in precipitation are altered under global warming, we will not be able to make any sensible predictions about consequent changes in terrestrial ecosystems.

What we can do, however, is to improve our understanding of how specific changes in climate would affect the structure and function of different terrestrial ecosystems. We may not be able to say which particular changes in climate are likely to occur in a given habitat, but we will be better placed to make contingency plans if we understand what would happen if a particular kind of climate change were to come about.

Data from Recent Plant Invasions

More recent plant invasions are less likely to provide information on the consequences of global climate change. Although there is a superficially appealing pattern of greater numbers of alien plants in northern temperate environments coming from warmer rather than colder countries of origin (Crawley, 1986b, 1987), this pattern is confounded by two overriding processes. The first is that the trade in plants, and human traffic generally, is greatly biased towards importation of species from south to north. The second is that the pool of potential immigrant

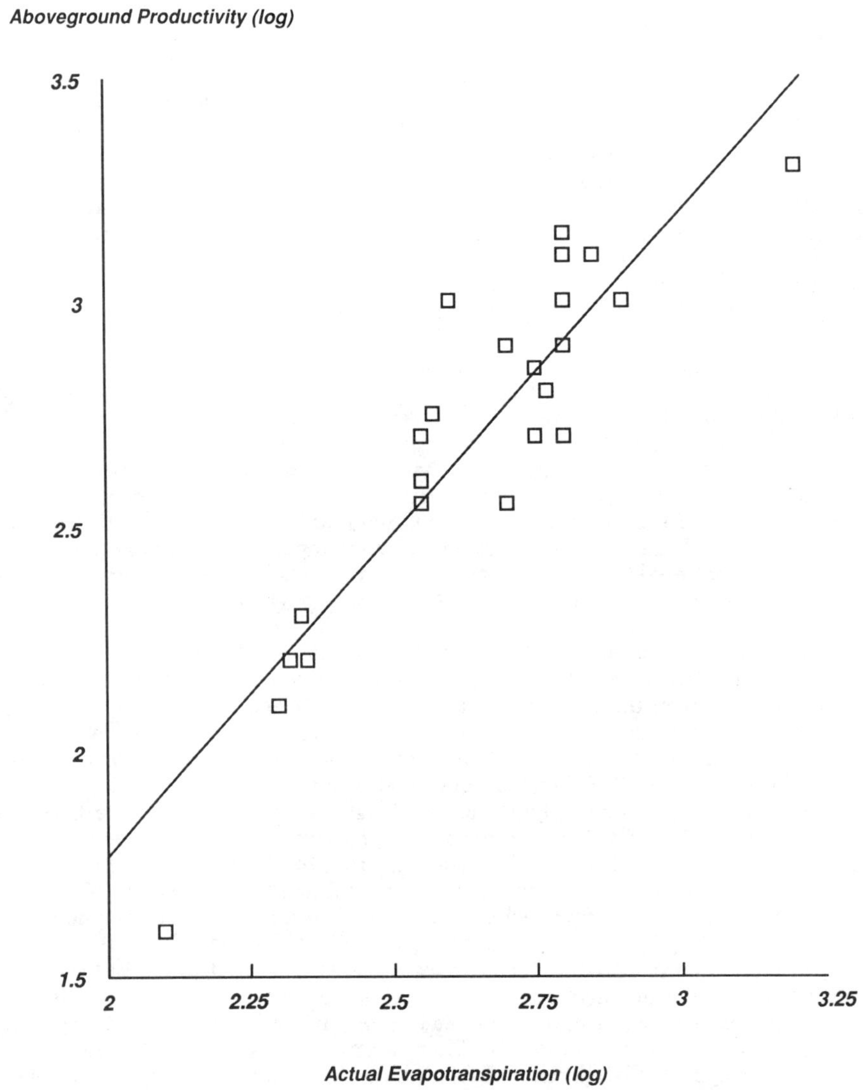

Fig. 3 Rosenzweig's rule: the empirical relationship between primary productivity and actual evapotranspiration (note the logarithmic scales). From Rosenzweig (1968).

species is substantially greater in southern (e.g., Mediterranean) than in northern (e.g., Arctic) communities. Thus, the pattern of more plants invading from warmer rather than colder places can not be interpreted as a response to global warming. Also, a great many of the important plant invasions from the south occurred before the present CO^2-driven episode of global warming began and must, therefore, be due to other causes.

It is inevitable, however, that the new plant species that will eventually come to dominance following climate change are currently classed as aliens. Plant speciation is unlikely in the time periods under consideration, so new plant communities will be assembled from a spatial redistribution of species that are already growing in other places. Some of the plants that will become important components of semi-natural vegetation following climate change may be resident aliens already; some may be naturalized (e.g., Turkey oak, *Quercus cerris*, and sweet chestnut, *Castanea sativa*), while others are casuals, currently requiring repeated reintroduction for their persistence (e.g., most garden plants and many of the coniferous trees of commerce).

Thus, while the plant invasions that have occurred over recent decades may tell us little about how ecosystems will change under global warming, it should be recognized that one of the first consequences of global climate change will be an increase in the rate of establishment of alien organisms, and an increase in the rate of spread of currently established aliens. Alien species will supplant native ones, because climate change is almost certain to reduce the competitiveness of resident, native species, thus making natural habitats more prone to invasion by alien organisms. In addition, the individuals of surviving native species are likely to be less vigorous following climate change, and hence to be more prone to debilitating attack by parasites and diseases. The vulnerability of native plants may further increase the probability that they would be replaced by invading alien species, because the invaders are likely to have left their own natural enemies behind in their places of origin.

The plants that invade naturally following climate change will probably not be those that we would have selected for ourselves, given the opportunity. They will be a ragbag of garden escapees (e.g., *Robinia pseudoacacia*), weeds (e.g., *Conyza canadensis*), and botanical curiosities (e.g., *Heracleum mantegazzianum*). A great many decisions about the importation of alien organisms will be faced by environmental legislators in the coming years, and it is clear that blanket policies banning the importation and release of foreign species would be totally counterproductive. Given the opportunities for invasion, however, the need for strict quarantine is greater than ever. In an ideal world, we would exert strict selection criteria in deciding which species to allow in to replace debilitated local flora and fauna, and which species to exclude. The problem, however, is that we don't know how to build ecosystems. We urgently need to obtain a fundamental understanding of how ecosystems are assembled from different kinds of component species, and how the different components interact and evolve with one another.

RESEARCH PRIORITIES

There are several different kinds of questions concerning the impact of global climate change on the structure and function of terrestrial ecosystems. In this section, I shall highlight five such questions, and discuss the kinds of research that need to be done and the difficulties that will be encountered in tackling the pertinent issues.

1) Given that changes in rainfall are likely to have greater consequences on terrestrial ecosystems than direct changes in temperature will have, what will be the impact of altered patterns of precipitation on the structure and function of terrestrial ecosystems?

The relationship between primary productivity and the turnover rate of the soil-based part of the hydrological cycle is well established (Fig. 3). Using this relationship, we can reasonably predict the amount of carbon fixed by terrestrial plants for a certain level of evapotranspiration. The predictions are robust for variations in annual evapotranspiration between places with different mean rainfalls, as well as between years of differing mean rainfall at a given location. Note, however, that: 1) the axes are on a log scale, so the standard errors are large on an untransformed scale; and 2) the relationship takes no account of how the precipitation occurs (e.g., as a constant drizzle or in a few heavy storms).

Given the consensus that global warming will lead to increased variance in precipitation, and to an increase in the frequency and severity of extreme events, it remains to be seen how well this empirical relationship holds up. At present, it is the best predictive relationship we have, and it can be used with other, more recently established empirical patterns (e.g., McNaughton et al., 1989) to make ecosystem-level predictions. If actual evapotranspiration were to fall as a result of reduced rainfall and/or reduced temperature, then primary productivity would decline. Such a decline would lead to reductions in the biomass of herbivorous animals and in secondary productivity. These reductions, in turn, would cause a reduction in the biomass of parasites and top predators. The consequences of these changes for ecosystem dynamics are more difficult to anticipate, but it is almost certain that a reduction in the total biomass of herbivores and predators would be associated with extinctions at both these trophic levels (it is hard to imagine that all the losses would be from the most abundant species). The consequences of a reduced diversity of herbivores might be compensated to some degree by an increase in the diversity of plant species (since each new plant species could support one or more specialist herbivores).

Plant diversity might increase somewhat as a consequence of reduced primary productivity, because over most of the range of productivity values, a negative relationship exists between productivity and diversity (the likelihood of competitive exclusion by fast-growing dominants is reduced; Grime, 1979). Only in the most extremely unproductive environments do we expect plant diversity to fall with declining total productivity. In these places, diversity is low simply because the pool of plant species that is adapted to survive at such low rates of resource supply is quite small.

In short, reduced rainfall might cause: reduced plant productivity but increased diversity; a change in herbivore species composition (fewer of the original species, but new herbivores on the extra plant species); and reduced natural enemy abundance. The net result of all these changes is likely to be increased instability at all trophic levels, due mainly to a reduction in mean population sizes.

2) What are the consequences of species extinctions resulting from climate change on ecosystem-level functions such as primary productivity, biomass, herbivore productivity, and so on (i.e., what, if any, are the ecosystem functions of diversity, and how much species loss can be compensated by replacement or alteration in the relative abundance of the surviving species)?

The most commonly cited model for the relationship between productivity and diversity was proposed by Oksanen et al. (1981). Their model is based on the central assumption that there are threshold levels of productivity in each trophic level, below which the ecosystem cannot support a higher trophic level. Thus, the most unproductive systems are taken as having only plants in them, because there is too little primary production to support herbivorous animals. Moderately productive systems have herbivores, but herbivore productivity is too low to support predator populations. Only in the most productive systems are natural enemies found. This argument leads to predictions about the abundance (i.e., standing crop) of the different kinds of organisms. Herbivores will be more abundant in systems of medium rather than high primary productivity, because at high productivity their standing crop is reduced by natural enemies. Unfortunately, there is no evidence at all to support any of the elements of this otherwise attractive logical edifice.

The loss of species from different trophic levels (plant, herbivore, predator, decomposer) is likely to have different consequences at the ecosystem level. In the first place, the degree of niche specificity exhibited by species in different trophic levels exhibits characteristic patterns. This aspect makes the scope for replacement of function by surviving species different in each trophic level. With plants, for example, there is considerable scope for replacement, because many species within most communities share similar niches (Tilman, 1988). Thus, should one plant species be lost, several others can probably replace its function.

Herbivorous animals, in contrast, are typically quite specialized in their feeding requirements, and many of the invertebrate herbivores are strictly monophagous (they feed from just a single host plant species). Because an extinct herbivore species is much less likely to be replaced by a surviving resident herbivore, the plant species upon which the extinct herbivore fed could experience a considerable increase in its realized growth rate. With its competitive ability altered, the plant could eventually become much more abundant. An increase in the abundance of the herbivore-free plant could be accompanied by the extinction of other, less competitive plant species.

Decomposer organisms tend to be rather generalized in the species identity of the substrates they feed upon (tending, instead, to specialize on particular chemical classes of dead organic matter such as proteins, cellulose, sugars, etc.). We know virtually nothing about the degree to which the function of different decomposer organisms can be substituted by others, but the object lesson of the Australian cattle farmers serves as a salutary reminder of the limits to which decomposer function can be stretched. After a number of years of intensive cattle production in Australia, it became clear that the native dung-feeding species was unable to dispose of cattle dung at anywhere near the necessary speed; undecomposed cattle droppings were accumulating at an alarming rate. The farmers were forced to conduct an intensive search for decomposer organisms in Africa and other places where ungulate herbivores occurred naturally at high densities. The limited ability of the decomposer flora and fauna to deal with novel substrates might hinder niche shifts in the feeding habits of other resident decomposers after existing decomposer species become extinct. Any effect on the rate of nutrient-cycling could have profound effects on ecosystem function.

Predators and parasites tend to be intermediate in their feeding specificity; they are somewhat more polyphagous than herbivores, but more specialized than decomposers (they are said to be oligophagous in their feeding). The consequence of the loss of a natural enemy species as a result of climate change will depend upon whether or not it contributes towards the population regulation of its prey species. If it does, then loss of the enemy will lead to increased population density of the herbivore and reduced abundance of the plant(s) on which that herbivore feeds (see the examples under Keystone Predators, above). Conversely, if the predator does not influence the abundance of its prey, predator extinction might have little ecosystem-level effect. This might occur if the predator was habitat limited (e.g., its numbers were determined by the abundance of suitable, safe nesting sites), or if herbivore abundance determined predator numbers without any reciprocal effect (e.g., the predators took only sick and young animals, which would have died in any case).

What is immediately clear from these thumbnail sketches is that it is impossible to say what will happen following species loss unless a good deal is known about the ecology of the species in question, its food species, and the natural enemies that prey upon it. Extinction may have no measurable impact on ecosystem function in one case, while in another it could set in motion a cascade of changes that results in a fundamental alteration of ecosystem structure and function (e.g., leading to a different dominant plant species, wholly different herbivore communities, and a new pattern of natural enemy dynamics). The present lack of information on the dynamics of most ecosystems severely restricts our predictive ability.

3) How do temperature changes influence the dynamics of coupled predator-prey and insect-plant systems (e.g., for temperature-conforming invertebrates, warming will mean alterations in development rates, searching efficiencies, relative susceptibility to natural enemies, and so on)?

A key component in the interaction between arthropod predator and prey populations is the relative magnitude of their developmental temperature thresholds. Few arthropods can develop below about 5°C, but herbivorous species typically have lower developmental thresholds than their natural enemies. Herbivorous arthropods are often sedentary, at least during their feeding stages, and they have little need to move about. Again, their host plants are typically in prime nutritional condition during spring, when the leaves are young and tender and air temperatures can be low. From the natural enemy's perspective, the problems of making a living are somewhat different; they must not emerge from their overwintering dormancy until there is some prospect of finding hosts in sufficient numbers to allow them to grow and reproduce. On the other hand, they must not delay longer than is absolutely necessary, lest other predators emerge earlier than they do and consume all their prey.

In these kinds of systems, therefore, the impact of the predator on prey population density is determined largely by spring weather conditions. In a cool spring, prey numbers build up over a long period before the predators emerge. When the predators do begin feeding, they make relatively little impact on prey abundance, and a pest outbreak can result. If the spring is warm, predators arrive earlier, while prey numbers are still relatively low. Predator feeding then keeps prey numbers in check, and no outbreak occurs.

The temperature thresholds exhibited by natural enemy species are almost certain to have been under intense natural selection for millenia. This being so, sudden changes in mean temperature, and in the frequency and intensity of extreme events, are likely to have profound effects on the behavior of arthropod predator-prey systems. The example just described suggests that warming would lead to lower densities of herbivorous insects and hence to reduced pest problems. On the other hand, lower average prey densities would probably lead to a reduction in mean predator numbers. Lower predator numbers, coupled with an increase in the variance of year-to-year spring temperature regimes, would tend to increase the probability of pest outbreaks, because the lower the predator population, the less likely it is that the predators could nip a potential outbreak in the bud. As before, the limited information available to us does not allow us to distinguish between equally plausible alternative predictions.

It is clear that global warming will affect predator-prey dynamics, but it is not at all clear how it will affect them. What does seem clear, however, is that if the variance in spring weather increases, then so will the variance in the effectiveness of pest control. Damaging pest outbreaks are therefore likely to become more, rather than less, frequent.

4) How are broad-scale, empirical patterns of ecosystem dynamics (e.g., correlations between rainfall and primary productivity and between primary and secondary productivities) to be understood in terms of mechanistic processes, and how can these processes be described in few-parameter models of ecosystem function?

The problem of incorporating mechanistic understanding into simple, high-level models is not new, and it is certainly not unique to questions of global warming. Although it would obviously be good to obtain a fundamental appreciation of all the processes underlying the response of terrestrial ecosystems to global climate change, there is simply no prospect of this happening on the time scale under which we are working. We shall have to be content with a mixture of empirical data, laced with whatever mechanistic information we can pick up along the way.

The above is not a pronouncement of hopelessness. There has been very little incentive in the past to attempt to build simple models of ecosystems, and there is clearly a great deal that can be done. Although it would be unrealistic to expect simple ecosystem models to make accurate predictions about particular cases, it would at least be possible to analyze the general consequences of a whole range of potential climate-induced changes. Such a modeling approach is one of the only approaches open to us, because any field experiments we might initiate will take about the same time to come to fruition as the predictions we are attempting to make. Any such experiments would also have their results irretrievably confounded by any concurrent changes in climate.

The aims of modeling are clear. We need to know how changes in climate affect changes in soil moisture, and how changes in soil moisture lead to changes in vegetation. These changes in vegetation may have two sorts of effects: 1) they may influence the albedo, surface roughness, and actual evapotranspiration that could then influence subsequent climate changes; and 2) they may affect herbivore and natural enemy populations, thus exacerbating future changes in vegetation and, hence, subsequent climate change.

Models of these processes will require new experimentation in order to parameterize their central relationships (e.g., irrigation and rain-exclusion trials in a variety of different ecosystems), and monitoring will be needed to discover the rate at which vegetation changes actually occur. The predictive component of any such work, however, is almost certain to be centered on theoretical models.

5) How can large-scale monitoring of change within ecosystems (e.g., through satellite image analysis) be translated, using simple models, into an understanding of the ecological processes that cause the observed changes?

The problem with monitoring schemes is that they can become a bottomless pit for resources, generating data at a far greater rate than that at which it is possible to assimilate and interpret them. On the other hand, without monitoring there will be no way to assess whether global climate change is having any impact on terrestrial ecosystems, because at present we do not gather data that would enable such trends to be detected, let alone understood.

Selection of which ecosystems to monitor (e.g., pristine or disturbed, semi-natural or agricultural), the location of the study sites (e.g., well away from sources of industrial pollution that might vary over the same time scale as climate changes), how to collect the data (e.g., the relative importance of satellite image analysis versus ground-level validation), how frequently to take measurements (e.g., weekly, monthly, or seasonally), and which parameters to measure (e.g., by species, or by trophic level) will no doubt exercise the next generation of fund-raisers for environmental research.

CONCLUSIONS

We know that greenhouse gases are accumulating exponentially in the atmosphere. We also know that the greenhouse effect works, because, if it did not, temperatures on earth would be too low for life to exist. What we do not yet know is how increases in the concentrations of greenhouse gases lead to changes in global mean temperatures. Models of world climate predict 1.5° to 4.0°C increases in global mean temperature in only a few decades, but precise changes in temperature are extremely difficult to anticipate. Furthermore, we have virtually no idea how a given change in global mean temperature will affect local patterns of rainfall, and hence the potential for primary productivity. There is thus a cascade of uncertainty; we are sure about the atmospheric composition, somewhat uncertain about the relationship between climate change and atmospheric composition, and completely unsure as to how changes in global mean temperature will affect the structure and function of terrestrial ecosystems.

A major difficulty in predicting the direct ecological consequences of global warming is that there is such massive local heterogeneity in temperatures within the same ecosystem at any given time (e.g., summer daytime temperatures may differ by 30°C between the upper and lower surfaces of the same stone; soil-surface temperatures in a fleeting sun-fleck on the forest floor may be 20°C higher than in the shade only a few centimeters away). It follows that predicting the consequences of a small change in overall mean temperature for whole-ecosystem function is going to be very difficult.

Because so many ecological processes involve temperature-conforming organisms (cold-blooded species), a great many vital ecological processes are directly temperature dependent (photosynthesis and decomposition are important examples). But temperature changes are also likely to have myriad indirect effects. Many ecological interactions such as parasitism and predation work in different ways at different temperatures (see the preceding section).

It is clear that global warming will have profound and complex effects on the interactions between insect herbivores and their insect enemies, and hence on the host plants on which the herbivores feed. It is not at all clear, however, whether general warming will lead to higher or lower average pest numbers, although it does seem likely that pest outbreaks will occur more frequently than at present.

A further consequence of alteration in the dynamics of competition and predation will be changes in the relative abundance of species. Formerly dominant species will decline, and some formerly rare species will increase. A great many rare species, however, are likely to be lost altogether, and it is almost certain that the extinction rate of plants and animals will rise during the next 100 years. It will rise as a direct result of habitat destruction, which will be exacerbated in Third World countries by increasing internal pressure from an ever-expanding human population. The extinction rate will also increase as a result of habitat pollution from all kinds of domestic and industrial processes. Extinctions resulting from natural disasters will increase as a result of changes in the timing, frequency, and intensity of climatic extremes such as droughts, hurricanes, floods, and winter cold. Finally, extinction rates will increase worldwide as a result of climate change; species unable to recruit in their present locations will be lost before they are able to disperse to suitable habitats elsewhere. Species diversity is declining now, and will decline at an increasing rate for the foreseeable future.

We do not understand the function of species diversity. Although it is clear that many ecological systems are remarkably resilient and can respond through species replacement to the loss of a considerable number of their dominant species, it is equally clear that there are limits to this resilience. Eventually, during the progression of extinctions, species will be lost whose ecological roles are irreplaceable, and catastrophic loss of ecosystem function will follow (forest will degenerate to scrub, scrub to desert, and so on). In order to ascertain the tolerance limits of ecosystems, we need to answer many questions: What is the minimum number of species in each trophic level that is necessary to maintain ecosystem function? How does this number vary with latitude, and with different soil types, altitudes, exposures, and weather patterns? Unfortunately, we have neither the time, nor the resources, to answer these questions by experiment; the consequences of global climate change will be with us before a single experimental result could be gathered.

In the face of global climate change, the political priorities are clear: 1) control human population growth; 2) reduce per capita energy use in the developed world, both by increasing the efficiency of vital energy use and reducing non-vital use; 3) substitute low-carbon, energy-efficient fuels; 4) reduce the emissions of all greenhouse gases (methane, chlorofluorocarbons, etc.); and 5) conserve the maximum number of species. Species conservation can be accomplished only by protecting large fragments of natural habitat. The notion that we can preserve diversity through the widespread adoption of gene banks is absurd. Gene banks are an expensive red herring that preserve a tiny and entirely arbitrary fraction of the gene pool in a time warp, where the genes are isolated from the continuous action of natural selection.

In terms of ecological research priorities, we need to begin monitoring selected terrestrial ecosystems in sufficient detail to permit detection and interpretation of changes attributable to global climate change. Which ecosystems to monitor, where to locate the studies, and which parameters to measure are major policy decisions yet to be addressed (e.g., how much should we rely on satellite image analysis, and how frequently should we calibrate this against field work). At the same time, we need to build simple, analytical models of ecosystem dynamics in an effort to determine the limits to which the resilience of ecosystems can be stretched before they change, perhaps irreversibly. In this endeavor, we should resist the temptation of building complicated computer simulation models of terrestrial ecosystems and their responses to climate change. We should learn from the experience of such earlier modeling efforts as the International Biological Programme (IBP 1964-74), where it became evident that complex simulation models are as hard to understand as real-world systems, but have the added disadvantage of being unrealistic.

We also shall need to draw on every possible source of information on how ecosystems have changed in the past (the fossil record, pollen records in peats and sediments, dendrochronology, historical archives, archaeological data, etc.) and use fully all those episodes of pollution, biocontrol releases, and invasions of alien organisms that can be treated as unplanned experiments.

Numerous difficulties must be faced in predicting the consequences of climate change on terrestrial ecosystems and in forecasting the effects of ecosystem changes on subsequent climate change (e.g., we do not even know whether the feedback to climate through changes in vegetation would be positive or negative). To compound the problems, we must now add the fact that it may be impossible to make medium- or long-term predictions if ecological systems are chaotic in their dynamics. If chaos is the rule, then the extreme dependence on initial conditions limits the possibility of precise prediction to no more than two or three time periods into the future. Even such a minimal level of predictive ability is better than none at all, and an informed ecological input is required for sensible contingency planning. We have a great deal to learn, and little time to learn it.

REFERENCES

Belsky, A. J., 1986, Does herbivory benefit plants? A review of the evidence, *American Naturalist*, **127**:870-892.

Bergerud, A. T., 1988, Caribou, wolves and man, *in*: "Trends in Ecology and Evolution," **3**:68-72.

Boucher, D. H., James, S., and Kesler, K., 1984, The ecology of mutualism, *Annual Review of Ecology and Systematics*, **13**:315-347.

Brokaw, N. V. L., 1985, Gap-phase regeneration in a tropical forest, *Ecology*, **66**:682-687.

Chesson, P. L., 1986, Environmental variation and the coexistence of species, *in*: "Community Ecology," Harper and Row, New York.

Clements, F. E., 1928, *Plant Succession and Indicators*, H.W. Wilson, New York.

Crawley, M. J., 1983, *Herbivory*, Blackwell Scientific Publications, Oxford.

Crawley, M. J., 1986a, *Plant Ecology*, Blackwell Scientific Publications, Oxford.

Crawley, M. J., 1986b, The population biology of invaders, *Philosophical Transactions of the Royal Society of London, Series B*, **314**:711-731.

Crawley, M. J., 1987, What makes a community invasible?, *in*: "Colonization, Succession and Stability," A. Gray, M. J. Crawley and P. J. Edwards, eds., Blackwell Scientific Publications, Oxford.

Crawley, M. J. and Gillman, M. P., 1989, Population dynamics of cinnabar moth and ragwort in grassland, *J. Animal Ecol.*, **58**:1,035-1,050.

Crawley, M. J. and May, R. M., 1987, Population dynamics and plant community structure: competition between annuals and perennials, *J. Theoretical Biol.*, **125**:475-489.

Davis, M. B., 1981, Quaternary history and the stability of plant communities, *in*: "Forest Succession: Concepts and Application," D. B. Botkin, H. H. Shugart and D. C. West, eds., Springer-Verlag, Berlin.

Dawkins, R., 1987, *The Blind Watchmaker*, Oxford University Press, Oxford.

Elton, C. S., 1958, *The Ecology of Invasions of Animals and Plants*, Methuen, London.

Gleason, H. A., 1926, The individualistic concept of plant association, *Bulletin of the Torrey Botanical Club*, **53**:7-26.

Grime, J. P., 1979, *Plant Strategies and Vegetation Processes*, John Wiley, Chichester.

Hassell, M. P., 1978, *The Dynamics of Arthropod Predator-Prey Systems*, Princeton University Press, Princeton, New Jersey.

Holling, C. S., 1959, The components of predation as revealed by a study of small mammal predation of the European Pine Sawfly, *Canadian Entomologist*, **91**:293-320.

Lack, D., 1954, *The Natural Regulation of Animal Numbers*, Clarendon Press, Oxford.

Lawton, J. H. and Hassell, M. P., 1981, Asymmetrical competition in insects, *Nature*, **289**:793-795.

Lotka, A. J., 1925, *Elements of Physical Biology*, Williams and Wilkins, Baltimore.

Lovelock, J. E., 1979, *Gaia: A New Look at Life on Earth*, Oxford University Press, Oxford.

Marrs, R. H., Roberts, R. D., Skeffington R. A., and Bradshaw, A. D., 1983, Nitrogen and the development of ecosystems, *in*: "Nitrogen as an Ecological Factor," J. A. Lee, S. McNeill, and I. H. Rorison, eds., Blackwell Scientific Publications, Oxford.

May R. M., 1973, *Stability and Complexity in Model Ecosystems*, Princeton University Press, Princeton, New Jersey.

May, R. M., 1981, *Theoretical Ecology* (2nd ed), Blackwell Scientific Publications, Oxford.

May, R. M. and Oster, G. F., 1976, Bifurcations and dynamic complexity in simple ecological models, *American Naturalist*, **110**:573-599.

McNaughton, S. J., Oesterheld, M., Frank, D. A., and Williams, K. J., 1989, Ecosystem-level patterns of primary productivity and herbivory in terrestrial habitats, *Nature*, **341**:142-144.

Oksanen, L., Fretwell, S. D., Aruud, J., and Niemala, P, 1981, Exploitation ecosystems in gradients of primary productivity, *American Naturalist*, **118**:240-261.

Paine, R., 1966, Food web complexity and species diversity, *American Naturalist*, **100**:65-75.

Rosenzweig, M. L., 1968, Net primary productivity of terrestrial communities: prediction from climatological data, *American Naturalist*, **102**:67-74.

Tilman, D., 1988, *Plant Strategies and the Dynamics and Structure of Plant Communities*, Princeton University Press, Princeton, New Jersey.

Volterra, V., 1926, Variation and fluctuations of the number of individuals in animal species living together, *in*: "Animal Ecology," R. N. Chapman, ed., (1931) McGraw-Hill, London.

Wolda, H., 1978, Fluctuations in abundance of tropical insects, *American Naturalist*, **112**:1,017-1,045.

THE IMPACT OF GLOBAL CLIMATE CHANGE ON MARINE ECOSYSTEMS

Jacqueline M. McGlade

Arbeitsgruppe Theoretische Ökologie
Forschungszentrum Jülich
Postfach 1917
D-5170 Jülich, FRG
and
Darwin College
Cambridge CB3 9EU, UK

INTRODUCTION

The world's climate is inextricably bound up with the dynamics of the oceans and their physical, chemical, and biological functioning. Changes in global climate will therefore affect not only the circulation patterns of the oceans, but also, by extension, the role of the oceans as a reservoir for heat and atmospheric gases. Elements such as carbon take approximately 100 years to be exchanged between the surface waters and deeper layers of the ocean: part of this process is intimately tied up with the production of phytoplankton in the surface layers of the ocean and coastal waters. The ocean, therefore, directly affects the rate at which atmospheric gases, such as carbon dioxide (CO_2), adjust to changes in their sources or sinks. To view climate change sensibly, it is therefore essential that the ocean be seen as an integral part of both the geosphere and the biosphere.

Foremost in the debate about climate change is whether or not the data and models currently used to make predictions about increasing atmospheric CO_2 are sufficient to determine the likely response time of the coupled atmosphere-ocean system. It is already known that alterations in certain coastal currents can lead to the catastrophic inundation of deltaic areas, with widespread damage and loss of life. Now there is a growing concern among the population at large that a worldwide ecological disaster awaits us, a concern driven by the apparent increase in severe storm events, excessively severe winters in the Antarctic, and a progression of uncharacteristic seasonal weather patterns across the globe.

In the following chapter, I trace the implications of these changes along several avenues, by first giving an overview of the main biogeographic events and processes in the ocean. In terms of geological time, the ocean basins are very recent and relatively short-lived structures. When the oceans opened up during the last 150 million years, they were invaded by the major extant groups of marine fish, including the modern teleosts. The structure of the continental shelf and littoral zone subsequently gave rise to many new niches, which were progressively colonized by invasions of marine fish. To a large extent, the life cycles of these species were determined by the oceanographic and bathymetric structures that arose during this period.

Next, I examine some of the major patterns of exploitation of marine resources, identifying those species and countries most likely to suffer from climate change, either directly through sea level rise, or indirectly through ecological displacement, loss of resources, or product substitution. Because the influence of the sea is felt all along its terrestrial boundaries, the discussion of biological impacts also includes the coastal region, particularly in areas bordered by mangroves, where the biota are intricately adapted to the tidal and sea level regimes. It is argued that fishing patterns along the world's coasts will be affected by changes in circulation such that within the exclusive economic zones (EEZs) established under the 1982 United Nations Law of the Sea Treaty, other biological and social problems will arise. Indeed, many of the artisanal fishing communities depend upon access to and harvesting of local fish and invertebrates for their survival. The unprecedented explosion of aquaculture in developing countries has brought fish products into the agricultural commodity market, where the incipient economic structures are quite different from those of the fishing industry, and rely on a steady product supply. Changes in sea level will therefore have an immediate effect upon these communities, because they often have no alternative sources of food, habitation, or employment, and might lose their aquacultural investments because of intermittent supply.

Beyond the EEZs, the problems of regulation and defining strategies for sustainable use of the oceans increase in the face of climate change. One option is to make the oceans equally accessible to all, but, paradoxically, the distant-water fleets that use the oceans are likely to be among those most affected by climate change because of their dependence on the pelagic resources of upwelling and polar regions. Fishing fleets will have to be controlled in the open ocean, but as yet there are no institutional structures in place to regulate or govern the activities of countries that might destroy these marine resources through climate-induced changes.

In the final section, I examine problems arising from the fact that the geopolitics of conservation and development of maritime ecosystems are generally at odds with each other. Most current management models for ecosystem exploitation are based on biological models of single species, with equilibrium/deterministic assumptions. Change is incorporated in hindsight, so that regulatory measures are constantly being altered. This inconstancy has led to a growing credibility gap between government and those involved in the commercial exploitation of the marine ecosystem. To overcome some of these problems, a management framework is suggested based on a multispecies, multiactor approach. The structure affords the inclusion of some of the issues arising from conflicts between commercial sectors, socio-economic factors, and instabilities within the ecosystem itself, that are all likely to arise as a result of climate change.

MARINE BIOGEOGRAPHY

The historical record of changes in the marine ecosystem in response to climate changes does not provide the level of detail needed to predict the effects of global warming precisely. One can, however, extrapolate probable future responses from past events. Therefore, this section focuses primarily on known alterations that have occurred in the past as guidance to possible future shifts (see Fig. 1).

Wandering Continents and Biological Invasion

Studies of continental drift show that the Pacific Ocean is approximately 175 million years old, that the Atlantic and Indian Oceans opened up 150 myr ago, and that 50 myr ago the structure of the oceans was similar to the present. On the other hand, the evolution of fish follows from the early jawless Agnathans in the Devonian (200 myr ago) and the teleosts in the Tertiary/early Quaternary (100 myr ago). We can conclude that colonization of the oceans by extant families of fish and invertebrates would have occurred as the oceans opened up; thus, the older the ocean basin, the more species it is likely to have. In addition, latitudinal differences exist between species of fish because of generation times and the time taken to switch from somatic to gonadic growth. Longitudinal differences also occur because of Coriolis forces and the physical structuring of the oceans, so that, for example, there are approximately 900 species in the western Atlantic compared to 300 in the east.

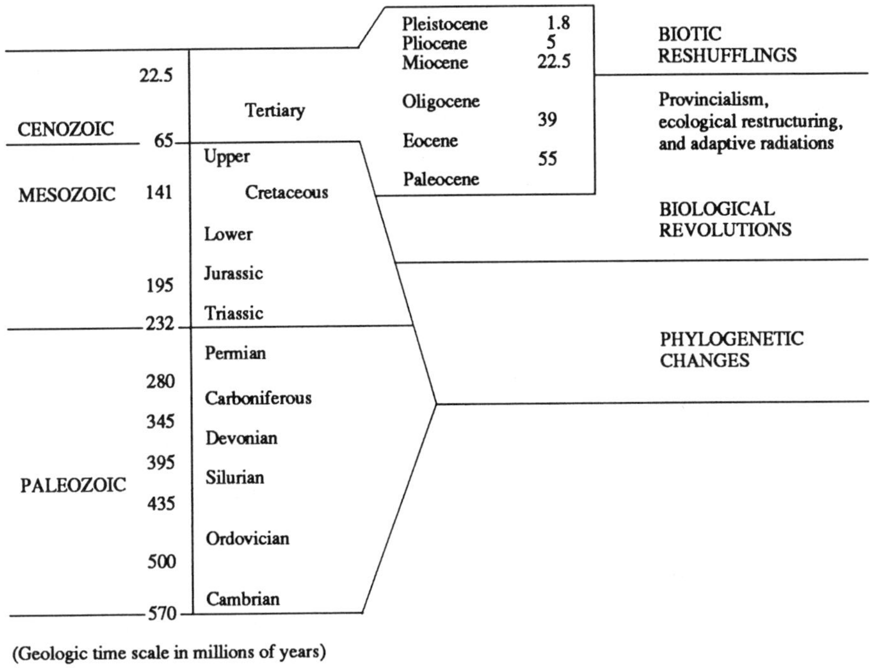

Figure 1. Changes in marine biogeographic processes throughout geologic time (millions of years) showing the emphasis on phylogenetic change during the Paleozoic, and biological revolutions and adaptive radiations in the Mesozoic and Cenozoic.

The idea of adaptive radiation is well known; for example, widespread extinctions opened up many new areas for colonization by new species on islands or whole continents, as when the dinosaurs were replaced by mammals. In the marine environment, there have been at least two major radiation events. Although each lasted 180 myr, they were qualitatively different from each other (Erwin et al., 1987). The first was an early Paleozoic explosion of marine invertebrates, which occurred from the Vendian to the late Ordovician (620 to 438 myr ago). During this radiation, the diversity of skeletonized marine animals rose from 1 family to 470. The second phase, from the Triassic to the Cretaceous (245 to 66 myr ago), brought the level from 200 to 600 families. However, at higher taxonomic levels, there were radical differences: the first radiation was dominated by the appearance of orders, classes, and phyla, while such new appearances were very rare in the second.

The first radiation occurred in two phases: during the Vendian-Cambrian interval (115 myr), 11 new skeletonized phyla appeared, including the Arthropoda, Brachiopoda, Mollusca, Echinodermata, and Chordata; in the second, Ordovician phase, many soft-bodied phyla appeared, such as the Nematoda, Sipuncula, and Nemertea. By the end of the Ordovician, more than half of the 62 living and fossil invertebrate and vertebrate examples of classes had appeared, and the number of orders (152) was twice that of the Triassic-Cretaceous interval (73). The second radiation in the Mesozoic was quite different: of the new post-Ordovician classes, nearly all were vertebrates—four classes of fish appeared during the Silurian and Devonian, followed by Reptilia in the Permian, and Mammalia in the Eocene, the latter two being introductions of terrestrial groups into the sea. The orders that arose during the Mesozoic radiation were primarily in the major groups of bivalves, gastropods, malacostracans, echinoids, fish, and reptiles, but were in total only half those that had appeared during the earlier Paleozoic radiation, and included no new skeletonized classes or phyla.

Erwin et al. (1987) associate these differences with the amount of adaptive novelty possible; in the first interval, between the Vendian to Ordovician, vast numbers of new niches and habitats opened up. These habitats were then subdivided; for example, in the benthos, burrowing and filter feeding from long stalks emerged, and new reefs were colonized. The second phase was triggered by the mass extinctions of the Permian; however, all of the phyla of the Paleozoic survived to colonize the new areas made available. Apparently, new classes of phyla were less able to establish themselves. Thus, the Mesozoic radiation, like the classic radiation of the mammals, was enabled by a preceding extinction event that largely emptied the marine habitats. These habitats were still sufficiently occupied to prevent the emergence of new body forms; instead, more competitive ability or progressive adaptation was required (Benton, 1987).

Given these aspects of marine radiation, it is clear that colonization and speciation are more complex than any prejudged idea of "adaptive" radiation. The appearance of new life forms versus diversification on a common body plan is of more concern to us, however, because it is quite likely that the effects of global climate change will tend to alter and remove habitats rather than create new colonizable ones. Increased extreme storm events in the coastal zones and sea level rise will tend to destroy existing areas without necessarily creating stable areas for re-colonization.

An example of habitat alteration can be seen in the Pleistocene hydrographic-climate history of the Mediterranean (Mars, 1963). The Strait of Gibraltar is the key point through which marine immigrants enter and seawater is exchanged with the Atlantic Ocean. It is postulated that, as the northern European ice sheet expanded, cloudiness and precipitation over the Mediterranean increased; this in turn increased precipitation relative to evaporation and reversed the flow pattern through the Strait. Surplus water flowed out from the Mediterranean into the Atlantic, with a reverse flow from the Atlantic over the sill at depth—a situation comparable to that in the Dardanelles at present. Species normally present in the cooler waters of the Northern Atlantic were able to colonize the Mediterranean, only to become locally extinct during the interglacial periods of warming and increased salinity. Continued warming led to the northern range extension of some shallow-water species along the northwestern African coast and up into the Mediterranean. In fact, all these changes can be seen today in areas that have been uplifted to expose marine sediments; in shallow-waters, assemblages have been found that are composed of species found off west Africa, while other assemblages are of species now found to the north, elevated tectonically in offshore areas, and some further assemblages are made up of elements that exist only in the Mediterranean. Similar faunal shifts, complicated by local hydrography and climatic changes, occurred throughout the world during the Pleistocene. Populations disappeared from some areas where they had once been dominant, and emerged in others.

Changes in biogeography would also have been associated with changes in patterns of predation and competition, yet very few invertebrate and vertebrate species became extinct during the Pleistocene. Indeed, despite the impressive fluctuations in environmental conditions, the Pleistocene extinction rates for invertebrates are not exceptional; no unusual effects on the structure or composition of the marine biosphere appear to have occurred. One suggestion is that the Pleistocene changes were variations on an already developed theme (Valentine, 1973). There was no major change in the environmental structure, but rather a series of oscillations containing the present within them. When the highest latitudes reached freezing, the water could get no colder, so further climatic deterioration would have meant a spread equatorwards of colder water. The climatic zones of lower latitudes would have been compressed, but no truly new climatic regimes would have been created. Sanctuaries would have been available, because even in the height of glaciations the tropics were sufficiently stable and warm to maintain many of the species in rich reef communities, despite local eliminations of reefs and shallow water communities due to sea level fluctuations.

These aspects of biogeography, which governed faunal changes during the Pleistocene, are important in estimating the likely impacts of climate change in the world today. They indicate that major changes in the marine biosphere probably occur only as a result of (i) fundamental alterations in environmental quality, (ii) the appearance of entirely new structures, or (iii) ecological repatternings, such as those following from continental dispersion, interconnection, assembly, and fragmentation.

The majority of the extant marine faunal groups present in the tropics seem to have been derived from Tethyan progenitors, although the distinctive fauna that developed in the higher-latitude Boreal province have contributed to important parts of the modern taxa. For example, the gastropod fauna, which were modernized in a very short time during the mid-Cretaceous, appear to have changed first in temperate latitudes and then spread into the tropics, because the latter assemblages retained their primitive aspects of herbivory and detrital feeding longest (Valentine, 1973). The increase in diverse feeding types at higher trophic levels coincides with a major transgression during the late Cretaceous, the establishment of the Circum-Antarctic Current, and the subsequent cooling of the high southern latitudes and the opening up of the North Atlantic. Although diversity did not significantly decline at the end of the late Mesozoic, extinction of the ammonites and many of the planktonic foraminifera gave the Cretaceous a higher rate of extinctions than average. The changes were dislocated both spatially and temporally, which virtually rules out a single event. This apparent enigma is more likely the outcome of climatic changes that led to isolation of species across their ranges and a temporary incapacity for diversification.

During the Cenozoic, provincialism and ecological structuring increased rapidly. In the tropics, environmental stability led to a high degree of diversity and specialization, while large fluctuations in solar energy and nutrient resources at high latitudes led to species that displayed a high adaptive tolerance to temperature distributions and salinities. It is important to note that during the Cenozoic, glacial changes brought about biotic shufflings rather than major phylogenetic changes. Deposits from the shallow seas of the continental margins show that species which lived during times of lowered sea levels are now generally associated with areas of colder water, corresponding to the fact that during glacial periods much of the ocean water was locked up as ice.

The scales of change that can be observed vary throughout the biosphere. For example, the marine fauna of the rocky shores of the two sides of the Atlantic Ocean, while similar in form, are quite different at the species level. The shores of Virginia and Brittany are separated by the deep Atlantic Ocean, which intertidal organisms cannot effectively traverse. Along the western coast of North America, we also see separation between areas such as Oregon and Alaska, where the shoreline communities are often composed of at least 50% endemic species. Here, as elsewhere in the marine biosphere, temperature acts as an effective barrier against widespread distribution. The discontinuities that arise are, in effect, boundaries for biogeographic areas called provinces, in which communities are similar in species composition from place to place. The establishment of provinces during the Cenozoic is thus characteristic of adaptive forms of radiation rather than of the more extreme form of biological revolutions that occurred in the Mesozoic.

There is significant evidence that the processes governing the assembly of species to form a community and hence, by extension, a province, are both ecological and evolutionary. In the case of fish, a classic example of endemic radiations is the species swarms of cichlids that populate the great lakes of Africa, such as Lake Tanganyika and Nyasa (Fryer and Iles, 1972). In areas of low nutrient supply that are otherwise stable, some of the genera have diversified into cohabiting flocks composed of up to 200 different species. The converse exists in areas of instability, such as the hot saline lakes, where another species, *Tilapia*, shows remarkable phenotypic plasticity but little overlap or cohabitation. Thus, the evolutionary strategies for this group are markedly different for stable and unstable conditions. In other very old lakes, such as Lake Baikal, similar phenomena are seen, in this case among the gammarid amphipods and gastropods. The extent to which these organisms have been unable to alter the largely oligotrophic conditions into a more unstable eutrophic state, together with a lack of appropriate immigrants, has led to the evolution of these flocks of closely allied species. The actual origin of this behavior may be tied to reproductive patterns in areas of low nutrient status, where populations of any one species become very thinly spread; however, such ideas as these can be easily misinterpreted via the effects of field sampling.

Similar phenomena can be observed in another oligotrophic environment, the deep ocean. These areas are characteristically regarded as regions of high stability but with low species representation, except in areas of deep ooze. Overall, deep ocean populations appear to be patchily distributed. Significant patterns of change occur along vertical transects with biogeographic divisions reflecting shallower areas, such as those on either side of the Laurentian Channel off eastern Canada (Fowler and McGlade, 1990). There is no apparent reason why the deep sea was not invaded during the late Precambrian, but it is not known whether the degree of diversity, and hence provinciality, is highest in areas such as the Pacific Ocean, where the Tethyan-Pacific separation is oldest, or in the Atlantic Ocean, where there have been many recent events in shallower areas.

Finally, for the pelagic realm, a model of lower diversity in areas of fluctuating or high productivity and vice versa has long held (Ryther, 1969). In the superoceans surrounding the Tethys continental mass, conditions would have been highly stable, but nutrients would have been scarce (removed as these oceans were from the supercontinents), in comparison to the highly disturbed coastal areas. As the continents divided up, the oceans became smaller; nevertheless, biotic change would have taken very different directions in the building of oceanic versus shelf communities. However, as the continents underwent changes, such as glaciation, the trends for change would have been in the same direction for both oceanic and continental shelf communities, because both areas would have suffered changes in nutrient loading and instabilities in sea conditions. From the little evidence that exists in the fossil record, changes in community structure reflect not only the age of the sea, but also the history of changes in nutrient loading and sea stability.

In conclusion, the impacts of change in the marine biosphere will differ according to location and biological productivity. These points must therefore be in the forefront of our minds when we try to predict the outcome of global climate change. Interpreting change in ecosystem structure resulting from large-scale fluctuations is not always easy. We can look, for example, at the effects of local warming of the sea on the apparently cyclical behavior of a kelp-urchin-lobster complex off Nova Scotia (Mohn and Miller, 1987; Mann, 1988). In this example, the system appears to exist in two states: one predominantly urchin/coralline, the other *Laminaria* kelp. Storms are known to destroy the kelp canopy and initiate a succession to an end point of *Laminaria*, but mass mortality of sea urchins caused by epizootics leads to a dramatic switch back to the kelp-dominated state. The rise of epizootics is probably triggered by physical factors that allow above-average water temperatures and/or the arrival of exotic water masses. The reverse change occurs when a population outbreak of sea urchins is brought about by particular wind, current, and temperature characteristics; the urchins can then graze the kelp to destruction. The interplay of physical and biological dynamics is such that, following physical disturbance, biological processes bring the system back to one of the two known states.

Population Regulation and Physical Oceanography

If we are to interpret change, we must be able to unravel the long- and short-term dynamics of marine communities. Having looked at the paleontological record, I shall turn my attention to the idea of population regulation. In the following section, I examine the role of life-cycle closure, vagrancy, and secondary invasions in lower marine vertebrates with respect to tidal mixing, retention areas, and bathymetry. Then I discuss the role of variance and physical oceanographic features in determining absolute levels of abundance, population richness, and ecological stability and diversity.

Five key factors are involved in regulating species structure and population: pattern, richness, abundance, temporal variability, and behavior/physiology. The first two act at the level of the species, the next two at the level of the population, and the last one at the level of the individual. Population regulation in a sense relies on individuals within each generation "closing" their life cycle: Sinclair and Iles (1988) and others have worked with this set of ideas in connection with fish populations. Briefly stated, the patterns and richness of populations of open ocean species are determined by the number and size of physical structures in the ocean, and population dynamics are determined by the ability of individuals to avoid vagrancy during their early life-history stages and thereafter to maintain population membership in various physical features until their life-cycle has been closed (Sinclair, 1988).

The first hypothesis is that a scaling relationship exists between the number of physical structures in the oceans and the number of populations of various fish species across all oceans (Sinclair, 1988). For example, the North Atlantic Gyre is the only physical feature of its kind within the distributional range of the European eel (*Anguilla anguilla*); hence, because this species occupies this feature for spawning, there is only one eel population. Atlantic salmon (*Salmo salar*), however, home and spawn in specific rivers, and thus there are large numbers of populations. Between the population-rich salmon and the population-poor eel, there is a wide variety of species that fits this general pattern.

A physical feature is defined as one which allows the life-cycle of an individual to be completed or closed. This definition places particular emphasis on the ability of individuals to complete their early life history stages successfully and "end up in the right place at the right time." The positioning of adult spawning is thus hypothesized to have to do with situating eggs and larvae in a regime that will retain them (Sinclair, 1988) rather than determining passive drift patterns into some later juvenile-stage placement, when individuals can take a more active role in migration. This second hypothesis is contained within Hjort's (1914) theory of a "critical period," in the "match-mismatch" hypothesis of Cushing (1973; 1982), and in the stability hypothesis of Lasker (1981); all three suggest that the collapse and rise in recruitment are directly related to the spatio-temporal matching of juvenile fish to their food resources above some threshold. Lasker, in particular, hypothesized that larval anchovy (*Engraulis mordax*) off California required a threshold concentration of prey particles in order to feed and survive. This concentration is substantially higher than normal background levels, so Lasker suggested that it could occur in small-scale layers, formed in inshore waters during periods of water column stability. Because these layers can be dispersed by storms and upwelling, anchovy recruitment and population strength depend upon the stability of the environment and the presence of the correct types of prey. This hypothesis has also been tested off the east coast of the United States on Georges Bank, for cod (*Gadus morhua*) and haddock (*Melanogrammus aeglefinus*), with some success.

For many marine species, population sizes vary over several orders of magnitude. It has been found that for herring (*Clupea harengus*), the occurrence of physical fronts corresponds very closely to the distribution of larvae in the continental shelf waters of the Atlantic. The residual circulation associated with a given system of fronts creates a retention zone of well-mixed water. A comparison between the horizontal area of the retention zones and population size led Iles and Sinclair (1982) to conclude that the population was directly determined by the zone's surface area. The retention of individuals within these areas results from a dynamic between vagrancy and individual behaviors that utilizes ocean currents and the microstructure of the water column, allowing individuals to stay in the zone longer than the residence time of the water itself. The relative separateness of stocks can then be attributable to the integrity of the retention areas during and shortly after spawning. This effect has also been noted by the author for pollock (*Pollachius virens*) off the east coast of Canada, where recruitment is correlated with the incidence of Gulf Stream eddies that pass over the spawning areas; the more numerous the eddies, the more physical disturbance to the potential retention area, and hence the lower the subsequent recruitment. The case of the pollock offers a prime example of the way in which the physical heterogeneity of the ocean directly impacts biological processes, and challenges the dependence of recruitment in fisheries solely on stock size.

It is possible to examine spatio-temporal variability in fish populations at a number of levels. All have a role to play, but in constructing models of marine ecosystems, the parametrization of different processes is sometimes very difficult. To give an idea of the types of factors involved, we can take five broad classes of processes derived from intraspecific and interspecific interactions, fishing, the physical environment, and anthropogenic effects. Intraspecific effects involve density-dependent changes in mortality, growth and reproductive rates, stock-recruitment relationships, and cannibalism. Interspecifc effects are the so-called stock replacement hypotheses, which suggest that an increase or decrease in one species causes an increase or decrease in another. Fishing involves the reduction of population size, which in turn causes the annual production of young fish to decrease. Concepts about the physical environment and its effects include hypotheses about retention areas, as discussed above, and relate climate- or weather-induced perturbations to changes in the spatio-temporal scales of specific physical structures. Anthropogenic effects relate to the impact of various pollutants on fish populations.

An example of the integration of the physical and biological processes in the ocean can be seen in the population dynamics of the flying fish (*Hirundichthys affinis*) in the Caribbean and western Atlantic (Mahon, 1985). A time series of monthly catches of this species since 1958 shows a regular seasonal pattern and significant interannual variability. However, there is no significant relationship between abundance of the parent stock and subsequent recruitment. In the Lesser Antilles region, strong eddy activity appears to be superimposed on the net westward current flows. A strong seasonal salinity signal, associated with the volume of Amazon discharge and plankton abundance, is related to turbulence and nutrient upwelling that coincide with the season of low salinity. Circulation is strongly influenced by the Southern and Northern Atlantic Oscillations. The timing of migration, spawning, and early life history (ELH) of the flying fish coincides with high plankton abundance and eddy activity—a large-scale eastward spawning migration with westward drift of the ELH stages is the most feasible explanation for this. Finally, recruitment is significantly correlated with Amazon runoff, temperature, and the Northern and Southern Oscillation Indices. This implies that any changes in Amazonian run-off, already of much concern in the climate debate because of deforestation effects, will influence the productivity of this species by altering not only local nutrient levels, but also the circulation patterns that arise from changes in the oceanic oscillation (Meade et al., 1979; Mazeika et al., 1980).

The physical and biological fields interact in the ocean over a wide variety of spatial and temporal scales (Horne and Platt, 1984). These interactions occur across a spectral window that ranges from the high frequencies of molecular diffusion to the low frequencies of the current systems, and between processes operating over similar scales. Thus, phytoplankton growth would be associated with turbulent diffusion, and abundance of pelagic fish populations would be associated with variations in physical fronts, upwelling, and eddies occurring over a temporal scale of seasons and a spatial scale of hundreds of kilometers. In particular, Simpson and Hunter (1974) showed that tidal mixing was important in coastal areas, and that the boundary between well-mixed and stratified waters was defined by a critical value of the buoyancy flux due to solar heating (B), the mean water depth (H), and the amplitude of the tidal current (BH/U^3). Subsequent work by Garrett et al. (1978) and Pingree and Griffiths (1978) showed that this parameter picks out all the most biologically active areas in the Gulf of Maine and the North Sea. The same process defines oceanic fronts, which occur over time scales of hours to seasons at continental shelf breaks (Horne, 1978), and small eddies, which occur over a time scale of minutes and seconds. In the range where the Richardson number is less than 1 (i.e., in internal waves), the scale of ocean variability is reduced to the order of millimeters (Garrett, 1979). This type of variability directly affects much of the biological data taken during oceanographic and resource surveys, but it generally remains unexplained. Thus, patterns of variability in the biosphere are likely to be highly influenced by changes in climate and hence ocean circulation. Estimates of biomass changes should therefore be viewed with much caution, unless some attempt has been made to analyze the effects of physical forcing.

Global Climate Models, Coupled Ocean-Atmosphere Models, and the Prediction of Climate-Induced Sea Level Change

During the past two decades, there has been a rapid increase in the complexity of the models used to examine climate circulation and the climatic effects of the ocean and atmosphere. Despite this increase in complexity, global-scale models are generally concerned with long-term average conditions, which are spatially averaged over latitudes. The loss of detail may in some circumstances be very important, but only now can we begin to undertake computations that allow us to assess the effects of these simplifying assumptions. It is not the intent in this sub-section to review all the climate models that are currently in use, but rather to point to their generic features and highlight the areas of greatest uncertainty in predicting changes. Indeed, prediction is particularly dangerous in the context of climate change, because the scenarios for various input parameters, such as CO_2 emissions, are sufficiently different from each other to generate significantly different results in such phenomena as sea level rise.

The most important variables of any climate model are radiative processes; the movement of energy around the globe, including convection; surface processes, such as changes in albedo; emissivity; energy exchanges between the land, ocean, and ice; and spatio-temporal resolution. The four basic types of models are Energy Balance Models, one-dimensional Radiative-Convective models, two-dimensional Statistical Dynamical models, and General Circulation Climate Models. The last type are also called Global Climate Models, the distinction being between oceanic (OGCMs) and atmospheric (AGCMs) general circulation models.

As stated throughout this book, there is a growing awareness that the CO_2 concentration in the atmosphere is increasing; earlier work by Keeling et al. (1982), showed a 7% increase in 21 years at Mauna Loa, Hawaii. A variety of direct and indirect measurements of CO_2 have been used to reconstruct the increases since 1860 (Rotty, 1983), and from these a number of global scenarios of future concentration increases have been derived. Nordhaus and Yohe (1983) have put together these scenarios to give the probabilistic view of an 80% chance that the pre-industrial CO_2 concentration will have been doubled by 2100. Recent model results support the idea that a significant coherence exists between average global temperature and CO_2 concentration from 1958 to 1988 (Kuo et al., 1990). Using a hierarchy of models, this increase in atmospheric CO_2 has been simulated to give a warming of 1.3° to 4.2°C in the global-mean surface air temperature (Manabe and Stouffer, 1980; Schlesinger, 1984; Hansen et al., 1984). Because a warming of this magnitude is between 25% and 100% of that which is estimated to have occurred during the 10,000-year transition from the last ice age to the present interglacial, the impacts are likely to affect all segments of human society.

With respect to equilibrating any changes in energy exchange, the ocean and the atmosphere operate on different time scales. Manabe and Bryan (1969) and Bryan (1984) have taken this into account by decoupling fast and slow time-scale processes in the numerical integration. Work in the area of linking atmospheric results to the ocean includes a "switch on" experimental procedure in which equilibrium climates are calculated for a coupled model corresponding to a normal concentration of atmospheric CO_2 and for the case of four times the normal concentration. The "switch on" time is the period that elapses after the climate corresponding to a normal level of CO_2 is instantaneously perturbed by quadrupling the CO_2 in the model atmosphere. The results show that: (1) at the surface of the earth, the warming is larger over land than over the ocean, but there could be considerable warming at depth in the vicinity of the Ross Ice Shelf in the Antarctic (Schlesinger et al., 1985); (2) because the ocean has a memory of many decades, it may take 75 years to reach 63% of its equilibrium warming (Schlesinger et al., 1985); (3) the response of the climate system continues to fall further and further behind its equilibrium warming for the exponentially increasing CO_2, with direct implications for the detection of CO_2-induced warming (Wigley and Schlesinger, 1985); and (4) cloud feedback may greatly increase climate sensitivity to increasing CO_2. If these and other findings are correct, the time scale of climatic response will be heavily influenced by the spatial inhomogeneities in the atmosphere and ocean, which are therefore likely to impede our ability to detect changes.

Since the publication of work by Mercer (1978) on the instability of the western Antarctic Ice Sheet, great concern has been expressed about the possible effects of global warming, and how fast sea levels will rise. Calculations to answer these concerns can be made in several ways. After dicussing the various calculations, Oerlemans (1989) concluded that for 2100, a value of 66 cm above the present-day stand is most likely. In addition to sea level rise per se, the effects of climate change will be (1) increased mean water temperatures, (2) increased vertical stability of the seas at the surface due to warming, (3) decreased latitudinal and seasonal sea-ice extent, (4) weakening of the poleward and seasonal shifts in storm tracks, surface turbulence, wind-driven surface currents, and coastal upwelling, and (5) a temperate decrease and high-latitude increase in net precipitation and runoff.

Predicting changes at the regional level is much harder. Some attempts have been made by Titus et al. (1985), Broadus et al. (1987), and MacCracken et al. (1987), and all conclude that local effects will be determined by regional differences in climate, landscape patterns, and regional subsidence.

The Effects of Sea Level Changes on Marine Fauna

Any increase in sea level will bring about many changes in localized areas, but if these changes proceed at a pace that is within the range of migration, growth, or evolutionary adaptive potential, then it is likely that the marine biosphere will only undergo some reshufflings. However, at the extremes of the globe, many of the processes that are tied up in primary productivity will clearly be at risk. For example, some of the world's largest fisheries occur at the boundaries of the arctic and sub-arctic oceans, in the marginal ice zone. Primary productivity is critical in these areas and depends upon the annual melting of the ice edge. The ice edge itself refracts and concentrates the sun's oblique rays, thereby creating massive blooms of algae on the underside of the edge. Because there are strong salinity gradients and it is extremely cold, the populations of aquatic invertebrates are too small to harvest this resource to any extent. Instead, much algae falls to the benthos, where squids and molluscs exploit the nutrients. They, in turn, are exploited by the large populations of mammals and birds that are characteristic of these polar areas. The spatial differences in projected global warming are such that the polar caps will become relatively warmer than other areas, especially during winter, so that icebound areas may eventually become ice free as albedo is reduced. If this occurs, the effects on the polar ecosystems would be catastrophic, simply because of their fragility and the lack of resistance to invasion by outside species.

In temperate areas, many fish, such as shad (*Alosa spp.*), use estuaries in which to spawn; in some other cases, such as the Pacific salmon (*Oncorhynchus spp.*), the trigger for spawning is a change in temperature. If sea levels rise, or temperatures alter, many areas will become unavailable for spawning, with the concomitant loss of populations of endemic or commercially important species of fish. Furthermore, the cues that precipitate migrations in many species are current and temperature changes, because they are proxies for seasonal change. Longer-term fluctuations in temperature are known to have caused poleward displacement and attraction of fish (Bardach and Santerre, 1981). However, as Sibley and Strickland (1985) point out, it is not only temperature that is important: global warming would lead to a shift in the ranges of many commercial species northward, and this would place populations in areas where wind speed, episodes of high light intensity, and the episodic nature of the spring phytoplankton blooms would be dramatically altered from current patterns. Thus, many species of fish, such as the oil sardine (*Sardinella longiceps*) in the waters off India, would be unable to match physiologically the conditions under which eggs normally develop, with obvious repercussions for recruitment.

A change in rainfall patterns will also influence river-fed fisheries. Examples of likely effects can be seen after the building of dams and barrages, such as the Aswan Dam in the Nile and the Kotri barrage in the Indus, where fisheries have declined by nearly 75% (Quraishee, 1975). River discharge, which can be controlled upstream, has a direct effect on the productivity of the adjacent coastal waters. An extension of this case occurs with the aquaculture of many salmonid species, where the brood stock is selected from one location. Such an approach creates a long-term drain on natural biodiversity, which would become a critical issue if sea level rise happened to eradicate the one river from which the brood stock had been taken, and up which the adults migrated to spawn.

The near-shore environment in tropical and sub-tropical regions of the world includes another ecosystem that is already being put at risk—the mangrove swamps. This highly important ecosystem is common to tidal and brackish waters, and generally contains a vast array of interdependencies between the coastal flora and the marine intertidal and coastal fauna. For example, in a typical mangrove swamp in Indonesia, plant litter from the mangroves and other terrestrial sources enters the system as leaves and sticks. As detritus, and with herbivorous excreta, it becomes a substrate for abundant bacterial flora, whereupon it passes through the bodies of detritus feeders (shellfish and invertebrates), then to primary or small carnivores, and finally to secondary carnivores (large fish and birds). Christensen (1978) estimated that for a stand of *Rhizophora* in southern Thailand, the annual production of organic matter above ground was equivalent to 7 tons of leaves and 20 tons of wood per hectare per year. This amounts to 10 tons of plant production not used by humans every year, which is almost five times the phytoplankton productivity of the same area of coastal zone without mangroves. Mangroves are also specifically associated worldwide with the reproduction and maintenance of fauna, such as birds (>200 species), crustacea (200 species), and fish (30 species) (see Garcia and Le Reste, 1981, for a review).

Unfortunately, before the biological and socio-economic importance of this component of the coastal ecosystem was realized, much of the mangrove forest in Southeast Asia had been damaged. By 1985, Java had lost 70% of its mangroves; Sulawesi, 49%; Sumatra, 36%; the Philippines, 45%; Thailand, 27%; and Peninsular Malaysia, 20%. Aquaculture, while not the only culprit, is one of the biggest (Csavas, 1988). In Thailand, the mangrove forests along the east coast have been over-utilized and heavily encroached upon. It is now likely that the loss of key parts of the indigenous flora and fauna in these areas will have changed the ecosystems sufficiently to negate the reconstitution of mangrove forest in any biological role. Indeed, much of the shoreline of southeast Asia has been radically altered to make way for semi-intensive and intensive shrimp farming, because the market value of shrimps has made large-scale pond construction a worthwhile investment. But if market dynamics change, thereby reducing local and foreign demand for shrimp and other aquacultural products, it is not clear that the local communities in these areas could survive the effects of climate-induced changes in the coastal seas, given that the highly productive mangrove systems have been replaced or potentially lost. In the context of sea level rise, the problems become even more critical, because construction of coastal lagoons and ponds has denuded the coastline of a natural physical barrier that could keep much of the deleterious effects of salt-water incursions and extreme storm events at bay.

Coral reefs are another part of the tropical system likely to suffer changes from a rise in sea level and an increase in sediment and nutrient loading in the coastal zone. The growth of reefs is intimately tied up with sea level and water quality, because the coral polyps depend upon a symbiotic relationship with photosynthetic zooxanthellae to provide them with energy. Nutrient enrichment through sewage and related discharges of nitrogen and phosphorus has already been shown to have a critical negative effect on coral survival (Walker and Osmond, 1982; Hawker and Connell, 1989). Add to this effect an increase in turbidity as a result of algal blooms and sediment loading, and it becomes clear that a number of coastal lagoons and coral reef islands would be badly affected by climate change. The picture looks even more bleak when estimates of sea level rise are compared to coral reef growth: Buddemeier and Smith (1988) estimate that the predicted rate is five times the present modal rate of vertical accretion on coral reef flats, and 50% greater than the maximum vertical accretion rates attained by coral reefs.

PATTERNS OF EXPLOITATION OF MARINE RESOURCES

Historical fluctuations in the marine ecosystem illustrate the mechanisms by which changes occur in response to climatic alterations. These past events provide the only guidance available as to possible responses to present or future global warming. Given current patterns of consumption, such changes can have devastating effects on those segments of the population that depend upon fish for their food or livelihood. This section outlines the effects which climate-induced changes could have on the sectors affected most directly: the fishing industry and the economies of coastal regions.

Throughout history, fishermen from around the world have been taking fish from coastal waters, but as we reach the end of the twentieth century, nearly 90% of the fish landed comes from the sea, and more than half of the world's fishery production comes from coastal and shallow waters stretching out to 30 mi. As a result, inland seas such as the Gulf of Thailand and the Gulf of Mexico have become vastly overexploited. To meet the demand in fish production, freshwater and marine aquaculture have grown and expanded in importance. But the growing operations for sessile molluscs, seaweeds, and shrimps usually occur in the coastal zone, which, as we have seen above, is the area most susceptible to changes in environmental quality resulting from climate change.

Of the fish that are taken by direct capture, the primary use is for human consumption, followed by production of fish oils and agriculture. In the developing world, one-third of animal protein for human consumption comes from fish. Given the increasing human population and consumption rates that have been observed over the past decade in less-developed areas, it is not clear whether fisheries production can keep up. In fact, current wisdom suggests that, by the turn of the century, the shortfall will be of the order of 29 million tonnes and will occur primarily in Asia.

Changes in oceanographic conditions brought about by such events as El Niño will definitely affect pelagic fish populations and will therefore exacerbate the effects of our dependence on fish. The geographical distribution of the world's major pelagic fisheries is characterized by the large volumes of fish off Latin America and Japan, where fish meal production dominates due to weak local demand, and off Africa and Asia, where the local demand is high and better technology is needed to extend the distribution and shelf life of those fish caught. All of the pelagic fisheries associated with eastern and western boundary current systems would be put at risk by climate changes. The situation in the demersal fisheries is somewhat different, although such features as artificial reefs would of course be affected adversely by sea level rise.

A case study of the Moroccan fishery, which is likely to suffer from climate change in the form of oceanic circulation alterations, will help to underline the complexity of the situation. Morocco has extremely rich fishing in her territorial waters, brought about by coastal upwelling along the coast between 12°N and 33°N. To the north of 25°N, upwelling occurs in the summer; to the south of 20°N, it occurs in winter and spring; between 20° and 25°N, it occurs all year round. The waters in the area are principally made up of the North and South Atlantic central waters. The mixing front is in the region of Cape Blanc at 21°N. The relative amounts of these two water masses alter throughout the season and have a tremendous impact on the fisheries and the actual species composition of the catches in the area. In 1986, the species that made up the major part of the fishery included the European pilchard (*Sardina pilchardus*), with 277,000 metric tonnes harvested; the chub mackerel (*Scomber japonicus*), 101,000 metric tonnes; octopods (*Octopodiae*), 22,000 metric tonnes; jack and horse mackerels (*Trachurus spp.*), 17,000 metric tonnes; and the European anchovy (*Engraulis encrasicolus*), 7,000 metric tonnes. The remainder of the 600,000 metric tonnes landed annually is made up of at least 50 other marine species; however, pelagic species that originate in the upwelling zone account for some 96% of the country's fish harvest. The exploitation of these resources depends heavily upon the market conditions for fishmeal and the price of fish oil extraction. In the past, stock collapses, such as that witnessed in the Peruvian anchovy, drove many animal-feed compounders to seek other sources of protein, and in more recent years, genetically engineered bacteria have been developed to take on this role. Even so, the present situation for Morocco is still one of heavy reliance on a resource whose abundance is determined by oceanic conditions. Climate change could thus have a tremendous impact upon the economic structure of Morocco as a whole.

Examples from "wild" fisheries, such as this, are all too easy to draw together. But some thought must also be given to the low-lying areas and mangrove forests, such as those in southeast Asia, which are rapidly filling up with coastal aquaculture ventures for agro-fisheries production. Humans use the mangrove system for timber (for fuel and construction materials), foodstuffs, folk medicines, and agricultural and other products (Saenger et al., 1983). Kapetsky (1985) estimates that there are approximately 460,000 artisanal fishermen living in and around the lagoons and mangroves of the world, and that these areas produce 925,000 t of fish and crustaceans. Mangroves also protect land against tropical storms, floods, and soil erosion (Christensen, 1978). Kapetsky (1985) has shown how valuable mangroves are in terms of the landings of fish and shrimp species associated with them. For example, in Khlung, Thailand, the value of fishery products captured inside the mangrove, including *Liza*, *Eleutheronema*, *Arios*, *Lates*, and *Ophichthus* was \$3,000/km^2 (1977), and for mangrove-associated *Penaeus* species caught elsewhere the value was \$10,000/km^2 (1977). In Sabah, Malaysia, the value of *Scylla serrata* was \$133/km^2 (1977), based on areal extent of mangroves; in the Peninsula, the value of *Penaeus*, *Stolephorus*, *Pampus*, *Polynemus*, and *Lutjanus* was \$277,235/km^2 (1979), based on the areal extent of mangroves, lagoons, and estuaries. In the Gulf Province in Papua, New Guinea, the fishery products value was \$476/km^2 (1977), based on shrimp caught outside mangroves, plus subsistence fishing and crabbing inside for *Penaeus*, *Metapenaeus*, *Scylla serrata*, *ambassids*, gobids, and catfish.

The communities in these areas are also committed to high levels of investment for shrimp aquaculture, so that even a relatively localized sea level rise or increase in storm events is likely to create immense hardships if ponds or living areas become inundated. Coastal-zone development, while attractive from the viewpoint of economic expansion, may soon become self-destructive as it hastens the erosion of beaches and coasts.

Patterns of offshore and coastal fisheries exploitation would vary significantly in response to climate change. The countries most likely to suffer, however, are those with distant-water fleets that exploit the large pelagic stocks, and those countries with economic infrastructures that either depend upon the revenues from offshore resources found in upwelling areas, or are based upon aquaculture of brackish- and fresh-water species in low-lying coastal areas.

THE GEOPOLITICS OF CONSERVATION AND DEVELOPMENT OF MARITIME ECOSYSTEMS

Because alterations in the marine ecosystem will affect the world's food supply, they will also alter the patterns of employment and habitation. Given the types of problems described above, governments must be prepared to deal with the hardships and displacements that would result from climate-induced changes in the marine ecosystem, particularly in the coastal zone. For it is the human and economic dimensions of climate-induced change that will motivate governments, international organizations, and individuals to address the changes that will accompany global warming.

To create an atmosphere in which to discuss conservation, economic and social concerns must be addressed, and the interactions between different actors in the system laid out. Issue analysis as a means for conflict resolution provides one such approach. Using a common currency, the effects of actions on various spatial, temporal, and social levels can be discussed interchangeably, thus giving a framework that is more embedded in the real exigencies of the system. By replacing a single decision-making function with the concept of passive and active players who condition the rate at which the removal of natural resources is occurring, incentive structures can be established to effect change, if change is in fact needed. This section outlines issues that affect marine ecosystems and lists those organizations and international agreements that relate to them. A brief description follows of some of the international mechanisms under which policy may be made concerning marine conservation, as well as an outline of the characteristics of a useful management structure to deal with this problem.

Coastal Implications

Of the world's 25 largest cities, 16 are situated on the coast. They include Tokyo (3rd largest), Calcutta (4), Greater Bombay (5), New York/New Jersey (6), Shanghai (8), Rio de Janeiro (9), Buenos Aires (11), Cairo (12), Jakarta (13), Karachi (14), Istanbul (16), Los Angeles (18), Dacca (19), Manila (20), Bangkok (23), and Tianjiin (24) (United Nations, 1985; World Port Index, 1986). Outside these conurbations lie many kilometers of coastline where the ecosystems are at risk, either directly from sea level rise, or indirectly through development and coastal migration of human populations. In the at-risk category are salt marshes, mangroves, salt ponds, ice-covered shores, reefs, areas that have been reclaimed from the sea for agriculture, and intertidal zones, where many coastal fisheries exist. Any plans to cope with changes in these areas will require not only foresight, but also political pressure, because in the event of sea level rise, government attention will be directed towards safeguarding large economic zones such as ports and cities.

The coastal zone is one of the more complex environments in which to set policy and management strategies, owing to conflicts between the many users. In a typical coastal area, a planner would have to consider such elements as port facilities, urban functions, waste treatment and disposal, coastal fisheries, production of oil, gas, and salt, extraction of minerals, sand, and gravels, natural conservation areas, and recreational activities. In the case of sea salt production, for example, much of the land area given over to solar evaporation would be lost with only a small increase in sea level (Bardach, 1989). Similarly, a small rise in sea level would remove much of the beach area for recreation in places such as Hawaii, where over 80% of the annual tourist population of 4.3 million uses the beach for swimming and recreation (Bardach, 1989).

It is also clear that a climate-induced sea level rise will be accompanied by an increase in extreme events such as storm surges. For example, in the last millennium there has been an increasing number of severe sea floods along the English Channel and North Sea coasts, with up to 15 in the 13th century, and an increasing number in the closing decades of the 20th century (Lamb, 1972). Some of the worst disasters, however, have occurred in deltaic and low-lying areas such as Bangladesh, where tropical cyclones caused a death toll of 250,000 to 500,000 in 1970, and where flooding in the 1980s has led to terrible hardships in the Gangetic delta. But these are the noticeable products of climate change; perhaps more serious is the increased development in areas known to be potentially in danger.

Institutional Structures

One of the key differences between climate-related problems and other geopolitical issues is that, hitherto, countries have felt no obligation to compensate others for creating conditions of hardship. A prime example is the problem of parity in sharing out the problems caused by greenhouse gases and coastal degradation today to benefit future generations. These issues revolve around four basic population groupings: developed countries expecting improved climatic conditions (e.g., Canada and parts of the Soviet Union), developed countries expecting a deterioration (e.g., the United States and parts of Europe), developing countries expecting better conditions, and developing countries expecting worse conditions (Weiss, 1989). In each case, there will be some costs involved in adapting to change, but those countries in the last category, already burdened with an increasing gap between their economies and those of the rest of the world, will find themselves even more disadvantaged.

At present, the atmosphere is being altered at an unprecedented rate. Strategies to moderate and prevent further alterations include controlling both fossil fuel consumption and the use of CFCs and other ozone-layer destroying agents, and preventing deforestation, misuse of soils, and fertilizer abuse. Long-term strategies involve collecting materials to alleviate loss of biodiversity, acquiring an understanding of traditional harvesting methods that allow cultures to exist in harsh climates, situating waste and hazardous material deposits in areas that are unlikely to cause human and ecological damage, encouraging research into critical areas of climate dynamics, such as feedback processes and ocean-atmosphere interactions, and developing incentive structures that will be necessary to encourage or discourage population migrations.

Most of the relevant international agreements that exist today pertain to pollution control; they include the Vienna Convention for the Protection of the Ozone Layer (1985), the Montreal Protocol on Substances that Deplete the Ozone Layer (1987), the Economic Commission of Europe Convention on Long-Range Transboundary Air Pollution (1979), and the Protocols to reduce emissions of sulfur (1985) and nitrogen oxides (1988) in Europe. But with respect to climate change, few of these are capable of setting any priorities. The marine agreements, however, do offer some avenues for action. For example, the United Nations' Law of the Sea Convention (1982), the London Ocean Dumping Convention (1972), the Convention on the Conservation of Antarctic Marine Living Resources (1980), the Convention for the Conservation of Antarctic Seals (1972), and the Convention for the Protection of the Natural Resources and Environment of the South Pacific Region (1986) all provide frameworks to establish guidelines for minimizing the direct effects of global changes associated with coastal flooding and water pollution.

In 1986, the International Geosphere-Biosphere Program (IGBP) was launched in Berne under the auspices of the International Council of Scientific Unions, which has nearly 70 member nations and represents 20 member scientific unions. Its function is to "describe and understand the interactive physical, chemical, and biological processes that regulate the total earth system, the unique environment that it provides for life, the changes that are occurring in this system, and the manner in which they are influenced by human actions." A document incorporating these ideas was circulated and used to set up a special committee, chaired by James McCarthy, a biological oceanographer, to carry out the first phase of activities, which is due to last until 1990.

The committee has established panels to assess research needs, develop implementation plans, and execute pilot experiments in four areas: terrestrial biosphere-atmosphere chemistry interactions, marine biosphere-atmosphere interactions, biospheric aspects of the hydrological cycle, and the effects of climate change on terrestrial ecosystems. There are working groups for technical needs in global geosphere-biosphere modeling, data and information systems, techniques for extracting environmental data from the past, and geosphere-biosphere observatories. For the marine biosphere in particular, activities include characterization of the oceanic carbon cycle and associated links with other elements known to limit biological productivity, studies of effects likely to have an impact on coastal and estuarine systems, and examination of the linkages among biogeochemical cycles and the physical climate system, in order to anticipate the effects of global change and climate feedbacks. Links are also being made between these and other programs, such as the Joint Global Ocean Flux Study and the World Ocean Circulation Experiment, to ensure world data coverage.

Already it is clear that cross-disciplinary activities are difficult to establish. As ecologists are called upon to develop models of the biological consequences of the changing climate, the lacunae in our knowledge are exposed. Further research into the actual effects that climate change will have on the marine ecosystem is an obvious priority. Also important, however, is the development of management models that will allow individuals, governments, and international organizations to create policies for minimizing damage and for coping with the changes that do occur.

An Integrated Management Assessment Framework for Marine Ecosystems

To prevent the development of risk-prone ventures, governments and planners must establish integrated management frameworks that allow the issues to be identified and evaluated. In general order of importance, the three principal management objectives for marine ecosystems are: (1) ecological stability and resilience through conservation, (2) economic efficiency and social well-being, and (3) conflict resolution. Each of the associated elements must be assessed in relation to the impacts of sea level rise, storm events, and altered patterns of ecosystemic functioning. The issues related to climate change must be assessed with respect to resource factors, harvesting and culturing factors, economic and resource use factors (including both market and nonmarket economics), and socio-cultural factors. The complexity of each component and the interactions between them determine the degree of complexity in the overall management of conservation and development.

Once the goals and complexity of the problem have been assessed, the possible solutions will be constrained by the institutional structures in place. The performance of systems for managing living marine resources must be assessed in terms of goals, systemic structures, and institutional constraints. In the context of global climate change, it is clear that parity in controls on human activities will be impossible. Thus, management options and institutional structures will only partially reflect the externalities induced by other parts of the globe. Especially important then are national and regional governments, because at this level, legislation and organizations have already made possible effective control of many environmental practices through such mechanisms as property use rights, zoning, licensing quotas, technology transfer, and financial inducements.

CONCLUSIONS

In summary, historical evidence indicates that the effects to be expected as a result of global warming will vary significantly depending upon the type of marine life affected, their geographic location, and the speed and severity of any changes. The key factors in species recruitment will include changes in currents, tidal patterns, temperature, and sea level. For the many marine species that live in coastal waters, changes in sea level will alter habitat; for some of these species, erosion of coastline or the disappearance of certain environments such as coral reefs could have a devastating effect.

Beyond these direct consequences, changes in human actions, such as destruction of mangroves for aquaculture and overfishing of pelagic fish stocks, may, in turn, influence the abundance, and indeed the survival, of many marine species. These effects will further influence human activity, given the heavy dependence of many people on the marine ecosystem as a primary source of food and/or livelihood. Thus, human interaction with the marine ecosystem must be considered in any strategies for dealing with the impact of climate change.

REFERENCES

Bardach, J. E., 1989, Global warming and the coastal zone, *Climatic Change*, **15**:117-150.

Bardach, J. E., and Santerre, R. M., 1981, Climate and aquatic food production, *in*: "Food Climate Interactions," W. Bach, I. Pankrath, and S. H. Schneider, eds., Kluwer Academic Publishers, Dordrecht.

Benton, M. J., 1987, Progress and competition in macroevolution, *Biol. Rev.*, **62**:305-338.

Broadus, J. M., Milliman, J. D., Edwards, S. E., Aubrey, D. G., and Gable, F., 1987, Rising sea levels and damming of rivers: Possible effects in Egypt and Bangladesh, Woods Hole Oceanography Institution (internal mimeograph).

Bryan, K., 1984, Accelerating the convergence to equilibrium of ocean-climate models, *J. Phys. Oceanogr.*, **14**:666-673.

Buddemeier, R.W., and Smith, S.V., 1988, Coral reef growth in an era of rapidly rising sea level: predictions and suggestions for long-term research, *Coral Reefs*, **7**:51-56.

Christensen, B., 1978, Primary production of mangrove forests, *in*: "Proceedings of the International Workshop for Mangrove and Estuarine Area Development for the Indo-Pacific Region," Philippine Council for Agriculture and Resources Research, Los Baños.

Convention on the Conservation of Antarctic Marine Living Resources, May 20, 1980, T.I.A.S. No. 10,240.

Convention on the Conservation of Antarctic Seals, June 1, 1972, T.I.A.S. No. 8,826.

Convention on Long Range Transboundary Air Pollution, November 13, 1979, *Int. Law M.*, **18**:1,440.

Convention on the Prevention of Marine Pollution by Dumping of Wastes and Other Matter (London Ocean Dumping Convention), December 29, 1972, T.I.A.S. No. 8,165.

Convention for the Protection of the Natural Resources and Environment of the South Pacific Region, November 25, 1986, 1987, *Int. Law M.*, **26**:38.

Csavas, I., 1988, Shrimp farming development in Asia, INFOFISH, **2**, 88:11-16.

Cushing, D. H., 1973, *Recruitment and Parent Stock in Fish*, University of Washington, Seattle.

Cushing, D. H., 1982, *Climate and Fisheries*, Academic Press, London.

Erwin, D. H., Valentine, J. W., and Sepkoski, J. J., 1987, A comparative study of diversification events - the early Paleozoic versus the Mesozoic, *Evolution*, **41**:1,177-1,186.

Fowler, M., and McGlade, J.M., 1990, Bathymetric and geographic distribution of demersal fish of the continental slopes of Nova Scotia and Newfoundland, Prog. Mar. Ecology (in press).

Fryer, G., and Iles, T. D., 1972, *The Cichlid Fish of the Great Lakes of Africa: Their Biology and Evolution,* T.F.H. Publications, Neptune City.

Garcia, S., and Le Reste, L., 1981, Life cycles, dynamics, exploitation and management of coastal penaeid shrimp stocks, *FAO Fisheries Technical Paper*, 203.

Garrett, C. J. R., 1979, Mixing in the ocean interior, *Dyn. Atmos. Oceans*, **3**:239-265.

Garrett, C. J. R., Keeley, J. R., and Greenberg, D. A., 1978, Tidal mixing versus thermal stratification in the Bay of Fundy and Gulf of Maine, *Atmosphere-Ocean*, **16**:403-423.

Hansen, J., Lacis, A. Rind, D., Russell, G., Stone, P., Fung, I., Ruedy, R., and Lerner, J., 1984, Climate sensitivity: Analysis of feedback mechanisms, *in*: "Climate Processes and Climate Sensitivity (Maurice Ewing Series 5)," J. E. Hansen and T. Takahashi, eds., Am. Geophys. Union, Washington, D.C.

Hawker, D. L., and Connell, D. W., 1989, An evaluation of the tolerance of corals to nutrients and related water quality characteristics, *Intern. J. Environmental Studies*, **34**:179-188.

Hjort, J., 1914, Fluctuations in the great fisheries of northern Europe viewed in the light of biological research, *Rapp. P.-v. Reun. Cons. Perm. Int. Explor. Mer*, **20**:1-228.

Horne, E.P., 1978, Interleaving at the subsurface front in the slope water off Nova Scotia, *J. Geophys. Res.*, **84**:7,707-7,725.

Horne, E.P., and Platt, T., 1984, The dominant space and time scales of variability in the physical and biological fields on continental shelves, *Rapp. P.-v. Reun. Cons. int. Explor. Mer.*, **183**:8-19.

Iles, T. D., and Sinclair, M., 1982, Stocks of Atlantic herring, their discreteness and abundance variability, *Science*, **215**:627-633.

Kapetsky, J. M., 1985, Mangroves, fisheries and aquaculture, *FAO Fisheries Report*, **338**:17-36.

Keeling, C. E., Bacastow, R. B., and Whorf, T. P., 1982, Measurements of the concentration of carbon dioxide at Mauna Loa Observatory, Hawaii, *in*: "Carbon Dioxide Review: 1982," W. C. Clark, ed., Oxford University Press, Oxford.

Kuo, C., Lindberg, C., and D. J. Thomson, 1990, Coherence established between atmospheric carbon dioxide and global temperature, *Nature*, **343**:709-714.

Lamb, H. H., 1972, *Climate Past, Present and Future,* Methuen, London.

Lasker, R., 1981, Factors contributing to variable recruitment of the northern anchovy (Engraulis mordax) in the California current: contrasting years, 1975 through 1978, *Rapp. P.-v. Reun. Cons. Int. Explor. Mer*, **178**:375-388.

MacCracken, M. C., Buddemeyer, R., and Knox, I., 1987, Water resources in the Sacramento-San Joaquin Valley and San Francisco Bay: A proposed case study of the impacts of changing climate, Lawrence Livermore National

Laboratory (internal mimeograph).

Mahon, R., 1985, personal communication.

Manabe, S. and Bryan, K., 1969, Climate calculations with a combined ocean-atmosphere model, *J. Atmos. Res.*, **26**:786-789. Manabe, S. and Stouffer, R. J., 1980, Sensitivity of a global climate model to an increase of CO_2 concentration in the atmosphere, *J. Geophys. Res.*, **85**:5,529-5,554.

Mann, K. H., 1988, Towards predictive models for coastal marine ecosystems, *in*: "Concepts of Ecology. A Comparative View," L.R. Pomeroy, and J. J. Alberts, eds., Ecological Studies 67, Springer-Verlag, New York.

Mars, P., 1963, Les Faunes et la stratigraphie du Quaternaire Mediterranean, *Recent Trav. St. Mar. End. Bull.*, **28**:61-97.

Mazeika, P. A., Burns, D. A., and Kinder, T. H., 1980, Mesoscale circulation east of the southern Lesser Antilles, *J. Geophys. Res.*, **85**:2,743-2,758.

Meade, R. H., Nordin, C. F., Curtis, W. F., Rodrigues, F. M. C., Do Vale, C. M., and Edmond, J. M., 1979, Sediment loads in the Amazon River, *Nature*, **278**:161-163.

Mercer, J. H., 1978, West Antarctic ice sheet and CO_2 greenhouse effect: A threat of disaster, *Nature*, **271**:321-325.

Mohn, R. K., and Miller, R. J., 1987, A ration-based model of a seaweed-sea urchin community, *Ecological Modeling*, **37**:249-267.

Montreal Protocol on Substances That Deplete the Ozone Layer, September 1987, Reference File, *Int. Environment Rep.* (BNA), **21**:3,151.

Nordhaus, W. D., and Yohe, G. W., 1983, Future paths of energy and carbon dioxide emissions, *in*: "Changing Climate," National Academy of Sciences, Washington, D.C.

Oerlemans, J., 1989, A projection of future sea level, *Climatic Change*, **15**:151-174.

Pingree, R. D., and Griffiths, D. K., 1978, Tidal fronts on the shelf seas around the British Isles, *J. Geophys. Res.*, **83**:4,615-4,622.

Protocol Concerning the Emissions of Nitrogen Oxides or Their Transboundary Fluxes, 1988, *Environ. Policy and Law*, **18**:228.

Protocol on the Reduction of Sulphur Emissions on Their Transboundary Fluxes by at Least 30%, July 6, 1985, Reference File, *Int. Environment. Rep.* (BNA), **21**:3,021.

Quraishee, G. S., 1975, Influence of the Indus River on the Marine Environment, *in*: "Proceedings of the Internal Conference on Management of Environment," Pakistan Academy of Science, Islamabad.

Rotty, R. M., 1983, Distribution of and changes in industrial carbon dioxide production, *J. Geophys. Res.*, **88**:1,301-1,308.

Ryther, J. H., 1969, Photosynthesis and fish production in the sea, *Science*, **166**:72-76.

Saenger, P., Hegerl, E. J., and Davie, J. D. S., eds., 1983, *Global Status of Mangrove Ecosystems*, International Union for the Conservation of Nature and Natural Resources Committee Ecological Papers 3, Geneva.

Schlesinger, M. E., 1984, Climate model simulations of CO_2-induced climate change, *in*: "Advances in Geophysics 26," B. Saltzman, ed., Academic Press, New York.

Schlesinger, M. E., Gates, W. L., and Han, Y. J., 1985, The role of the ocean in CO_2-induced climate change: Preliminary results from the OSU coupled atmosphere-ocean general circulation model, *in*: "Coupled Ocean-Atmosphere Models," J. C. J. Nihoul, ed., Elsevier Oceanography Series, Amsterdam.

Schwalb, A., 1978, The TIROS-N/NOAA A-G Satellite Series, *NOAA Technical Memorandum NESS 95*.

Sibley, T. H., and Strickland, R. M., 1985, Fisheries: some relationships to climate change and marine environmental factors, *in*: "Characterization of Information Requirements for Studies of CO_2 Effects: Water Resources, Agriculture, Fisheries, Forests and Human Health," M. R. White, ed., United States Department of Energy, Carbon Dioxide Research Division, Washington, D.C.

Simpson, J. H., and Hunter, J. R., 1974, Fronts in the Irish Sea, *Nature*, **250**:404-406.

Sinclair, M., 1988, *Marine Populations: An Essay on Population Regulation and Speciation*, Washington Sea Grant Program, Univ. of Washington Press, Seattle.

Sinclair, M., and Iles, T.D., 1988, Population richness of marine fish species, *Aquat. Living Resour.*, **1**:71-83.

Titus, J. G., Leatherman, S. P., Everts, C. H., and Kriebel, R. J. D., 1985, *Potential Impacts of Sea Level Rise on the Beach at Ocean City, Maryland*, Environmental Protection Agency, EPA 230-10-85-0113.

United Nations, 1985, *Estimates and Projections of Urban, Rural and City Populations, 1950-2025: The 1982 Assessment,*, United Nations Department of International and Social Affairs, United Nations, New York.

United Nations Convention of the Law of the Sea, December 10, 1982, *Int. Law M.*, **21**:1,261.

Valentine, J. W., 1973, *Evolutionary Paleoecology of the Marine Biosphere*, Prentice-Hall, New Jersey.

Vienna Convention for the Protection of the Ozone Layer, March 22, 1985, 1987, *Int. Law M.*, **26**:1,516.

Walker, D. I., and Osmond, R. A. G., 1982, Coral death and phosphate pollution at Aqaba, Red Sea, *Mar. Poll. Bull.*, **13**:21-25.

Weiss, E. B., 1989, Climate change, intergenerational equity and international law: an introductory note, *Climate Change*, **15**:327-335.

Wigley, T. M. L., and Schlesinger, M. E., 1985, Analytical solution for the effect of increasing CO_2 on global mean temperature, *Nature*, **315**:649-652.

World Port Index, 1986, (10th edition) Defense Mapping Agency Hydrographic Topographic Center Publication 150, Washington, D.C.

PREDICTION IN CHAOTIC NONLINEAR SYSTEMS: TIME SERIES ANALYSIS FOR APERIODIC EVOLUTION

Henry D. I. Abarbanel

Marine Physical Laboratory
Scripps Institution of Oceanography
and
Department of Physics
University of California
San Diego, CA 92014, USA

INTRODUCTION

This chapter deals with time series analysis for observations taken on systems that exhibit aperiodic or chaotic behavior. Such systems, which must be nonlinear, abound in physical and biological applications, and have been studied for nearly a century (Poincaré, 1899) in contexts ranging from classical mechanics and fluid dynamics to epidemiology. For climatic systems, the lessons of chaos for fluid dynamics are of special relevance.

The following discussion is meant to be pedagogical. I wish the reader to develop a sense of how to ask and answer questions about chaotic time series that will reveal interesting, important, and relevant information about the dynamics that give rise to the time series in the first place. To me, time series analysis is a set of methods for revealing structure in apparently structureless data. Underlying the data from a chaotic, deterministic system is geometric structure that has fractional dimension in general, but regularity in the appropriate time domain. The structure is an attractor, in the sense that a large set of initial conditions for the system asymptotically lie on this same geometrical structure. Because it has fractional dimension, the structure is called a strange attractor (Eckman and Ruelle, 1985). Signal processing on strange attractors is the focus of this chapter.

When one looks at a chaotic orbit, one sees no regularity to the time evolution. When one looks at its Fourier spectrum, one sees a broadband collection of frequencies. The first task, then, is to discover in what space one should be looking at the time series. This starting step in the analysis of chaotic data goes under the jargon-like name of *phase-space reconstruction.* In working out how the phase space is reconstructed, we will encounter several ideas about *invariants* of the motion on strange attractors. These invariants will help us classify the motion we are trying to "process," and we will return to them further on, because they play the same role for nonlinear systems as the characteristic frequencies of a linear system enjoy for linear time series analysis. Because the individual orbits of nonlinear systems can differ wildly with even small changes in initial conditions (Lorenz, 1984), the orbits cannot be a reliable subject for our study. This, of course, is a major departure from the Newtonian view that we should study in detail all specific orbits of evolving systems. This same view was probably held by Poincaré (1899), though it did not enter the textbooks.

Lastly, I will discuss what one can do once the proper space for seeing dynamics has been established, and one has classified the system at hand in terms of its invariants. This generalization of linear prediction theory to the domain of nonlinear systems is, in many ways, the most speculative, bold, and fruitful aspect touched on here. We will study how to make parametrized models of the evolution of the system through the reconstructed phase space, and what information about invariants of the motion we can believably glean from the data.

The plan then is

- to establish the domain in which to *visualize* the motion and carry out the analysis of chaotic time series
- to *classify* the system that produces the time series in terms of the invariants of the chaotic motion
- to make models of the motion in the visualization space that respect the invariants and allow us to *predict* future behavior of the system

The full program outlined here is not a completed one; many computational and conceptual issues remain unaddressed. One of the most interesting is that of *extrinsic noise*, which inevitably contaminates any data (laboratory-, field-, or computer-generated) that we might want to process with these methods. External noise is broadband and thus spectrally indistinguishable from the chaotic signal we want to analyze. Conventional methods for separating the signal from the noise rely on their differing spectral characteristics. If we tried to use those methods, we would, in effect, "throw the baby out with the bathwater." I will not comment on noise reduction in chaos, though this might be a major engineering goal of this and other related work. Noise reduction is an ongoing and challenging research topic, and should be addressed separately. Another set of issues about which no comment will be made is that of control of nonlinear chaotic systems. If one is able to predict, then control methodologies can be established. A detailed study of the optimum methods and their stability is yet to be made. Finally, there is the matter of the application of all this technology to matters of immediate and long-range interest—the weather and climate. Given the present state of ignorance about climate or weather as a strange attractor, it may be that the methods described will allow a few steps of progress to be made in understanding these complex systems, and in predicting their behaviors.

Is there any indication that the climatic system is chaotic, so that the kind of material I promise here might prove relevant? The answer is a strong "Yes!," as has been discussed in Chapter I. Part of the answer is theoretical and is based on the work of Lorenz (1984, 1989a) and many others, and on both simple and complex models that exhibit chaos. As Lorenz (1989a) notes (emphasis added):

> The climate system is unquestionably chaotic. It possesses obvious annual and diurnal periods, but, when these and their overtones and all other established or suspected periodic variations are subtracted out, *a strong signal remains.* Further supporting evidence is the sensitive dependence on initial conditions exhibited by virtually all reasonably realistic atmospheric models.

Also supporting this point of view is an analysis by Fraedrich (1986), which analyzes several climatological data sets by the methods we will discuss in the next section, and concludes rather persuasively that the state of observed climatic variables is fractional-dimensional. These remarks made, we are ready to describe the powerful analytical tools that can be applied to climate and other dynamical systems.

GETTING INTO THE RIGHT SPACE: PHASE-SPACE RECONSTRUCTION

Saturation of Invariants Method

As indicated above, the first step in making an analysis of a chaotic time series is to locate the space in which the processing is to take place. The now-conventional method for determining this space used in the study of dynamical systems makes several assumptions that are more or less defendable in a rigorous fashion, and that appear both plausible and correct about the actual physical systems with which we work.

The first assumption is that the time series we observe has a deterministic, dynamical system underlying its evolution. This seems to be something we must assume, but it excludes the possibility that the data arise from random noise, which has no structure of any deterministic nature.

The second assumption we make is that the data represent the evolution of the system after all transients have died out, and that this evolution is taking place on a geometric structure that is finite-dimensional (typically a small dimension, though that is not critical) and remains within a bounded region of the finite-dimensional space in which it is located. The technical idea here is that the system orbits emanating from various initial conditions enter some bounded region of state-space and remain within that region for all further time. Within this region is a set of points visited by the orbits, which is called the *attractor.* This attractor may be simple and regular, as, for example, motion that evolves to a fixed point in the space and then remains there. It may be periodic, which means that all orbits move on, or asymptotically in time on, a recurring set in the space. The motion may consist of a single frequency and its harmonics, or of a set of incommensurate frequencies and their harmonics. In the latter case, the time-domain picture of the orbit is quite complicated, but the Fourier spectrum is simple to classify. Finally, the attractor may be complex and a set of non-integer dimension (we will make clear what such an outlandish statement means quite soon) on which orbits move in an aperiodic fashion. Such orbits are very sensitive to the initial conditions, or, equivalently, to the way they enter the bounded region in which they stay permanently. Slightly differing initial conditions can give rise to very different detailed orbits, which, nonetheless, visit the same attractor.

The third assumption is that even though we have a finite sample of the orbit—no infinitely long orbits are available to us—this segment of the orbit is long enough and well enough sampled in time that we can reconstruct a good representation of the attractor and have not missed any pieces of the geometry. This requirement is less fundamental than the others and, in principle, can be met by increasing the amount of data and the sampling rate. It is good to keep in mind, however, that we are assuming the geometric structure on which the orbits evolve has been well sampled.

From the first assumption about an underlying differential equation, we imagine we ought to be able to "see" the nature of any orbit by plotting the independent degrees of freedom of the system against each other. Recall that any set of differential equations can be written as a collection of first-order equations, and that the independent degrees of freedom are those required to describe these first-order equations. Basically, this is because one always requires the number of first-order equations as the number of initial conditions, and to unfold the evolution of the system one must follow it in that same number of dimensions. The simple harmonic oscillator makes the point quite clearly. The variable $X(t)$, satisfying

$$\ddot{X}(t) + \omega^2 X(t) = 0,$$

has evolution equations that can be written as

$$\dot{X}(t) = V(t),$$

and

$$\dot{V}(t) = -\omega^2 X(t).$$

In the first form, we would accept a one-dimensional solution $X(t) = A\sin(\omega t) + B\cos(\omega t)$, which is fine, while in the second form we would be inclined to look at plots of $X(t)$ against $V(t)$; namely, a two-dimensional plot. The latter reveals that the motion is circular; the former does not. That is, the *geometry* of the solution is revealed in the latter, while the analytic form revealed in the former is useful but less informative. Indeed, in the cases we shall come to shortly, there is no analytic solution known, and probably none achievable, so the second point of view is imperative.

Our first problem is to determine what kind of data is presented to us and what we are to do with it in order to extract the independent derivatives as degrees of freedom. We shall assume now and henceforth that we have measured or computed a data set of measured scalar values $x(n); n = 1, 2, \ldots, N$. There could be several such data sets corresponding to different measurements of the system, but we shall continue as if we had just one. The data are presented at discrete times, because that is how they are measured. To create the derivatives of this data set we must know the δ in time between measurements. We then form, in addition to $x(n)$,

$$\dot{x}(n) = \frac{x(n+1) - x(n)}{\delta},$$

and

$$\ddot{x}(n) = \frac{x(n+2) - 2x(n+1) + x(n)}{2\delta^2},$$

and so forth for higher derivatives. If δ is too large, this procedure is an uncertain, or at least inaccurate, way to measure the derivatives of $x(t)$. However, if we look at the various formulae for the numerically formed derivatives, we see that we are just adding information about $x(n+1)$, then $x(n+2)$, and so forth to the values of $x(n)$. The obvious way to skirt the issue of inaccurate evaluations of the derivatives is to take the d-lagged values of $x(n)$,

$$[x(n), x(n+\tau_1), x(n+\tau_2), \ldots, x(n+\tau_{d-1})], \tag{1}$$

as coordinates, instead of the derivatives themselves.[1] In this choice, we have allowed for the possibility that we would wish to evaluate the data at differing lagged values. If we are less general and slightly more practical, we can focus on a single lag τ and write at every time step the ***d*-dimensional vector of independent degrees of freedom**:

$$\mathbf{y}(n) = [x(n), x(n+\tau), x(n+2\tau), \ldots, x(n+(d-1)\tau)]. \tag{2}$$

If the dynamics described by the data $x(n)$ is evolving on an attractor of dimension d_a, which may be fractional, we should certainly have captured the structure of that attractor when d becomes sufficiently larger than d_a. If the motion were regular, for example, and described the surface of a sphere in three dimensions as the orbits evolved, then choosing $d = 3$ would be the minimum dimension to capture the full view of the attractor. Increasing d would embed the sphere in larger and larger spaces and reveal no further information about its structure. It would make any numerical calculations regarding the system more and more time consuming as the dimension got larger and one had to search for those small numbers of points comprising the dynamics in bigger and bigger volumes. Just how big must d be to reveal the geometric properties of the attractor without ambiguity? Mañé (1980) and Takens (1981) found that it is sufficient to take d as the next integer equal to or larger than $2d_a + 1$. For the sphere where $d_a = 2$, this criterion would require $d = 5$. So although this criterion

[1]This idea seems first to have been considered by David Ruelle and then taken up by numerous other authors. See Eckman and Ruelle (1985) for a discussion.

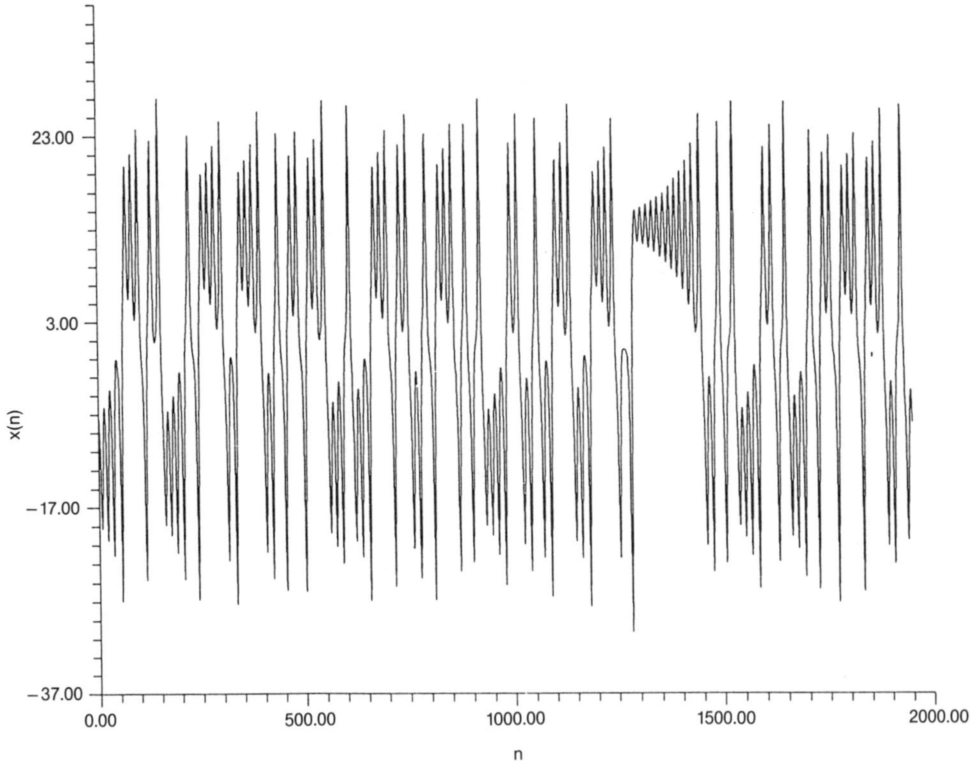

Fig. 1. Time series for data from the Lorenz model; $x_1(n) = x_1(n\Delta t)$.

is sufficient, it may give a d that is too large for some attractors. The criterion $d > d_a + 1$ arises from a topological theorem that considers how one must view an arbitrary topological space by embedding it in a Euclidean space R^d. If the topological space is sufficiently smooth, the criterion is clearly overkill. It is very useful, however, to have a sufficient upper limit on d in hand in order to confine one's search for this embedding dimension.

Before going on to more practical and therefore less rigorous criteria for choosing d, let us pause and look at data from a computer-generated dynamical system, just to get a sense of how the preceding fits into something concrete. The system we look at is Lorenz's original model (Lorenz, 1963):

$$\begin{aligned} \frac{dx_1(t)}{dt} &= \sigma(x_2(t) - x_1(t)) \\ \frac{dx_2(t)}{dt} &= -x_1(t)x_3(t) + rx_1(t) - x_2(t) \\ \frac{dx_3(t)}{dt} &= x_1(t)x_2(t) - bx_3(t), \end{aligned} \tag{3}$$

with parameter values $\sigma = 16$, $b = 4$, and $r = 45.92$. Figure 1 shows the time series $x_1(n\delta t) = x_1(n)$ for an initial condition chosen at random. The equations of motion were solved with a simple fourth-order Runge-Kutta solver that has a time step $\delta t = 0.0025$. We determined that with the same initial conditions, the orbit was unchanged for values of δt that were somewhat larger or any amount smaller than this. I think the reader will agree that the orbit is irregular. The Fourier spectrum of this orbit is shown in Fig. 2. We take 1,024 points of the orbit,

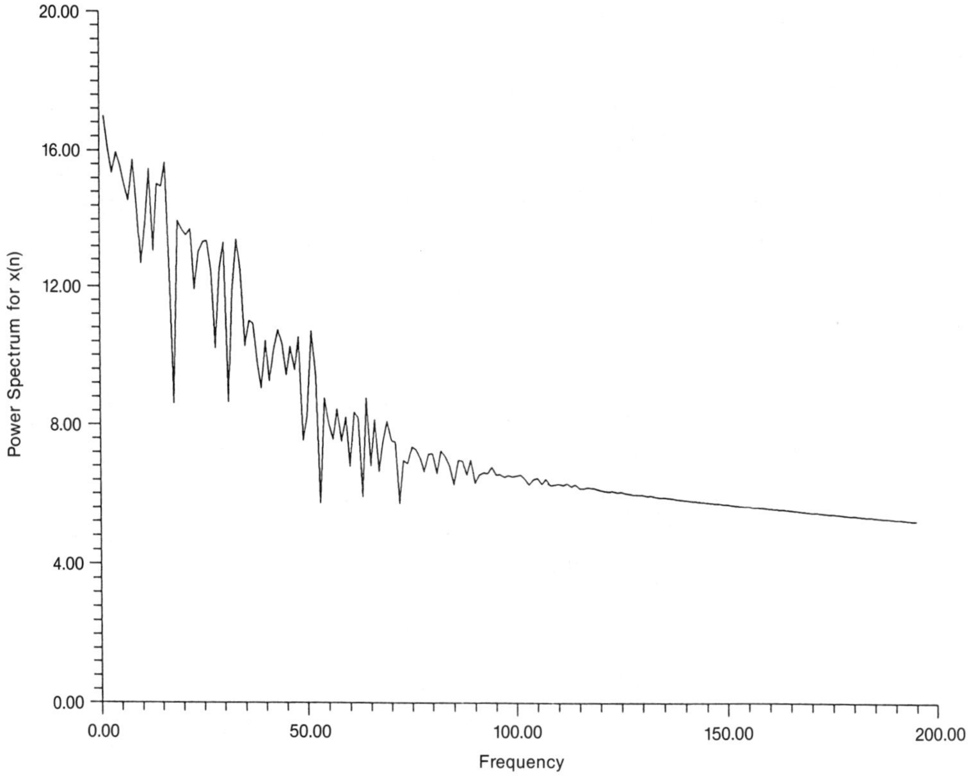

Fig. 2. Fourier power spectrum from $x_1(n)$ data.

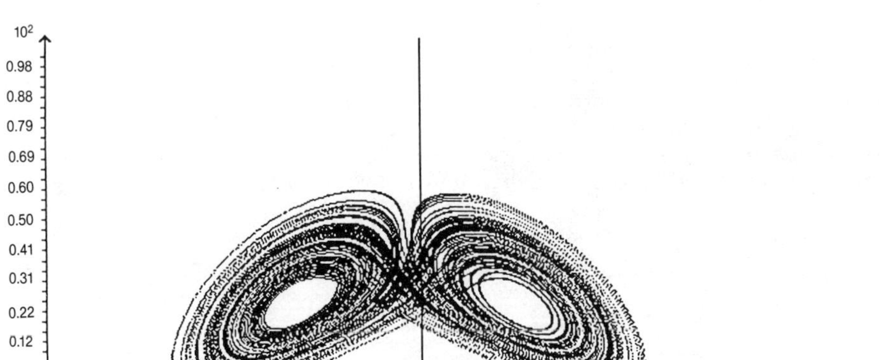

Fig. 3 Phase portrait of the Lorenz attractor. The attractor has been projected onto a plane through the x_3 axis. The parameters of the Lorenz model are $r = 45.92$, $\sigma = 16.0$, and $b = 4.0$.

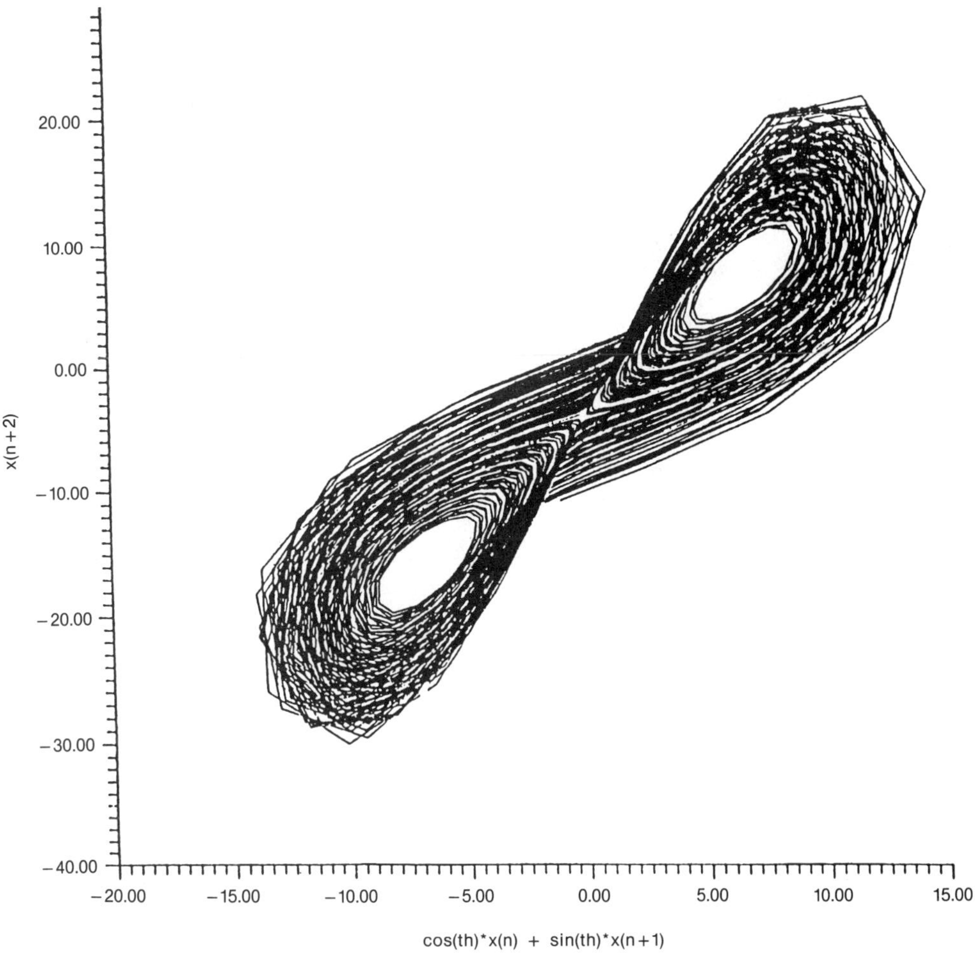

Fig. 4. The phase portrait of the Lorenz attractor as reconstructed from $x_1(n)$ data only. The vectors $[x_i(n), x_1(n+1), x_1(n+2)]$ are plotted against one another as a function of n. The figure is projected on the $x_1(n+2)$ axis.

transform them, and display the logarithm of the magnitude of $x(f)$, the Fourier transform of $x(n)$. Beyond the value of about $f = 125$ in the arbitrary units displayed, the spectrum exhibits no structure and is not shown. For lower frequencies, we see a rich broadband structure with no dominant sharp lines. Fig. 3 displays the full orbit in $(x_1(n), x_2(n), x_3(n))$ coordinates projected onto a plane passing through the x_3 axis and making an angle of 1.46 radians with the x_1 axis. Although the probably familiar geometric structure of the Lorenz attractor is now visible, recall how it was hidden in Figs. 1 and 2. Finally, we display a plot from this example in Fig. 4, with $x_1(n+2)$ on the vertical axis versus $x_1(n)\cos(\theta) + x_1(n+1)\sin(\theta)$ on the horizontal axis. In this picture, $\theta = 1.31$ radians. The geometric structure visible in the original phase space or state-space plot of (x_1, x_2, x_3) is revealed through the vectors $\mathbf{y}(n) = [x_1(n), x_1(n+1), x_1(n+2)]$. Our choice of $d = 3$ was, of course, motivated by our knowledge of the original system, but the point should be clear. The dynamics is in the $\mathbf{y}(n)$ and is not revealed in the $x(n)$.

Before leaving this example, it is good to remember that the Lorenz model came from a severe truncation of the fluid dynamic equations for convection in the lower atmosphere. They were not an invention of Lorenz's imagination. In the equations, σ is the Prandtl number of the fluid, b is related to the scale on which the first linear instability from pure conduction takes place, and r is the Rayleigh number in units of the critical Rayleigh number for the fluid's first instability. The peculiar behaviors shown in these figures are clearly connected with the issues considered in Chapters 1 and 2.

Lorenz (1984) introduced another model, which is of the driven, damped, general circulation of the atmosphere; you may read about details of its properties and interpretation in Lorenz (1984; 1989a). The model reads

$$\begin{aligned} \frac{dx_1(t)}{dt} &= -x_2(t)^2 - x_3(t)^2 - ax_1(t) + aF, \\ \frac{dx_2(t)}{dt} &= x_1(t)x_2(t) - bx_1(t)x_3(t) - x_2(t) + G, \\ \frac{dx_3(t)}{dt} &= bx_1(t)x_2(t) + x_1(t)x_3(t) - x_3(t); \end{aligned} \tag{4}$$

its parameters are explained in the original articles. "Winter" in this model is chaotic and is determined by the parameters $a = 0.25, b = 4.0, F = 8.0$, and $G = 1.0$. A typical solution for $x_1(n)$ is shown in Fig. 5; its Fourier power

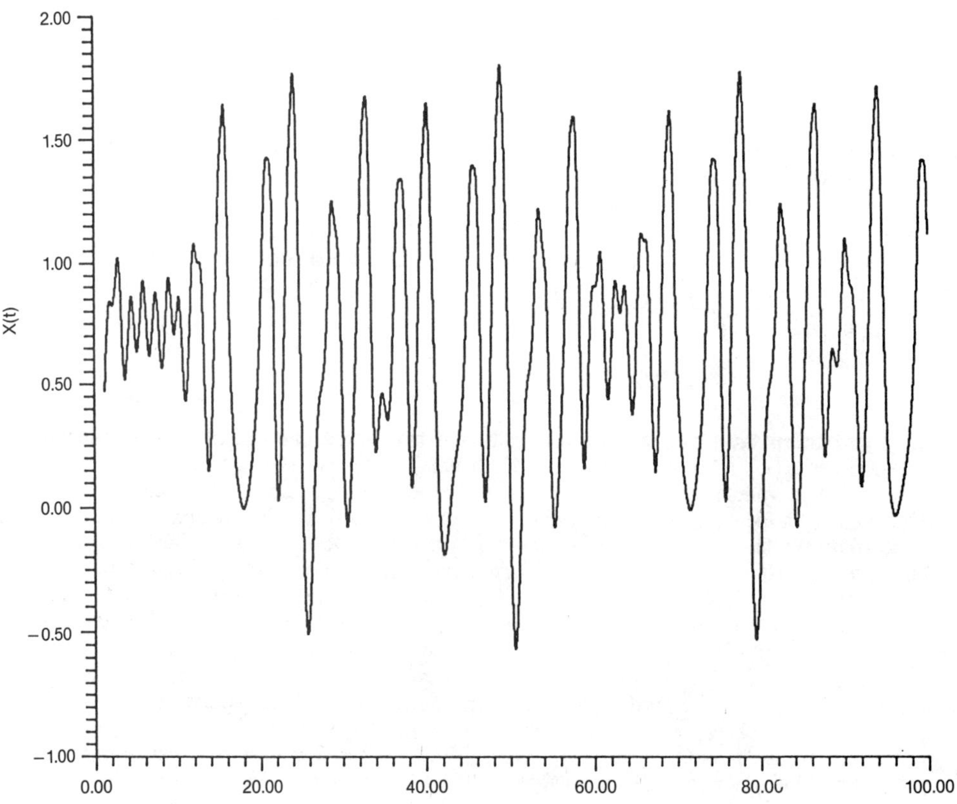

Fig. 5. X(*t*) for the new Lorenz model. Parameters for the model are dt =0.025, a =0.25, b =4.0, F=8.0, and G=1.0. The data is shown for $1.25 < t < 100$.

spectrum is shown in Fig. 6. This simple model is another example from geophysical fluid dynamics that is chaotic. The model's attractor has a dimension of about 2.5 and is twisted enough to require an embedding dimension larger than 3 in order to describe it in terms of the lagged vectors $\mathbf{y}(n)$ (Lorenz, 1989b).

If one is convinced that a dimension d can be made from lagged, scalar data, then two questions naturally arise:

- Is the use of scalar pieces of information to reconstruct motion in R^d a sleight of hand?
- What way can d be determined so it is the minimum necessary for the unfolding of the attractor from $x(n)$ data?

We can answer the first question because the system is nonlinear, and all the actual variables $\dot{x}, \ddot{x}, \ldots$ are interconnected; they are never uncoupled as in the diagonalization of a linear system. The spectral or other dynamical information in any variable is in all variables, so lags, which are complicated nonlinear combinations of the "true" variables, are perfectly good coordinates to reveal the

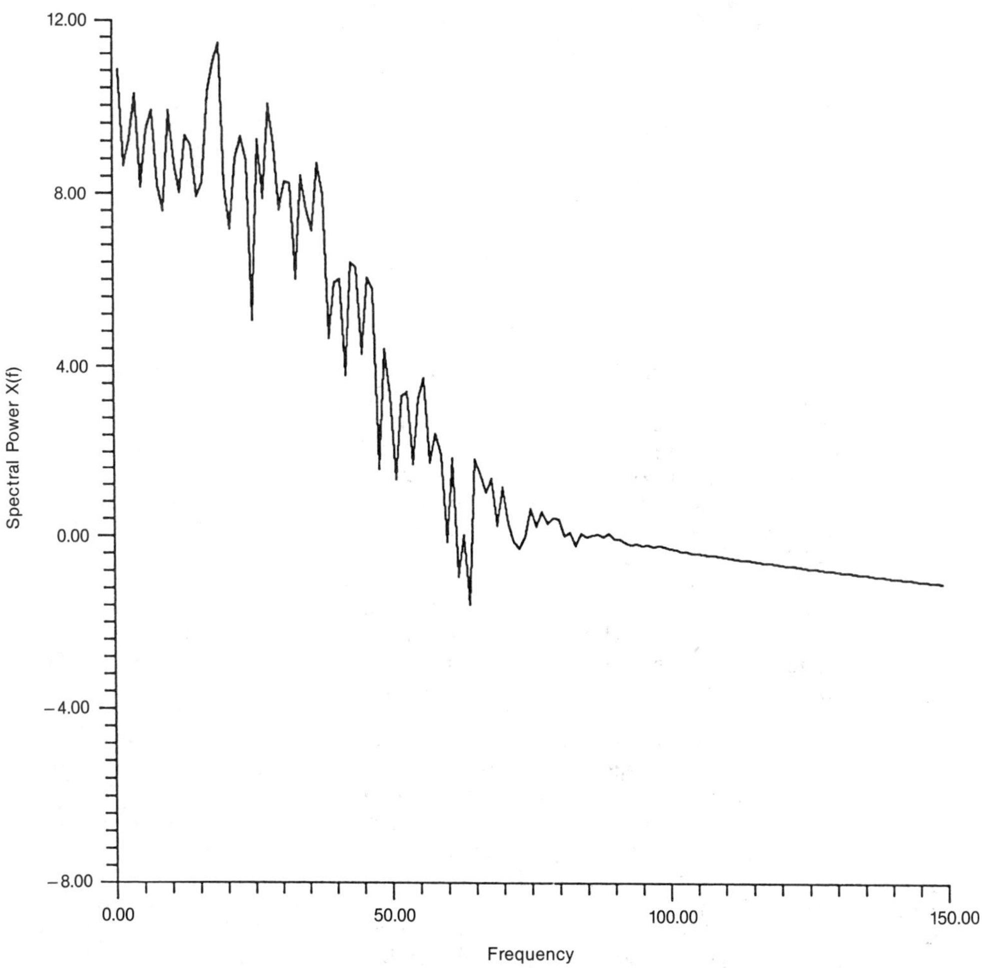

Fig. 6. Fourier power spectrum from *X(t)* data from the new Lorenz model.

dynamics. The Mañé (1980) and Takens (1981) results are for much more general constructions of d-vectors $\mathbf{y}(n)$ than simple lagged terms from $x(n)$, but it is the essential coupling of the degrees of freedom in nonlinear systems that underlies this construction. Once one chooses lagged variables as coordinates for the dynamics, the choice of lags τ becomes an important practical issue. I recommend the discussions of Fraser (1988; 1989) and Fraser and Swinney (1986) on this issue. If one uses for a common lag τ about 5% to 10% of the time associated with the first local minimum of the autocorrelation function of the data, all seems to work well (Abarbanel, 1990; Abarbanel et al., 1989a, 1989b; Brown, 1989). This amount of time is not precisely that chosen by Fraser, but it is of the same general magnitude.

As to the second question, I do not yet know the answer; I have several possible answers and a preference, but not a strong opinion. There are two general frameworks in which an answer is attempted to our second question:

1. Identify a quantity that varies with embedding dimension d until a critical dimension d_E is reached, after which it becomes independent of d. Compute this quantity for a d set that is increasing and determine d_E by the value at which independence sets in. The dimension of the attractor d_a should be one such quantity, because if we view the attractor in too low a dimension, we will underestimate d_a. At the critical dimension d_E and above, our determination of d_a will give the same number. We will consider several candidate quantities that "saturate" at some d_E.

2. Choose a large embedding dimension, call it D, and make it larger than one would expect the "best" embedding dimension d_E to be. Use one of the previous tests to determine a large enough embedding dimension, or guess right. Compute the $D \times D$ sample covariance matrix of the data

$$\sum_{l=1}^{N} \mathbf{y}^T(l)\mathbf{y}(l)$$

and determine its singular value spectrum (Press et al., 1986; Broomhead and King, 1986; Broomhead and Jones, 1989). There will be D singular values, of course, but only d_E of them will be "relevant." The criterion of relevance then becomes the issue. The idea is that $D - d_E$ components of $\mathbf{y}(n)$ are contaminated by noise from experimental error, or rounded off in the computing machine or whatever source. The full D-dimensional space is occupied only because of this noise. The singular value spectrum will reveal a set of "large" singular values above the noise floor set by the contamination, and a set of singular values of more or less equal magnitude, which are the noise. The number of "large" singular values is d_E. We will describe a new idea for determining d_E using theoretic criteria for the determination of the relevance of the singular value spectrum of the sample covariance matrix (Abarbanel and Kadtke, 1989).

Let us start with the first matter. We will discuss a type of invariant that is related to phase-space averages over the ***invariant density*** $\rho(\mathbf{y})$ on the attractor. This density gives us a measure of the number of times an orbit visits some neighborhood of the R^d space in which we have embedded it. The evolution of the orbit $\mathbf{y}(n) \to \mathbf{y}(n+1) \to \mathbf{y}(n+2)$, and so on may be viewed as a *mapping* from R^d to itself. Points in R^d at "time" n are mapped into points of R^d at time $n+1$, and so on throughout the orbit. The locations of the points $\mathbf{y}(n)$ define the attractor. The distribution of points on the attractor need not be, and generically is not, uniform (Theiler, 1989; Paladin and Vulpiani, 1987). The distribution is defined directly in terms of the measurements $\mathbf{y}(n)$ as

$$\rho(\mathbf{y}) = \lim_{N\to\infty} \frac{1}{N} \sum_{k=1}^{N} \delta^d\{\mathbf{y} - \mathbf{y}(k)\}. \tag{5}$$

Recalling that there is a map in R^d, which we call $\mathbf{F}(\mathbf{y})$ and which takes $\mathbf{y}(n)$ to $\mathbf{y}(n+1)$, the invariant density can be written in the equivalent form

$$\rho(\mathbf{y}) = \lim_{N\to\infty} \frac{1}{N} \sum_{k=1}^{N} \delta^d\{\mathbf{y} - \mathbf{F}^k[y(0)]\}, \tag{6}$$

where $\mathbf{F}^k(\mathbf{y})$ is the kth iterate of the map:

$$\mathbf{F}^2(\mathbf{y}) = \mathbf{F}[\mathbf{F}(\mathbf{y})],$$

and so forth.

The invariance of this density is best stated in terms of its moments. For this purpose, we consider an arbitrary function $g(\mathbf{y})$ in the phase space R^d and integrate it with $\rho(\mathbf{y})$ to form $< g >$:

$$< g >= \int d^d y \rho(\mathbf{y}) g(\mathbf{y}), \tag{7}$$

which is also equal to

$$< g >= \lim_{N\to\infty} \frac{1}{N} \sum_{k=1}^{N} g(\mathbf{F}^k[\mathbf{y}(0)]), \tag{8}$$

by the definition of $\rho(\mathbf{y})$. This is an ergodic theorem statement of the equality of spatial and temporal averages. The definition of $< g >$ shows plainly that it is independent of initial conditions on the orbit $\mathbf{y}(0)$, making it a candidate for something that is *not sensitive to initial conditions.* The invariance of $< g >$ is in the statement

$$< g >= \int d^d y \rho(\mathbf{y}) g(\mathbf{y}) = \int d^d y \rho(\mathbf{y}) g(\mathbf{F}(\mathbf{y})),$$

which is easily demonstrated from the definition of $\rho(\mathbf{y})$.

We will take as the function $g(\mathbf{y})$ powers of the local average number of points on the attractor (i.e., in the data $\mathbf{y}(l)$) in a d-dimensional sphere of radius r with its center at $\mathbf{y}$:

$$n(r, \mathbf{y}) = \lim_{M\to\infty} \frac{1}{M-1} \sum_{j=1}^{M-1} \Theta(r - |\mathbf{y} - \mathbf{y}(j)|), \tag{9}$$

where $\Theta(x)$ is the Heaviside function: $\Theta(x) = 1$ for $x > 0$ and $\Theta(x) = 0$ for $x < 0$. The Euclidean length of vectors in R^d is denoted by $| \bullet |$. This function $n(r, \mathbf{y})$ clearly counts up and averages the number of points in the ball of radius r. It is an integral over the invariant density of the two-point function

$$n(r, \mathbf{y}, \mathbf{w}) = \Theta(r - |\mathbf{y} - \mathbf{w}|), \tag{10}$$

because

$$n(r, \mathbf{y}) = \int d^d w \rho(\mathbf{w}) n(r, \mathbf{y}, \mathbf{w}). \tag{11}$$

Now we average the powers of $n(r, \mathbf{y})$ over the attractor to arrive at the invariant quantity

$$C(r, q) = \int d^d y \rho(\mathbf{y}) [n(r, \mathbf{y})]^q. \tag{12}$$

The attractor has an inhomogeneous invariant density; this quantity is sensitive to various local inhomogeneities on different spatial scales corresponding to the power q chosen. Arguments cited by Paladin and Vulpiani (1987)[2] suggest that for r that is smaller than the natural size of the attractor and larger than the minimum distance between points on the attractor, the $(q+1)^{th}$ correlation function behaves as

$$C(r,q) \approx r^{\phi(q)}. \tag{13}$$

Each of the $\phi(q)$ is an invariant of the attractor, as should be clear from its definition.

The point to be made about the $\phi(q)$ is that they are easy to compute, and they depend on the dimension d of the embedding space—the dimension of the lagged vectors $\mathbf{y}(n)$. Because they measure geometric properties of the attractor, the $\phi(q)$ should become unchanging above a critical dimension d_E, which captures the full structure of the attractor. For $d < d_E$, they should change in some complicated fashion, depending on the space and the attractor. Actually, this statement could be made for the entire function $C(r,q)$, but the $\phi(q)$ have an additional interpretation we want to explore in a moment. For now, our emphasis is on the $C(r,q)$, or the scaling exponents $\phi(q)$, as indicators of the minimum embedding dimension d_E needed for viewing the attractor and providing the space in which we are to do our signal processing.

The case $q=1$ was suggested some years ago by Grassberger and Procaccia (1983a, 1983b) as a way to achieve a simple characterization of an attractor. Many others have investigated and generalized the $q=1$ case to the $C(r,q)$ and applied the results to the interpretation of the structure of strange attractors, but the basic ideas seem to have been uncovered by Renyi (Renyi, 1970a; 1970b; Paladin and Vulpiani, 1987) more than 20 years ago.

We will return shortly to the case $q=1$ and show how $C(r,q)$ or other invariants are used to establish d_E, but for now we look at the value of $\phi(q)$ when $q=-1$. In this case, we are averaging the inverse of the local number of points on the attractor. We can interpret this as the number of little boxes of size r (each of which contains an average order unity number of points) needed to contain all the points. If we ask instead how many boxes of size r, $N_B(r)$, are needed to cover all the points on the attractor as $r \to 0$, then we would come up with the approximate rule

$$N_B(r) \approx \frac{\text{Constant}}{r^{d_a}}, \tag{14}$$

where d_a is the dimension of the set that we are trying to cover with the $N_B(r)$ boxes. Testing this claim on a part of the real line or on an area in the plane will give $d_a = 1$ or 2 for those cases; this generalization to objects of arbitrary dimension was conceived by Hausdorff and Besicovitch 100 years ago. The dimension of an attractor d_a is thus defined in an operational way that makes a clear connection with the idea that dimension is related to the occupancy rate of the points in space.

Any of the $\phi(q)$ for any q, integer or fractional, would do as indicators of the saturation of our invariants as we compute them for various embedding dimensions d. The most popular choice among the values of q is $q=1$, but I suspect this is based more on historical precedent than on some sure insight that 1 is the optimum value. In principle, it should not matter which $\phi(q)$ one chooses to evaluate when testing its independence of d; except for finite N, which we always have, some may be more sensitive than others. In any case, we will report here on the calculation of $C(r,1)$:

[2]This discussion of the inhomogeneity of points distributed on an attractor provides a clear picture of how one determines quantitative measures of this inhomogeneity and how one looks for effects in physics of this phenomenon.

$$C(r,1) = \lim_{N\to\infty} \frac{1}{N(N-1)} \sum_{i,j;i\neq j}^{N} \Theta(r - |\mathbf{y}(i) - \mathbf{y}(j)|)$$

as a function of the embedding dimension of the vectors

$$\mathbf{y}(i) = [x(i), x(i+1), x(i+2), \ldots, x(i+(d-1))].$$

The index $\phi(1)$ is also commonly known as ν in the literature and is known to formally to satisfy $\nu \le d_a$ (Paladin and Vulpiani, 1987), while in practice the two numbers seem to agree to about 1%. It is also now known that the minimum number of points N in the data set required to compute ν to 5% accuracy is (Smith, 1988)

$$N_{\text{MIN}} \ge 42^M, \tag{15}$$

with M the greatest integer *less* than the dimension of the set; namely, d_a.

We now report the calculation on two attractors to show the good and bad aspects of trying to establish d_E from this method. First, in Fig. 7, using $x_1(n)$ data for the Lorenz model, we display $\log C(r,1)$ versus $\log r$ for $d = 1, 2, \ldots, 5$. It is known from many other calculations that $d_a \approx 2.06$ for this attractor, so at least 1,800 or so points are required to get an accurate evaluation of ν. We use 4,500 points in order to exceed that limit. Indeed, the number 4,500 was chosen on the criterion that the orbit being sampled visits the attractor rather completely. We felt that visiting only part of the attractor would give a wrong answer for both ν

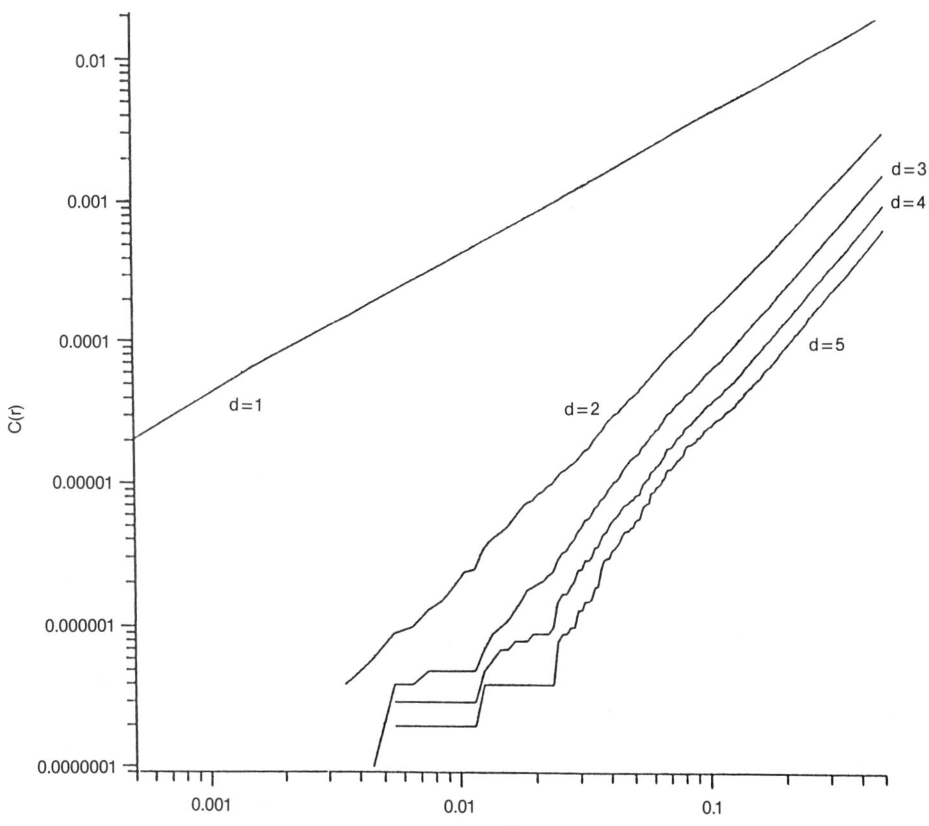

Fig. 7. $q=1$ correlation function for $x_1(n)$ Lorenz data. Log $C(r)$ is plotted versus log (r).

and other $\phi(q)$. Our actual interest here is not the specific number ν, but only a dimension d_E where $C(r,1)$ becomes invariant. Determining $C(r,q)$ using an orbit that does not visit the entire attractor is sure to be an error, but we have only a numerical test for this, which is to evaluate $C(r,q)$ or other invariants for varying N and d, and see when they both saturate. The figure shows reasonably clearly that for a large range of intermediate values of r, $C(r,1)$ is unchanged. For very small r, the function changes quite a lot, as expected, because there are no data there for finite sample size, and for large r we are somehow sampling the entire attractor.

The good news from Fig. 7 is that it was quite easy to calculate. Using a code created by James Theiler, who instructed us on its use, we were able to compute the data used in Fig. 7 in less than a minute on a Sun workstation. Because it was so easy, we also computed the same quantity using $x_2(n)$ and $x_3(n)$ data; the results are shown in Figs. 8 and 9. The information in each figure should be the same, because all variables are coupled nonlinearly among themselves. The less-than-good news is that in any *objective* way, it is really not possible to decide whether d_E is 2, 3, or 4, though 2 or 3 are most attractive. We know from other calculations that $d_a = 2.06$, making $d_E = 2$ a bad choice. Choosing 3 would be safer, and 4 safer still. In fact, from the reconstruction of the Lorenz attractor shown in Fig. 4, we can infer that $d_E = 3$ will do nicely. Just as a warning, the fact that the Lorenz differential equations have three degrees of freedom does not imply that $d_E = 3$ for the embedding space capturing the attractor for the data that is discrete. Recall the Mañé and Takens results about this.

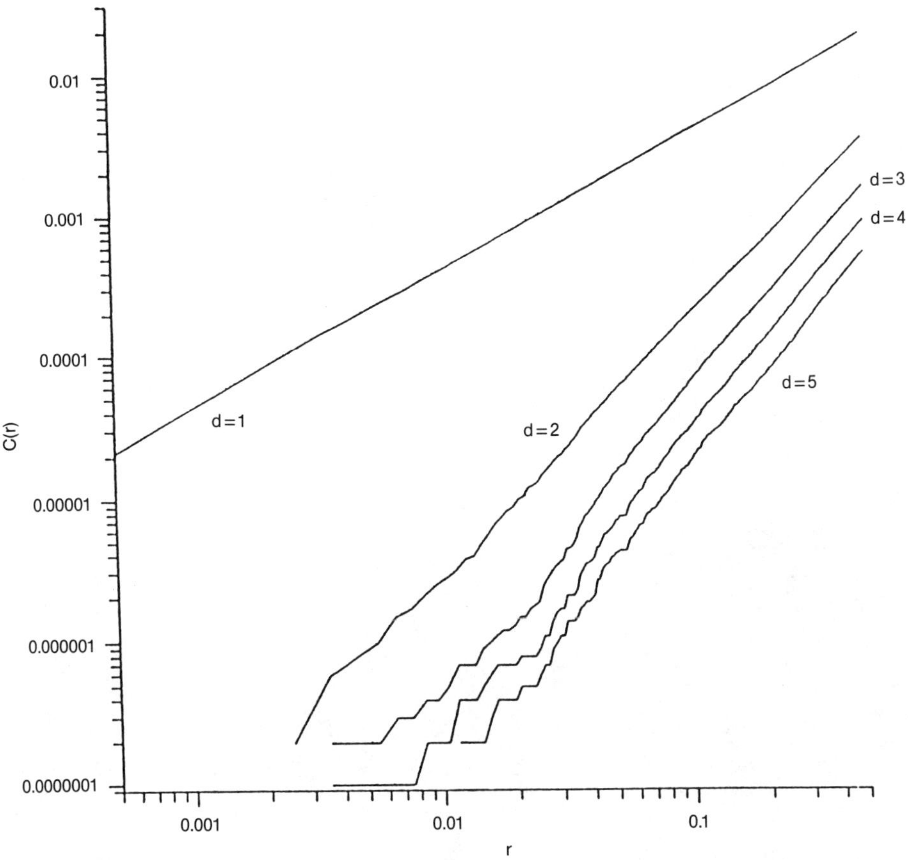

Fig. 8. $q=1$ correlation function for $x_2(n)$ Lorenz data. Log $C(r)$ is plotted versus log (r).

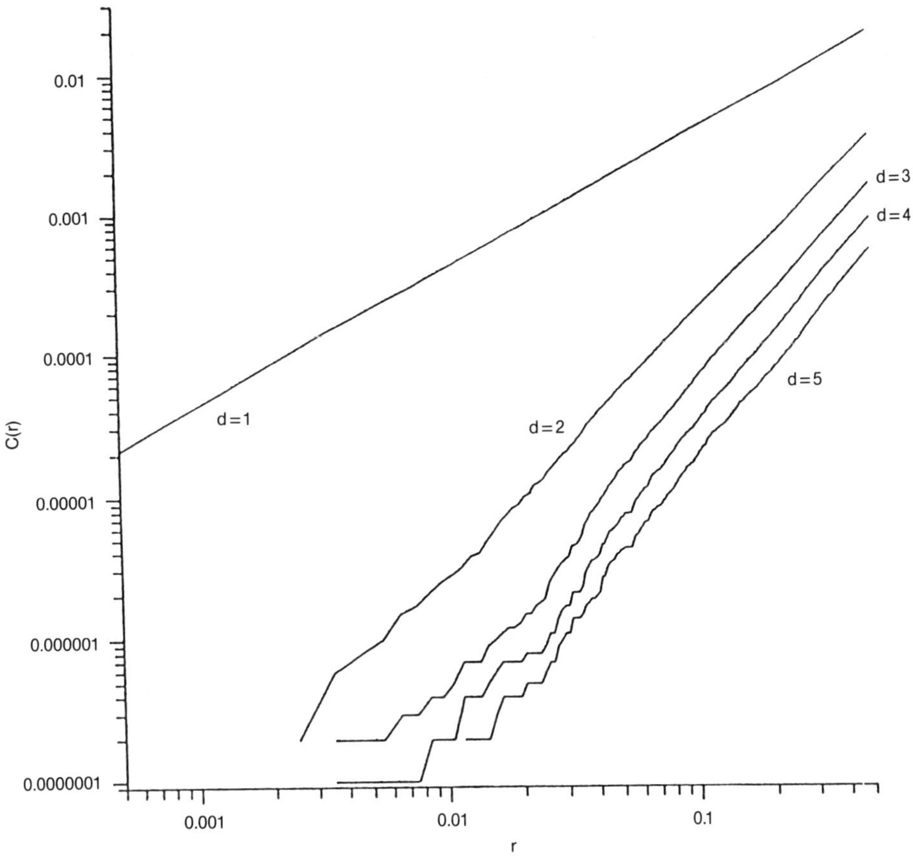

Fig. 9. $q=1$ correlation function for $x_3(n)$ Lorenz data. Log $C(r)$ is plotted versus log (r).

Our second data set comes from a map from the plane to itself devised by Hénon (1976):

$$\begin{aligned} x_1(n+1) &= 1.0 - ax_1(n)^2 + x_2(n) \\ x_2(n+1) &= bx_1(n), \end{aligned} \tag{16}$$

with conventional parameter values $a = 1.4$ and $b = 0.3$. The attractor for this map is shown in Fig. 10; it has $d_a = 1.26$. The function $\log C(r,1)$ is plotted versus $\log r$ for $x_1(n)$ and $x_2(n)$ data in Figs. 11 and 12. Again, the computation is easy, and here the determination that $d_E = 2$ is less ambiguous. One could safely choose 3, but there is no temptation to do so.

In my opinion, the conclusion should be that this method of computing a quantity such as $C(r,q)$ for various numbers of data points N, and in various dimensions d, is acceptable. Although we cannot conclude with assurance that the d_E we choose will be the minimum, it seems unlikely to be too small. Thus, we can confidently proceed, knowing that the embedding dimension in which we choose to work will do the job. Still, one would like an independent check, especially if the data happen to be noisy, in which case the other dimensions of d-space will be filled up rapidly by the noise.

Singular Value Spectrum

The singular value spectrum method is widely used in areas such as radar

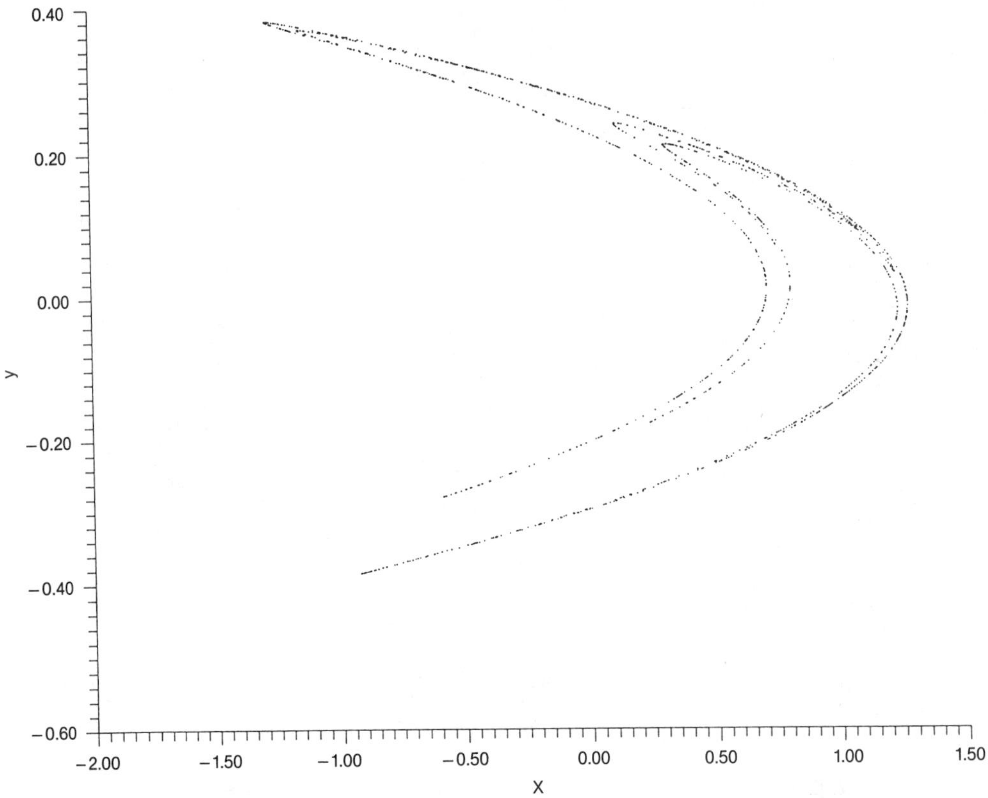

Fig. 10. Attractor for Hénon map. Fifteen hundred points on the orbit are plotted. The attractor dimension is $d_a = 1.26$.

engineering (Wax and Kailath, 1985; Orfanidis, 1988; Schmidt, 1979), where one has a noisy signal measured at d receivers and wishes to know how many independent degrees of freedom S (we would call this number d_E) are present in the signal. The S-dimensional signal subspace is found, and then various questions are asked of those signals—which one is Air France Flight 107, and which one is a Cessna carrying the local flight school, for example. The method measures d components of an overall signal vector, which we will continue to call $\mathbf{y}(l)$, at a sequence of times $l = 1, 2, \ldots, N$. It is presumed that the signal measured is composed of the "true" incoming signal $\mathbf{Y}$, which has S degrees of freedom in it (though it is a d-vector because of the measurement process) and is contaminated by some noise $n(l)$. It is necessary to make some assumption about this noise, so we will take it to be Gaussian, white, zero-mean, and additive, which gives it the properties $< n(l) >= 0$ and

$$< n_a(l) n_b(l') >= \sigma^2 \delta_{ab} \delta_{ll'}, \tag{17}$$

where $< \bullet >$ indicates an average over the realizations of the noise, and the indices run from 1 to d. If this is the case, then the sample covariance matrix

$$R_{ab} = \frac{1}{N} \sum_{l=1}^{N} y_a(l) y_b(l) \tag{18}$$

has the form

$$R_{ab} = \frac{1}{N} \sum_{l=1}^{N} Y_a(l) Y_b(l) + \sigma^2 \delta_{ab}, \tag{19}$$

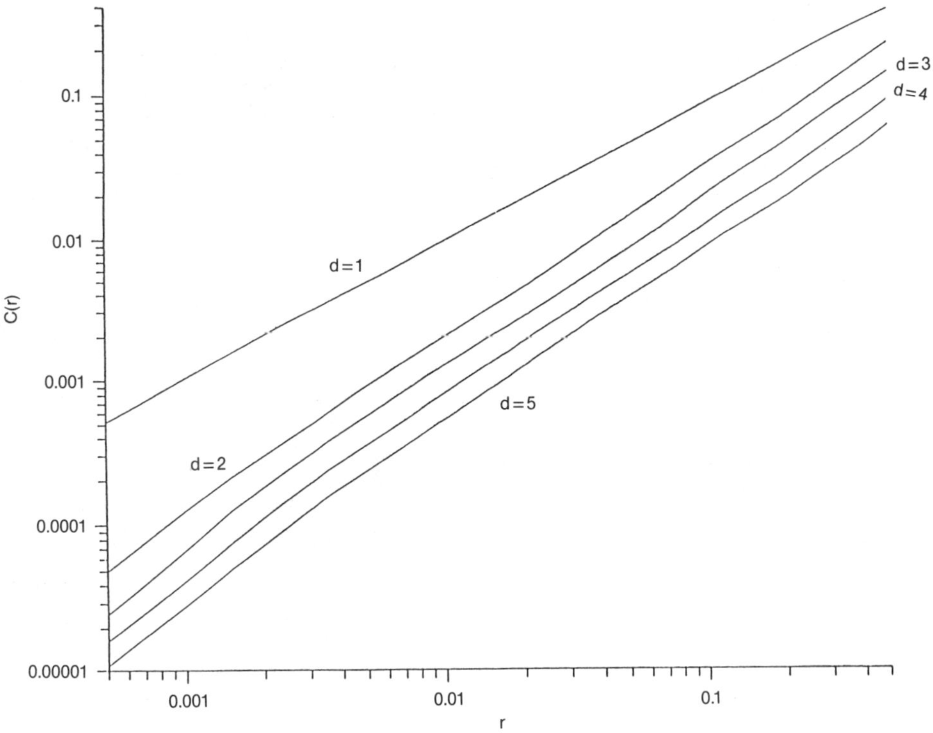

Fig. 11. $q = 1$ correlation function for $x_1(n)$ Hénon data. Log $C(r)$ is plotted versus log (r); 4,500 points of the orbit are used.

and the goal is to estimate the rank of the $\mathbf{Y}$ sample covariance matrix, which is supposed to be S.

The traditional way to do this is to compute $\mathbf{R}$ and look at its singular values. If the assumption about the additive noise is correct, then S singular values will sit above the noise floor at σ^2, and $d - S$ will be lost in the noise. Various sophisticated approaches to this separation of signal and noise are possible (Orfanidis, 1988; Lawley, 1956), but two examples show that this is still a somewhat subjective exercise.

The first example comes from Wax and Kailath (1985), who place two sinusoids ($S = 2$) with random phases in a $d = 7$ dimensional space with signal-to-noise ratio (defined as $10 \log[\frac{1}{2\sigma^2}]$) taken to be 10 dB. With $N = 100$, the seven singular values were 21.24, 2.17, 1.43, 1.10, 1.05, 0.94, and 0.73. It seems clear that the separation of the two "large" singular values from the five "small" ones is problematic.

Another example is taken from the work of Broomhead and Jones (1989). They generate data from a nonlinear, electrical oscillator and then look at the singular values of $\mathbf{R}$ computed from the time series of voltage lagged. The singular value spectrum for $d = 35$ from their work is reproduced in Fig. 13, along with an indication of the estimated noise arising from the analog-to-digital conversion of the signal. Alongside the singular value spectrum are the components of the first 12 singular vectors. The determination of d_E is again an unsure matter. Indeed, in another section of their paper, Broomhead and Jones indicate that the dimension d_a of the attractor in the oscillator data is order 2!

Information Theoretic Methods

The preceding discussion leads naturally to our last topic concerning the determination of the minimum or best dimension of embedding space in which to view strange attractors. The method here derives from work by statisticians (Akaike, 1974; Schwartz, 1978; Rissanen, 1978; Stone, 1979) on the question we have been addressing: for a data series in some space of dimension d, how does one determine the number of dynamical degrees of freedom in the data, and what portion of the data is noise contamination? Statisticians put the matter this way: if one observes N data vectors $\mathbf{y}(1), \mathbf{y}(2), \ldots, \mathbf{y}(N)$ and has a family of models that characterize features of these observations, how does one choose the model that best fits the data? The "best description" of the data is formulated in terms of a parametrized family of joint distribution functions of the $\mathbf{y}(l)$:

$$f(\mathbf{y}(1), \mathbf{y}(2), \ldots, \mathbf{y}(N) | \mathbf{a}). \tag{20}$$

Because the data vectors $\mathbf{y}(l)$ are presumed to be contaminated by measurement errors, this kind of probabilistic description is appropriate. The parameters $\mathbf{a}$ are chosen to minimize the *information* required to describe the data. In order to choose the minimum number of parameters required for describing the actual data, first make a maximum likelihood estimate of the parameters already in the model; call this $\hat{\mathbf{a}}$. Let S be an integer that indicates the number of parameters in $\mathbf{a}$—for our choice of model below, it will be precisely the dimension of the signal subspace—and then form the data description length (DDL) function:

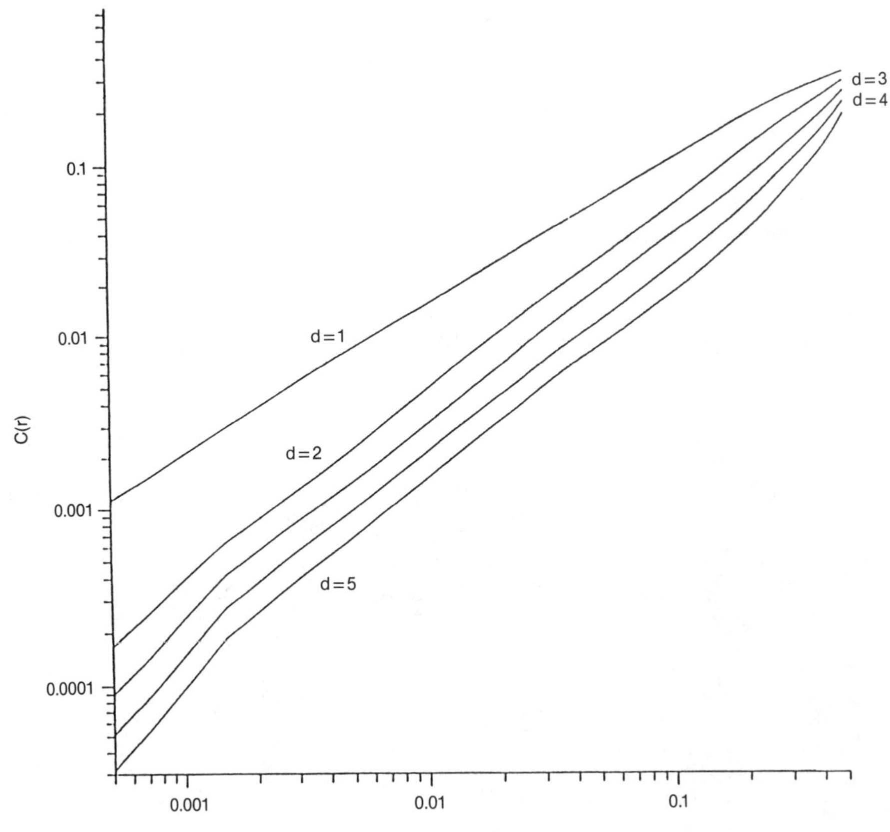

Fig. 12. $q=1$ correlation function for $x_2(n)$ Hénon data. Log $C(r)$ is plotted versus log (r); 4,500 points of the orbit are used.

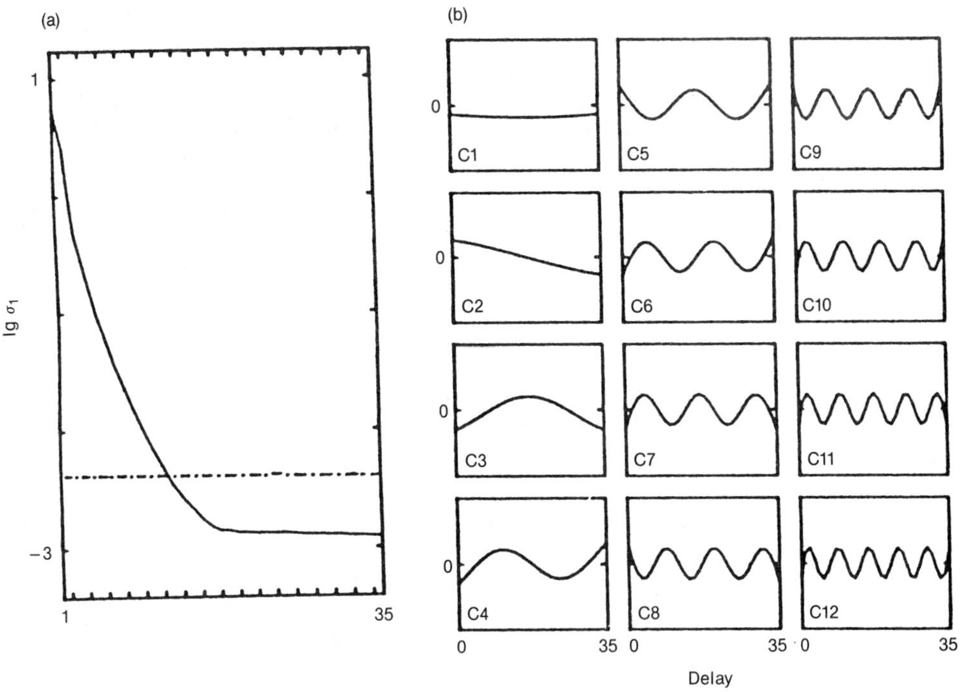

Fig. 13. Singular value spectrum for data taken from Broomhead and King (1986). $d = 35$ for the decomposition into singular values.

$$DDL(S) = -\log f(\mathbf{y}(1), \ldots, \mathbf{y}(N)|\hat{\mathbf{a}}) + \frac{\log N}{2}(\text{degrees of freedom}), \tag{21}$$

where "degrees of freedom" is the number of free parameters in the set $\mathbf{a}$.[3]

The idea is to (1) choose a model $f(\mathbf{y}(1), \ldots, \mathbf{y}(N)|\mathbf{a})$, which describes known features of the data, (2) find $\hat{\mathbf{a}}$ by the usual minimization of $f(\mathbf{y}(l)|\mathbf{a})$ with respect to the $\mathbf{a}$, and then (3) find that k which minimizes the DDL(S). The two terms in DDL(S) represent a balance between the "best fit parameters," as determined by maximum likelihood, and the information contained in those parameters. The choice of model is dictated by several considerations, the primary one being the information one has about the data set. Next in importance is the ease of manipulation of the model. As more information about the data is used, the model may become more complicated, and the likelihood of analytical work is lessened.

The first task then is to make a model of the distribution of the data vectors $\mathbf{y}$ in R^d. The simplest piece of interesting information is the sample covariance matrix $\mathbf{R}$. (The mean of the vectors $\mathbf{y}(l)$ is not very useful and can systematically be subtracted out without further discussion.) If this is the only information we have, then the best we can do in describing the distribution of the data is to identify the best ellipse in R^d, which is determined by the eigenvectors and eigenvalues of $\mathbf{R}$ that "fit" the data. A set of model distribution functions that are functions of quadratic forms in the $\mathbf{y}(l)$ are all we can utilize. The simplest function to work with is Gaussian. Recalling the form of the sample covariance matrix

[3]This is actually the criterion of Schwartz and Rissanen. Wax and Kailath (1985) discuss the Akaike criterion as well.

$$R_{ab} = \frac{1}{N}\sum_{l=1}^{N} Y_a(l)Y_b(l) + \sigma^2\delta_{ab},$$

when the signal is contaminated by additive, Gaussian, white noise, we choose models that are parametrized by a covariance matrix of this form; namely, models of the form

$$f(\mathbf{y}(1),\mathbf{y}(2),\ldots,\mathbf{y}(N)|\mathbf{a}) = \mathcal{N}\exp-[N\,\mathrm{trace}(\mathbf{R}\,\mathbf{R}(S)^{-1})], \tag{22}$$

where $\mathcal{N}$ is the normalization factor

$$\mathcal{N} = \frac{1}{[(2\pi)^d\det(\mathbf{R}(S))]^{N/2}}, \tag{23}$$

and $\mathbf{R}(S)$ is the $d \times d$ diagonalized sample covariance that has S degrees of freedom and is contaminated by noise:

$$\mathbf{R}(S) = \sum_{m=1}^{S}(\lambda_m - \sigma^2)|V_m><V_m| + \sigma^2 I, \tag{24}$$

with the λ_m and $|V_m>$ being the eigenvalues and eigenvectors of $\mathbf{R}(S)$. The parameters $\mathbf{a}$ for our model are then the λ_m, σ^2 and the independent components of the $|V_m>$. These eigenvectors are orthonormal, so the number of independent parameters in this kind of model is $S+1$ for the λ_m and σ^2 and $dS - S - \frac{S(S-1)}{2}$ for the components of the eigenvectors. In this case, the term "degrees of freedom" in the DDL(S) formula is $S(d - \frac{S-1}{2}) + 1$. Note that there is no noise whatsoever; the matrix $\mathbf{R}(S)$ is singular and has rank S. We expect the covariance matrix of the noise-free data $\mathbf{Y}$ to have this property, which is why we chose parameters $\mathbf{a}$ to go in our model of the data.

The maximum likelihood values of the parameters in $\mathbf{R}(S)$ are easily found from the Gaussian $f(\mathbf{y}(1),\ldots,\mathbf{y}(N)|\mathbf{a})$ above to be

$$\hat{\lambda}_m = \ell_m, \quad m = 1,2,\ldots,S, \tag{25}$$

$$\hat{\sigma}^2 = \frac{1}{d-S}\sum_{i=S+1}^{d}\ell_i, \tag{26}$$

and

$$|\hat{V}_m> = |C_m> \quad m = 1,2,\ldots,S, \tag{27}$$

where the ℓ_m and the $|C_m>$ are the eigenvalues and eigenvectors of the sample covariance matrix. The eigenvalues have been ordered, so $\ell_1\,\ell_2 > \ldots > \ell_d$. Combining these values with the determination of the number of free parameters in $\mathbf{R}(S)$ given above, we arrive at the formula for DDL(S):

$$DDL(S) = -\log[\prod_{i=S+1}^{d}\ell_i{}^N] + \log[\hat{\sigma}^{2(d-S)N}] + \frac{\log N}{2}(S(d - \frac{S-1}{2}) + 1). \tag{28}$$

The minimum embedding dimension d_E is the value at which DDL(S) is a minimum for fixed $d > S$ and fixed N. As Wax and Kailath (1985) show, when this formula is applied to their example cited above, the minimum of DDL(S) is at $S = 2$, as it should be.

When we apply Eq. 28 to data from a dynamical system, we must check that the singular value spectrum of $\mathbf{R}$ is stabilized by varying the sampling time with which we make the lags, or equivalently, in this case of computer-generated data, the time step in the solution of the differential equations. If we choose this

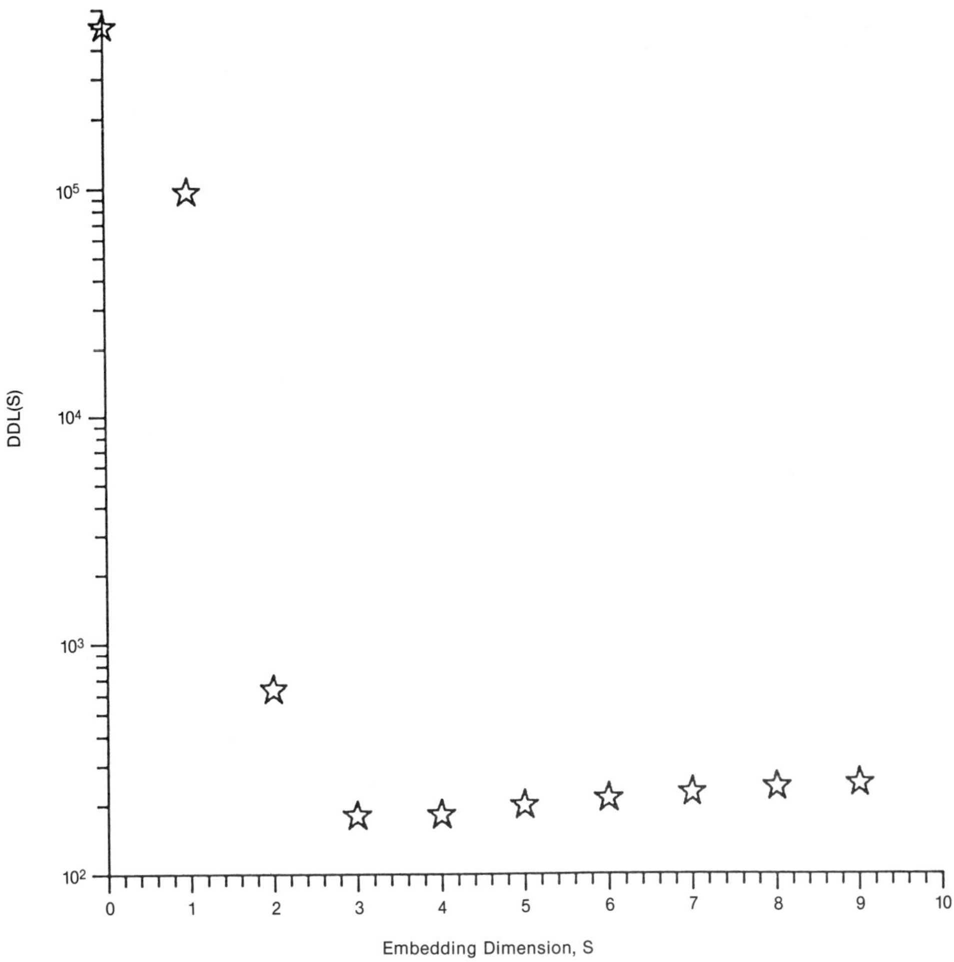

Fig. 14. Data description length function for data from the Lorenz model. The attractor dimension is $d_a = 2.06$, and the minimum of *DDL(S)* at $S = 3$ shows that the minimum embedding dimension to capture this attractor is 3.

time so that the ratio of singular values is unchanged for a wide range of sampling times about our choice, and if the number of samples is large enough (probably about 4,000, but we used 10,000), the DDL(S) function has a minimum at $S = 3$ for the original Lorenz model equation, as shown in Fig. 14. The eigenvalue spectrum of the sample covariance matrix in our case, $d = 10$, includes 0.993, $5.55x10^{-3}$, $2.34x10^{-4}$, $1.81x10^{-4}$, and six eigenvalues around $1.7x10^{-4}$. Choosing the d_E on the basis of the eigenvalues would be problematic at best. Indeed, one would almost certainly be led to choose $d_E = 2$ on the basis of the eigenvalues, and this certainly reflects the known dimension of the Lorenz attractor, $d_a \approx 2.06$. The DDL(S) minimum chooses that value we know to be correct from our earlier demonstration of the reconstruction of the attractor in $d = 3$ dimensions. It is worth noting that in constructing this DDL(S) function, it was important for numerical stability to *add noise* to the computer data in order to make the sample covariance matrix reliably invertible. Adding too much noise would swamp the answer we are looking for, but having too little noise is also treacherous. We added white, Gaussian, zero-mean noise at a level roughly 1% that of the signal.

For the more recent Lorenz model equation shown in Eq. (4), we find that we consistently have a minimum embedding dimension of 4 for δt, the integration

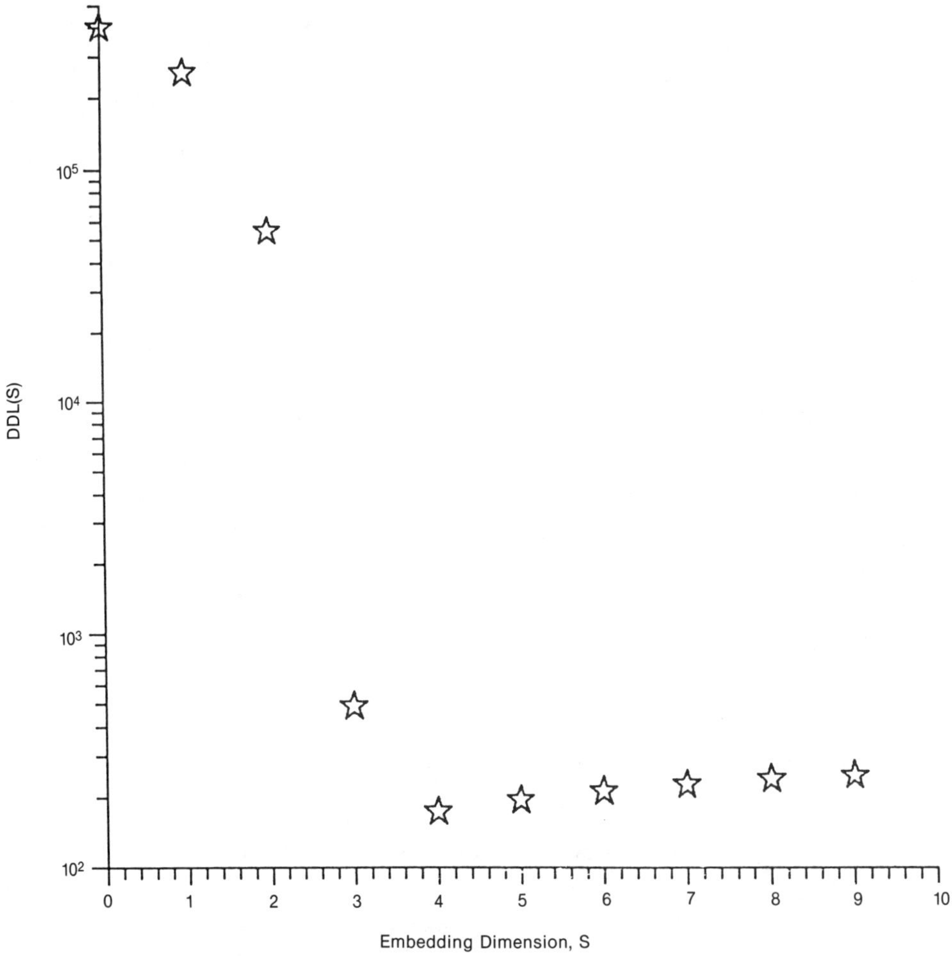

Fig. 15. Data description length function for data from the new Lorenz model. The attractor dimension is $d_a = 2.5$, and the minimum of *DDL(S)* at $S = 4$ shows that the minimum embedding dimension to capture this attractor is 4.

step and sampling time, because all lags τ are equal in the range $0.025 \leq \delta t \leq 0.045$. To be more precise, let us take the value $\delta t = 0.030$, which is very close to that used by Lorenz, in a space of $d = 10$ dimensions. The eigenvalues for the data were 0.86, 0.13, $4.83x10^{-3}$, $3.84x10^{-4}$, and a cluster of six more around $3.0x10^{-4}$. Should one, two, or perhaps four of these be termed large? The DDL(S) function is shown in Fig. 15 and has a clear minimum at $d_E = 4$. This number is less than the $2d_a + 1 = 6$ that is the maximum required by the Mañé-Takens results and is consistent with observations by Lorenz on this model (Lorenz, 1989c). Contaminating the data from the solution of the differential equation as above by Gaussian, white, zero-mean random noise produced no degradation of the method when the magnitude of this noise was up to $\approx 1\%$ of the signal, which is order unity.

This DDL method requires a few comments. First of all, we are not actually implying that the distribution of the $\mathbf{y}(l)$ in R^d is Gaussian. Clearly it is not! Instead, we are working within the framework of the data description length, where selection of a model is required and the minimum parametrization of that model is then sought. In our case, we parametrize the information we have (the sample covariance matrix) and ask what minimum set of parameters are needed in the model. We select the $\mathbf{R}(S)$ matrix, which has S true eigenvalues and $d-S$ noise-filled elements, to describe the data as represented by $\mathbf{R}$. The method, as we are using it here, poses the following question: How many of the singular values are needed to provide a minimum description of the data? Instead of answering by subjectively choosing the number d_E of "large" singular value, the method chooses, in an *objective* way, the number of singular values that gives a balance between the maximum likelihood value of the model distribution function and the amount of information required to describe the data.

The final word is certainly not in on this method, but if more information about the data is available, for example, the triple correlation function,

$$\frac{1}{N}\sum_{l=1}^{N} y_a(l)y_b(l)y_c(l),$$

or higher moments, then one can make a model distribution function that includes parametrized versions of these higher-order correlation functions, and can test that parametrization for its information content in the fashion we have outlined. The natural way to incorporate this additional information about the attractor is to make a cumulant representation of the characteristic function (Cramér, 1946) of the model $f(\mathbf{y}(1),\ldots,\mathbf{y}(N)|\mathbf{a})$ and use the parametrized cumulants within it. The minimization of the DDL(S) function for this model then incorporates more detailed features about the actual distribution of the data points in R^d. This and other ideas, such as including information in the model about the scaling moments $\phi(q)$, remain to be fully explored.

INVARIANTS, OPTIMUM MOMENTS, AND LYAPUNOV EXPONENTS

More on Invariants

In the previous section we discussed invariants of the attractor on which orbits move in either regular or chaotic ways. The invariants are very special quantities in the study of any dynamical system. The main use of these quantities, from a practical or applied point of view, is that of classifying dynamical systems. It is of some interest to be able to recognize a dynamical system from its time series. In the case of a chaotic system, such recognition is impossible from properties of the time domain view of the orbit itself. *That orbit is sensitive to initial conditions*, which means that a slight change in initial conditions will lead to a completely different time trace, due to the instabilities manifest in the dynamics. This sensitivity is present whether one views the orbit as a projection onto the measured one-dimensional time series $x(n)$, or has the opportunity to look at the orbit in an appropriate embedding space that captures the full attractor. What remains unchanged about the system—and can be used to recognize it on the next encounter—are the invariants.

To my knowledge, there are two kinds of invariants:

1. *Moments of the invariant distribution*, which have been discussed above. In this section we will discuss an *optimal* set of moments for classifying our knowledge about $\rho(\mathbf{y})$.

2. *Lyapunov exponents*, which are global quantities that express the way in which geometric constructs, such as line segments or small volumes, shrink and/or grow as they are transported about the attractor.

Because both of these quantities are independent of, and thus insensitive to, initial conditions, they are useful to consider. Let us begin with the Lyapunov exponents, since we have partially covered the invariants associated with the $\rho(\mathbf{y})$.

We will return to the optimal moments at the end of this section.

Lyapunov Exponents: Growth and Instability on an Attractor

The idea of Lyapunov exponents arises in the study of the *linear* instability of an orbit. We will focus our discussion on maps in R^d, because ultimately, these are what exists in a digital world. Suppose we have a map $\mathbf{F}(\mathbf{y})$ from $R^d \to Rd$; the orbits of this dynamical system are $\mathbf{y}(1), \mathbf{y}(2), \ldots$, and they emerge from an initial condition $\mathbf{y}(1)$ by iteration of the map

$$\mathbf{y}(n+1) = \mathbf{F}(\mathbf{y}(n)). \tag{29}$$

If we have found an orbit, either analytically or numerically, it is natural and actually imperative to inquire into the stability of this orbit under small perturbations in the initial condition: $\mathbf{y}(1) \to \mathbf{y}(1) + \Delta\mathbf{y}(1)$. This inquiry yields a new orbit $\mathbf{y}(n) + \Delta\mathbf{y}(n)$, which also is generated by the map $\mathbf{F}(\mathbf{y})$. If the deviation $\Delta\mathbf{y}$ remains small, it is governed by the linearization of the map as seen by

$$\begin{aligned} \mathbf{y}(n+1) + \Delta\mathbf{y}(n+1) &= \mathbf{F}(\mathbf{y}(n) + \Delta\mathbf{y}(n)), \\ &= \mathbf{M}(\mathbf{y}(n))\Delta\mathbf{y}(n) + O(\Delta\mathbf{y}(n)^2), \end{aligned} \tag{30}$$

so

$$\Delta y_l(n+1) = M(\mathbf{y}(n))_{kl}\Delta y_l(n), \tag{31}$$

where $\mathbf{M}(\mathbf{y})$ is the $d \times d$ matrix

$$M(\mathbf{y})_{kl} = \frac{\partial}{\partial y_l} F_k(\mathbf{y}), \tag{32}$$

and is evaluated along the orbit. The issue in the study of the linear stability of the orbit is whether the $\Delta\mathbf{y}$ evolving according to the linear map determined by $\mathbf{M}$ remains small or grows. The matter rests in the eigenvalues of $\mathbf{M}(\mathbf{y}(n))$.

It is useful to look at this issue when the orbit is a *fixed point* of the map $\mathbf{F}(\mathbf{y})$; namely, $\mathbf{y}(n+1) = \mathbf{y}(n) = \mathbf{y}_0$, independent of "time." In such a case, the linear map is

$$\Delta\mathbf{y}(n+1) = \mathbf{M}(\mathbf{y}_0)\Delta\mathbf{y}(n), \tag{33}$$

and has the solution

$$\Delta\mathbf{y}(n+1) = \mathbf{M}^n\Delta\mathbf{y}(1). \tag{34}$$

If the matrix $\mathbf{M}$ has eigenvalues $\lambda_1, \lambda_2, \ldots, \lambda_d$ ordered so that $\lambda_1 \geq \lambda_2 \geq \ldots \geq \lambda_d$, then for large n

$$\Delta\mathbf{y}(n+1) \approx \exp[n\lambda_1]|v_1 > c_1, \tag{35}$$

where $|v_1>$ is the eigenvector of $\mathbf{M}(\mathbf{y}_0)$ associated with λ_1, and c_1 is the projection of $\Delta\mathbf{y}(1)$ on this eigenvector. If $\lambda_1 < 0$, the orbit is linearly stable, because any perturbation $\Delta\mathbf{y}(1)$ decays back to the orbit itself. If $\lambda_1 > 0$, the orbit is linearly unstable. The eigenvalues λ_a are the Lyapunov exponents we are seeking, and we next want to generalize this definition to a time-dependent orbit. We accomplish this generalization in the natural way, by defining

$$\mathbf{M}^n(\mathbf{y}(1)) = \mathbf{M}(\mathbf{y}(n))\mathbf{M}(\mathbf{y}(n-1))\cdots\mathbf{M}(\mathbf{y}(1)), \tag{36}$$

which has information about each step of the orbit generated by $\mathbf{F}(\mathbf{y})$ from $\mathbf{y}(1)$. Only in 1968 was it proven by Oseledec (1968) (see also Eckmann and Ruelle, 1985) that the logarithm of the eigenvalues of the $d \times d$ matrix

$$\lim_{n\to\infty}[\mathbf{M}(\mathbf{y}(1))^{n\,T}\mathbf{M}(\mathbf{y}(1))^n]^{\frac{1}{2n}} \tag{37}$$

exists, and the eigenvalues are the Lyapunov exponents. Under the usual set of technical conditions, which always hold for physical systems and computations because of round-off errors or noise, this limit is independent of $\mathbf{y}(1)$. The Lyapunov exponents are thus interesting for chaotic systems as well.

If we know the map, calculating the Lyapunov exponents is difficult, but poses more of an algebraic challenge (Greene and Kim, 1987; Shimada and Nagashima, 1979; Benettin et al., 1979) than one requiring new inventions. The paper by Greene and Kim (1987) is especially clear and useful. If the system under study is a flow rather than a map, there is really only one difference in solving it, because digitizing its solution turns it into a map in any case. The difference is that for a flow—that is, for differential equations—there is always a Lyapunov exponent 0, which arises from making a displacement $\Delta\mathbf{y}$ exactly along the original orbit. The presence of the exponent 0 serves as a check for all programs that determine exponents.

The greater challenge, and one that is still unsolved in practice, is to extract from a time series the values of the Lyapunov exponents of the system generating the data. The problems connected with this extraction are twofold: (1) one must approximate with finite data the Jacobian matrix $\mathbf{M}(\mathbf{y})$, which involves derivatives in the phase space $\mathbf{y}$; and (2) after accurately approximating the Jacobian $\mathbf{M}$ along the orbit, one must extract the eigenvalues of an increasingly ill-conditioned matrix. This last statement means that the ratios of the terms in $\mathbf{M}$ grow without bound as $n \to \infty$ in the definition of the λ_a; even for finite but large n, the matrix is quite ill-conditioned. This problem can be treated, however, so let us focus on its first aspect.

There are many very good codes for determining properties of $\mathbf{M}(\mathbf{y}(l))$ along an orbit (Eckmann et al., 1986; Wolf et al., 1985; Sano and Sawada, 1985). Each code looks along an orbit. In the neighborhood in R^d of a point along the orbit $\mathbf{y}(l)$, the code seeks out neighbors that are sufficiently close to allow the derivatives needed in $\mathbf{M}$ to be constructed. Each method reported in the literature makes an approximation to the local tangent space of the attractor at $\mathbf{y}(l)$ by using properties of the neighbors of $\mathbf{y}(l)$. Each method is very good at producing the largest Lyapunov exponent λ_1, especially if it is positive, and each works with varying skill at determining the others (Abarbanel et al., 1989a).

The largest exponent, λ_1, is actually quite important, so the fact that it is easy to determine is quite helpful. A dynamical system is chaotic when small changes in an orbit amplify, yet the orbit evolves on a compact object. This evolution seems to be coupled with the appearance of at least one positive Lyapunov exponent, which dictates that nearby points on the attractor separate after time n as $\exp[\lambda_1 n]$ times their initial separation. Although complicated orbits do seem to appear when more than one exponent is zero, the situation is not generic, so chaos is often defined as having at least one positive λ_a.

For classifying the attractor, it would be useful to have as many of the λ_a as possible, so one needs methods for determining more than λ_1. All methods more or less do the same thing. Each follows an orbit $\mathbf{y}(l); l = 1, 2, \ldots, N$ and in R^d looks around at each time step for the neighbors $\mathbf{y}^\alpha(l)$ of $\mathbf{y}(l)$. A neighbor is a point that falls within some prescribed distance, which we will call σ. This means that $|\mathbf{y}(l) - \mathbf{y}^\alpha(l)|^2 \leq \sigma^2$ for each neighbor $\alpha = 1, 2, \ldots, N_b$, regardless of the actual time label on the neighbor. Call the difference vectors $\mathbf{y}(l) - \mathbf{y}^\alpha(l) = \mathbf{z}^\alpha(l)$. We are seeking the best local mapping from $\mathbf{z}^\alpha(l)$ to $\mathbf{z}^\alpha(l+1)$ in terms of the difference vectors, *as the distances* $\mathbf{z}^\alpha$ *go to zero.* This best fit is defined in terms of a *local* map at each step l:

$$\mathbf{z}^\alpha(l+1) = \mathbf{F}_l(\mathbf{z}^\alpha(l), \mathbf{a}), \tag{38}$$

with parameters $\mathbf{a}$ to be fit by requiring that the least squares difference

$$\sum_{\alpha=1}^{N_b} |\mathbf{z}^{\alpha}(l+1) - \mathbf{F}_l(\mathbf{z}^{\alpha}(l), \mathbf{a})|^2 \tag{39}$$

be minimized with respect to the $\mathbf{a}$. If we know the map $\mathbf{F}(\mathbf{z})$, then the Jacobian at the point $\mathbf{y}(l)$ on the orbit is the linear coefficient in the expansion of $\mathbf{F}_l(\mathbf{z})$ around $\mathbf{z} = 0$. A common fitting procedure is to take $\mathbf{F}(\mathbf{z})$ to be linear, so that the local map becomes

$$z_a^{\alpha}(l+1) = J_{ab} z_b^{\alpha}(l). \tag{40}$$

The least squares problem then is to determine the elements of the $d \times d$ matrix $\mathbf{J}$ at each l. These elements would then be the representative of $\mathbf{M}(\mathbf{y}(l))$.

As noted above, this procedure works rather well for the largest exponent λ_1, but extracting the other exponents in this fashion seems to be difficult. I do not know whether the problem resides in the often small number of neighbors and their finite distance from the orbit, which makes the finite differences entering the Jacobian derivatives uncertain, or if the problem is elsewhere.

Because the problem is extrapolation to $\mathbf{z} = 0$, it seems likely that enlarging the number of terms in the polynomial fit to the local map, or perhaps making the local map a rational $\mathbf{z}$ function might produce better results. Preliminary results by Bryant (1989) indicate that taking additional terms in the polynomial map,

$$z_a^{\alpha}(l+1) = J_{ab} z_b^{\alpha}(l) + K_{abc} z_b^{\alpha}(l) z_c^{\alpha}(l) + L_{abcd} z_b^{\alpha}(l) z_c^{\alpha}(l) z_d^{\alpha}(l) + \cdots, \tag{41}$$

results in excellent determinations of the λ_a in the correct embedding dimension. If the dimension d of the embedding space is too large, spurious exponents having no dynamical meaning will be associated with directions in R^d off the attractor. Bryant's techniques for dealing with this problem are reminiscent of the discussions of Broomhead and Jones (1989) on the identification of singular values of the sample covariance matrix that are not in the dynamical degrees of freedom of the attractor.

Before ending the commentary on Lyapunov exponents, it may be useful to remark why determining the largest exponent is so easy. It has to do with the severe ill-conditioning of the matrix $\mathbf{M}^n(\mathbf{y}(1))$. The largest term in the diagonalized matrix is $\exp[n\lambda_1]$, which dominates the entire matrix so completely that even a reasonably accurate approximation to the elements of the Jacobian all along the orbit will not disturb this value. The prevalence of this term is quantitatively apparent in the trace of the matrix,

$$\mathrm{trace}\mathbf{M}^n(\mathbf{y}(1)) = \sum_{a=1}^{d} \exp[n\lambda_a],$$

where it exponentially dominates the sum. To determine the next term in the trace, for example, one must determine the elements of $\mathbf{M}$ to the approximate accuracy $\exp[n(\lambda_2 - \lambda_1)]$ all along the orbit. The considerable difficulty of making such determinations clearly signals additional tricks in matching the known finite differences to the correct $\mathbf{z} \to 0$ extrapolation.

The "bottom line" here is that we can at present reliably determine only λ_1, and although there is some glimmer of being able to do better, the jury is still out. From the dynamical point of view, using the λ_a to classify the attractor is desirable, because the interpretation of these invariants is clear, and one can do quite a bit of good qualitative reasoning by knowing them. This may also be true of the scaling indices $\phi(q)$ mentioned in the previous section, if indeed one can determine them without ambiguity (Theiler, 1989). In any case, the need to ascertain a reliable, reasonably "user-friendly" black box algorithm for determining the Lyapunov exponents from a data set remains a challenge worth pursuing.

Optimal Moments: Recognizing the Attractor

In order to classify dynamical systems on the basis of information in the invariant density, we must select moments of that density which contain the "most" information. To do this we seek a set of "optimum" basis functions in R^d, which are concentrated on the attractor, so we can expand the density $\rho(\mathbf{y})$ in this set and characterize the attractor by the coefficients in this expansion. This is in many ways directly analogous to the identification of the invariant spectral frequencies for a linear system where the basis functions are sines and cosines.

The complete invariant density $\rho(\mathbf{y})$ has too much information in it for our purposes. Any finite sequence of N points producing

$$\rho_N(\mathbf{y}) = \frac{1}{N}\sum_{k=1}^{N}\delta^d\{\mathbf{y} - F^k[\mathbf{y}(1)]\} \tag{42}$$

has a finite resolution on the attractor. That resolution is approximately N^{-1/d_a} (it is exactly that if the attractor is uniform), which is the order of the mean distance of N points on a d_a-dimensional set. The finite number of points and the finite resolution on the attractor make our job easier than it might have been if the resolution had been very accurate down to all scales. We can be confident that an expansion in a good basis set will capture the major features of the attractor with only a few terms.

To work with this matter of finite resolution, we introduce a complete orthonormal set of functions $\psi_\mu(\mathbf{y})$ defined on $\mathbf{R^d}$, which can serve as a basis set. We then expand $\rho(\mathbf{y})$ in terms of this basis

$$\rho(\mathbf{y}) = \sum_{\mu=1}^{G} B_\mu \psi_\mu(\mathbf{y}). \tag{43}$$

Truncating this expansion at some finite order ($\mu = G$) provides a finite resolution representation corresponding to whatever information we have on $\rho(\mathbf{y})$. The coefficients B_μ will be the invariants of the dynamical process that characterize $\rho(\mathbf{y})$ within a given basis $\psi_\mu(\mathbf{y})$. After our discussion of how to select the $\psi_\mu(\mathbf{y})$, we will establish how one extracts B_μ, either from the data vectors $\mathbf{y}(n)$, or from a parametrized map $\mathbf{F}(\mathbf{y},\mathbf{a})$, which represents the evolution of points on the attractor. We will need this last item in the next section.

While any complete orthonormal set of functions $\psi_\mu(\mathbf{y})$ would do to determine our B_μ, some are more appealing than others. For example, Fourier series formed by taking

$$\psi_{\mathbf{m}}(\mathbf{y}) = e^{i\mathbf{m}\cdot\mathbf{y}} \quad \mathbf{m} = (m_1, m_2, \ldots m_\alpha)$$

are formally fine. However, because the attractor occupies only a small portion of R^d, most of the work performed by the Fourier representation of $\rho(\mathbf{y})$ is expended in making $\rho(\mathbf{y})$ vanish off the attractor. What we seek are orthonormal functions that are *concentrated on the attractor*, so all the work in the expansion of $\rho(\mathbf{y})$ is expended exhibiting structure where the attractor is located. This would also result in the need for many fewer B_μ than are required for Fourier series or other familiar choices for $\psi_\mu(\mathbf{y})$.

An *optimal* choice, using information in the data set, is constructed in the following manner (Rosenfeld and Kak, 1981; Root, 1987). Take the total data set $\mathbf{y}(n)$; $n = 1, 2, \ldots$, and divide it into two portions. We treat the first portion (of length N) as the data we are trying to model. The second portion of the data set (of length N') is used to construct orthonormal functions. The $\psi_\mu(\mathbf{y})$ orthonormal functions will be used in our expansion of $\rho(\mathbf{y})$, shown in Eq. (43). To construct these functions explicitly, we further divide the second portion of the data into G groups of length L ($N' = LG$). Each group is a sample of the invariant attractor. If L is large enough, each sample is a significant look at $\rho(\mathbf{y})$. Treat each of the G data sets as an independent sample of $\rho(\mathbf{y})$ and form the invariant distribution for the αth sample

$$\rho_\alpha(\mathbf{y}) = \frac{1}{L}\sum_{k=1}^{L} \delta^d[\mathbf{y} - \mathbf{y}(k,\alpha)],$$

with $\alpha = 1, 2, \ldots, G$. The data point $\mathbf{y}(k,\alpha)$ is the kth member of the αth sample. Of course, the mean density of the G samples is just the total invariant density of the a data set of length N':

$$\rho(\mathbf{y}) = \frac{1}{G}\sum_{\alpha=1}^{G} \rho_\alpha(\mathbf{y}). \tag{44}$$

From the G samples, $\rho_\alpha(\mathbf{y})$, we form the following *phase-space correlation function*

$$R(\mathbf{z},\mathbf{w}) = \frac{1}{G}\sum_{\alpha=1}^{G} \rho_\alpha(\mathbf{z})\rho_\alpha(\mathbf{w}). \tag{45}$$

It can be shown (Fukunaga, 1972; Rosenfeld and Kak, 1981) that the normalized eigenfunctions of this correlation function are the optimal eigenfunctions for expansion of functions localized on the attractor. By optimal, we mean that when these eigenfunctions are expressed as a finite series in an eigenbasis, they provide the best representation, in a least squares sense, of the information in $\rho(\mathbf{y})$. The label α is to be treated as a sampling index from a set of independent looks at the data, each of which is to be thought of as a selection from a uniform statistical distribution of invariant densities. The various averages over α then appear quite natural.

The requirement that $\psi_\mu(\mathbf{y})$ be an eigenfunction of $R(\mathbf{z},\mathbf{w})$ leads to

$$\int d^d z\, R(\mathbf{w},\mathbf{z})\psi_\mu(\mathbf{z}) = \mu\psi_\mu(\mathbf{w}). \tag{46}$$

The $\psi_\mu(\mathbf{y})$ are normalized as follows:

$$\int d^d w\, \psi_\mu(\mathbf{w})\psi_{\mu'}(\mathbf{w}) = \delta_{\mu\mu'}. \tag{47}$$

As the number of samples G becomes infinite, the set of eigenfunctions becomes complete in the usual least squares sense. If we insert Eq. (45) into Eq. (46), we see that for finite G, $R(\mathbf{w},\mathbf{z})$ becomes a finite sum of separable kernels. It is easily seen that in this case the eigenfunctions $\psi_\mu(\mathbf{y})$ must have the form

$$\psi_\mu(\mathbf{y}) = \sum_{\alpha=1}^{G} C_\alpha^\mu \rho_\alpha(\mathbf{y}). \tag{48}$$

The eigenfunctions defined in this fashion are localized near the attractor, just as we wished. This follows directly from Eq. (48), because $\psi_\mu(\mathbf{y})$ is made of the $\rho_\alpha(\mathbf{y})$'s, which vanish off the attractor.

Inserting Eqs. (45) and (48) back into Eq. (46) reduces the eigenvalue equation to a finite matrix problem. The coefficients C_α^μ are the G-vectors, which are eigenvectors of the $G \times G$ matrix:

$$A_{\alpha\beta} = \frac{1}{G}\int d^d z\, \rho_\alpha(\mathbf{z})\rho_\beta(\mathbf{z}), \tag{49}$$

i.e.,

$$\sum_{\beta=1}^{G} A_{\alpha\beta} C_{\beta}^{\mu} = \mu C_{\alpha}^{\mu}. \tag{50}$$

We now turn to the normalization condition, Eq. (47). Insertion of the representation for $\psi_\mu(\mathbf{y})$, given by Eq. (48) into Eq. (47), and use of the relationship between the C_α^μ's and $A_{\alpha\beta}$, given by Eqs. (49) and (50), dictates that the vectors C_α^μ obey the following normalization condition:

$$\sum_{\alpha=1}^{G} C_{\alpha}^{\mu} C_{\alpha}^{\mu'} = (\mu G)^{-1} \delta_{\mu\mu'}. \tag{51}$$

(Incidentally, this equation also shows that all the eigenvalues μ are positive.)

Formally, the elements of $\rho_\alpha(\mathbf{y})$ are delta functions. Hence, numerically speaking, computation with them is really not possible. We choose to replace $\delta^d(\mathbf{x})$ by

$$\delta^d(\mathbf{x}) \rightarrow \frac{1}{(\pi\varpi)^{d/2}} e^{-|\mathbf{x}|^2/\varpi} \equiv f_\varpi(|\mathbf{x}|),$$

which, when ϖ is small, represents only a small loss of resolution in calculating $\rho_\alpha(\mathbf{y})$. f_ϖ also has the same integral as the delta function it replaces. To this approximation,

$$\rho_\alpha(\mathbf{y}) = \frac{1}{L} \sum_{k=1}^{L} f_\varpi(|\mathbf{y} - \mathbf{y}(k, \alpha)|), \tag{52}$$

and Eq. (49) becomes

$$A_{\alpha\beta} = \left(\frac{1}{\pi\varpi}\right)^{d/2} \frac{1}{GL^2} \sum_{k,j=1}^{L} \exp\left[-|\mathbf{y}(k,\alpha) - \mathbf{y}(j,\beta)|^2/\varpi\right]. \tag{53}$$

We are now in a position to calculate our optimal eigenfunctions $\psi_\mu(\mathbf{y})$ from the G data sets. Use Eq. (53) to numerically calculate the $G \times G$ matrix $A_{\alpha\beta}$. Next calculate the eigenvalues μ and eigenvectors C_α^μ of this matrix, being sure to normalize them according to Eq. (51). We can then form the eigenfunctions $\psi_\mu(\mathbf{y})$ by using the normalized C_α^μ's and the $\rho_\alpha(\mathbf{y})$'s (in the form of Eq. (52)) in Eq. (48).

In Figs. 16-21, we show $\rho_1(\mathbf{y})$–$\rho_5(\mathbf{y})$ and $\psi_1(\mathbf{y})$ evaluated for the Hénon attractor from a data set of $G = 5$ samples that are each $L = 750$ steps in length. These data are displayed on a grid of 75 points in each coordinate direction. As we shall see in a moment, we do not need the actual functional form of $\rho_\alpha(\mathbf{y})$ and $\psi_\mu(\mathbf{y})$. The projection of $\rho(\mathbf{y})$ on $\psi_\mu(\mathbf{y})$ (the B_μ's) can be evaluated with only $A_{\alpha\beta}$, the C_α^μ's, and the μ's, so the time needed to extract the information necessary for constraining the fit will be far less in practice than it might appear.

We now have a set of G orthonormal functions $\psi_\mu(\mathbf{y})$, extracted from G samples $\rho_\alpha(\mathbf{y})$ of the invariant distribution. We can use the orthonormality condition in Eq. (47) to project a particular B_μ out of Eq. (43):

$$B_\mu = \int d^d y \, \rho(\mathbf{y}) \psi_\mu(\mathbf{y}).$$

This shows that the B_μ are invariants of the mapping, because they are integrals of ψ_μ and have the density $\rho(\mathbf{y})$. If we insert Eqs. (44) and (48) into this expression, we get

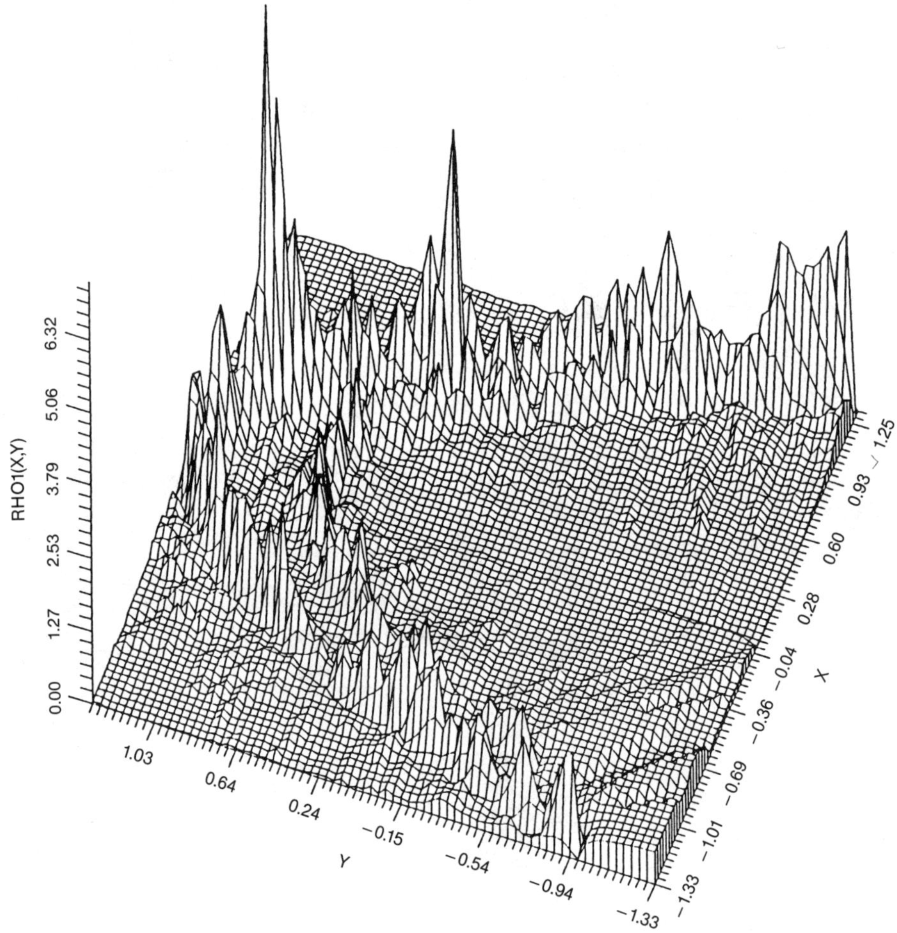

Fig. 16. Invariant density function for Hénon attractor that comes from the first 750 points of the 3,750-point data series. The grid size is 75×75 for the plot.

$$B_\mu = \sum_{\alpha,\beta=1}^{G} A_{\alpha\beta} C^\mu_\beta = \mu \sum_{\beta=1}^{G} C^\mu_\beta, \tag{54}$$

where we have used the eigenvector relationship between the C^μ_α's and $A_{\alpha\beta}$ expressed in Eq. (50). Equation (54) has been used to calculate the B_μ's numerically from the data. As the reader can see, only the eigenvalues and eigenvectors of the matrix $A_{\alpha\beta}$ were needed to determine the B_μ's completely. As we stated above, the $A_{\alpha\beta}$ matrix is numerically calculated through Eq. (53). An independent check of our calculation of the B_μ's can be found by inserting Eq. (48) into Eq. (43). The results are

$$\rho(\mathbf{y}) = \sum_{\mu=1}^{G} B_\mu \sum_{\alpha=1}^{G} C^\mu_\alpha \rho_\alpha(\mathbf{y}).$$

By definition, $\int d^d y\, \rho(\mathbf{y}) = \int d^d y\, \rho_\alpha(\mathbf{y}) = 1$. Using this fact, we can integrate the above expression for $\rho(\mathbf{y})$ to get

$$1 = \sum_{\mu=1}^{G} \frac{B_\mu{}^2}{\mu},$$

where we have used Eq. (54) to simplify our expression. We have used this equation to check all of our numerical calculations of B_μ.

The method for evaluating the B_μ from the $N' = GL$ data vectors $\mathbf{y}(k,\alpha)$ in R^d should now be clear. The B_μ are the G numbers characterizing the invariant density $\rho(\mathbf{y})$ by its projection on the optimum basis vectors $\psi_\mu(\mathbf{y})$. Now we wish to see how to evaluate them from a parametrized mapping $\mathbf{F}(\mathbf{y},\mathbf{a})$.

To determine B_μ from the map $\mathbf{F}(\mathbf{y})$—suppressing the parameters $\mathbf{a}$ for a moment—we return to the definition of the invariant density as expressed by Eq. (42). Call A_k the projection on $\psi_\mu(\mathbf{y})$ of each term in the sum in the following equation:

$$\begin{aligned} A_k(\mu) &= \int d^d y\, \psi_\mu(\mathbf{y}) \delta^d[\mathbf{y} - \mathbf{F}^k(\mathbf{y}(1))] \\ &= \psi_\mu[\mathbf{F}^k(\mathbf{y}(1))]. \end{aligned} \tag{55}$$

We interpret Eq. (55) as saying that $A_k(\mu)$ is the projection of $\delta^d\, [\mathbf{y} - \mathbf{F}^k(\mathbf{y}(1))]$ onto the orthonormal eigenfunctions $\psi_\mu(\mathbf{y})$. Using this interpretation, we expand the δ-function in terms of $\psi_\mu(\mathbf{y})$ to get

$$\delta^d[\mathbf{y} - \mathbf{F}^k(\mathbf{y}(1))] = \sum_{\mu=1}^{G} A_k(\mu) \psi_\mu(\mathbf{y}).$$

Equation (42) can now be written as

$$\rho(\mathbf{y}) = \sum_{\mu=1}^{G} \left[\frac{1}{N} \sum_{k=1}^{N} A_k(\mu) \right] \psi_\mu(\mathbf{y}).$$

Comparing the preceding equation to Eq. (43) indicates that

$$\begin{aligned} B_\mu &= \frac{1}{N} \sum_{k=1}^{N} A_k(\mu) \\ &= \frac{1}{N} \sum_{k=1}^{N} \psi_\mu[\mathbf{F}^k(\mathbf{y}(1))]. \end{aligned}$$

From Eq. (48) and the definition of $\rho_\alpha(\mathbf{y})$, we rewrite this equation as

$$B_\mu = \frac{1}{N} \sum_{k=1}^{N} \left[\sum_{j=1}^{L} \sum_{\alpha=1}^{G} \frac{C_\alpha^\mu}{L} \delta^d[\mathbf{y}(j,\alpha) - \mathbf{F}^k(\mathbf{y}(1))] \right].$$

This expression requires quite high powers of $\mathbf{F}(\mathbf{y},\mathbf{a})$ to be evaluated, and we cannot be confident that such high powers of the parametrized map are numerically accurate. When the map is near the correct, or optimum, map, then we are accurately reproducing $\mathbf{y}(k+1)$ from $\mathbf{y}(k)$ by a single application of

$\mathbf{F}(\mathbf{y},\mathbf{a})$. We utilize this to replace $\mathbf{F}^k(\mathbf{y}(1))$ in the last formula with $\mathbf{F}(\mathbf{y}(k))$. The expression for the B_μ's then becomes

$$\begin{aligned} B_\mu &= \frac{1}{N}\sum_{k=1}^{N}\left[\sum_{\alpha=1}^{G}\sum_{j=1}^{L}\frac{C_\alpha^\mu}{L}\delta^d[\mathbf{y}(j,\alpha)-\mathbf{F}(\mathbf{y}(k))]\right] \\ &= \frac{1}{N}\sum_{k=1}^{N}\left[\sum_{\alpha=1}^{G}\sum_{j=1}^{L}\frac{C_\alpha^\mu}{L(\pi\varpi)^{d/2}}\exp[-|\mathbf{y}(j,\alpha)-\mathbf{F}(\mathbf{y}(k))|^2/\varpi]\right], \end{aligned} \tag{56}$$

using our smoothed replacement for the delta functions in the invariant density.

Equation (56) is the one we would use to calculate the B_μ's from a map. To implement it, we take all N points in the first portion of the data set and iterate them once through the map $\mathbf{F}(\mathbf{y},\mathbf{a})$. When $\mathbf{F}(\mathbf{y},\mathbf{a})$ is near the correct map, iterating the data set once will result in points that are still on the attractor. We then evaluate the Gaussian and numerically sum all the contributions between the iterates of the data set and the points in the G samples.

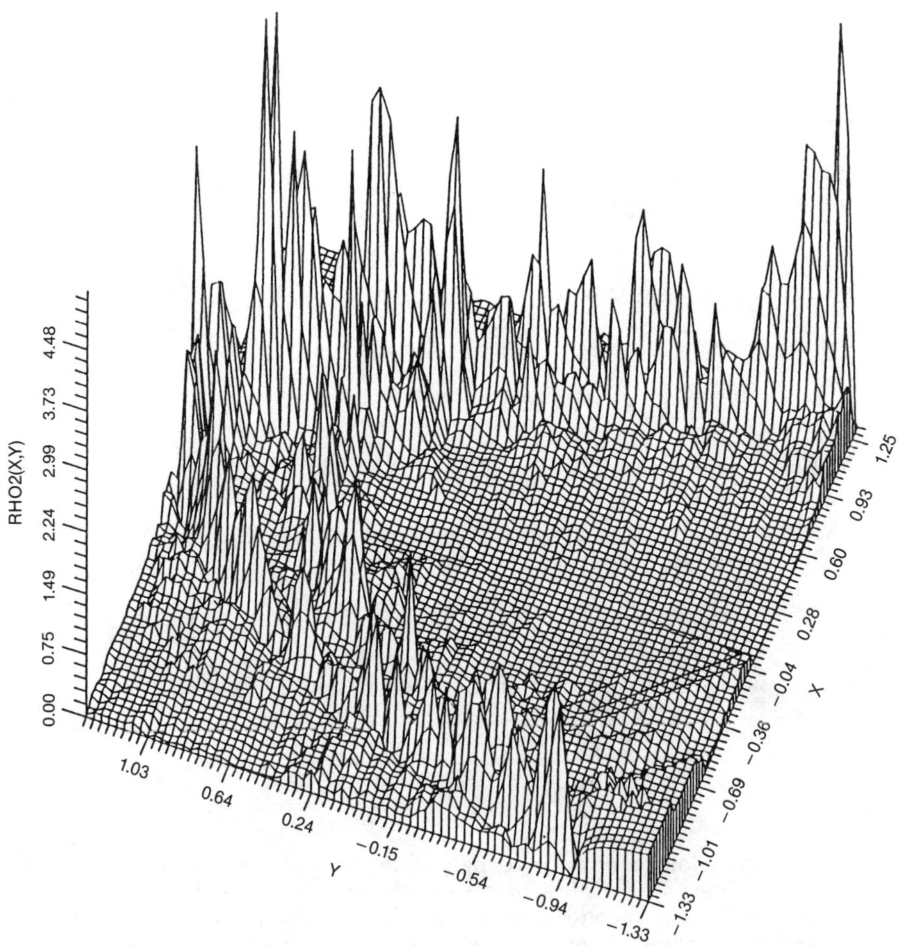

Fig. 17. Invariant density function for Hénon attractor that comes from the second 750 points of the 3,750-point data series. The grid size is 75×75 for the plot.

We close this section with an observation about the B_μ's. By combining the definition in Eq. (55) with the identity

$$\delta^d[\mathbf{y} - \mathbf{F}^{k+1}(\mathbf{y}(1))] = \int d^d w\, \delta^d[\mathbf{y} - \mathbf{F}(\mathbf{w})]\delta^d[\mathbf{w} - \mathbf{F}^k(\mathbf{y}(1))],$$

we can derive the recursion relation

$$A_{k+1}(\mu) = \sum_{\mu'} T_{\mu\mu'} A_k(\mu'),$$

in which the *transition matrix* T is given by

$$T_{\mu\mu'} = \int d^d y\, \psi_\mu(\mathbf{y})\psi_{\mu'}(\mathbf{F}(\mathbf{y})). \tag{57}$$

In a matrix notation, the recursion relation $A_{k+1} = TA_k$ leads to an expression for

$$\int d^d y\, \rho_L(\mathbf{y})\psi_\mu(\mathbf{y}) = \rho_L(\mu) = \frac{1}{L}\sum_{k=1}^{L} A_k(\mu),$$

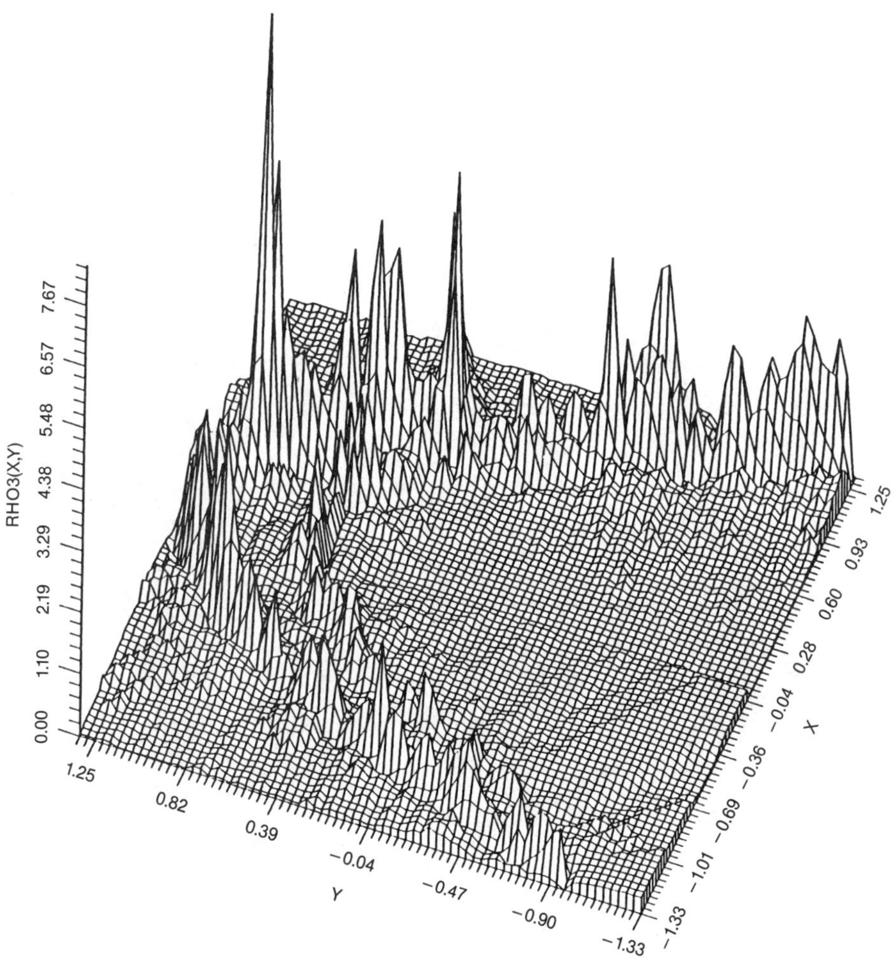

Fig. 18. Invariant density function for Hénon attractor that comes from the third 750 points of the 3,750-point data series. The grid size is 75×75 for the plot.

which is

$$(1-T)\rho_L(\mu) = \frac{1}{L}(1-T^L)A_1.$$

Because $\lim_{L\to\infty} \rho_L(\mu) = \int d^dy\, \rho(\mathbf{y})\psi_\mu(\mathbf{y}) = B_\mu$, this shows that the B_μ's are the components of the eigenvector of T, with eigenvalue unity, and further, that all other eigenvalues must lie within the unit circle if the expression for $\rho(\mathbf{y})$ converges. By our assumption that the $\rho(\mathbf{y})$ we observe is unique, we infer that the eigenvalue unity of T is nondegenerate. We tried to implement this observation about the B_μ's to yield a method for numerically determining them from Eq. (57) (Grossmann and Thomae, 1977; Abarbanel and Latham, 1982) but found that round-off error undermined our efforts.

We now have two kinds of invariants with which to establish that the attractor we are looking at is one we have or have not seen before. That is, we can classify the dynamical system so we can recognize it again. The invariants are of two kinds: (1) the Lyapunov exponents, and (2) moments of the invariant distribution $\rho(\mathbf{y})$. Determining the Lyapunov exponents is still a research issue, though I believe it is one that can be cleared up and automated. The moments of the invariant distribution on the attractor are easy to compute, and the discussion

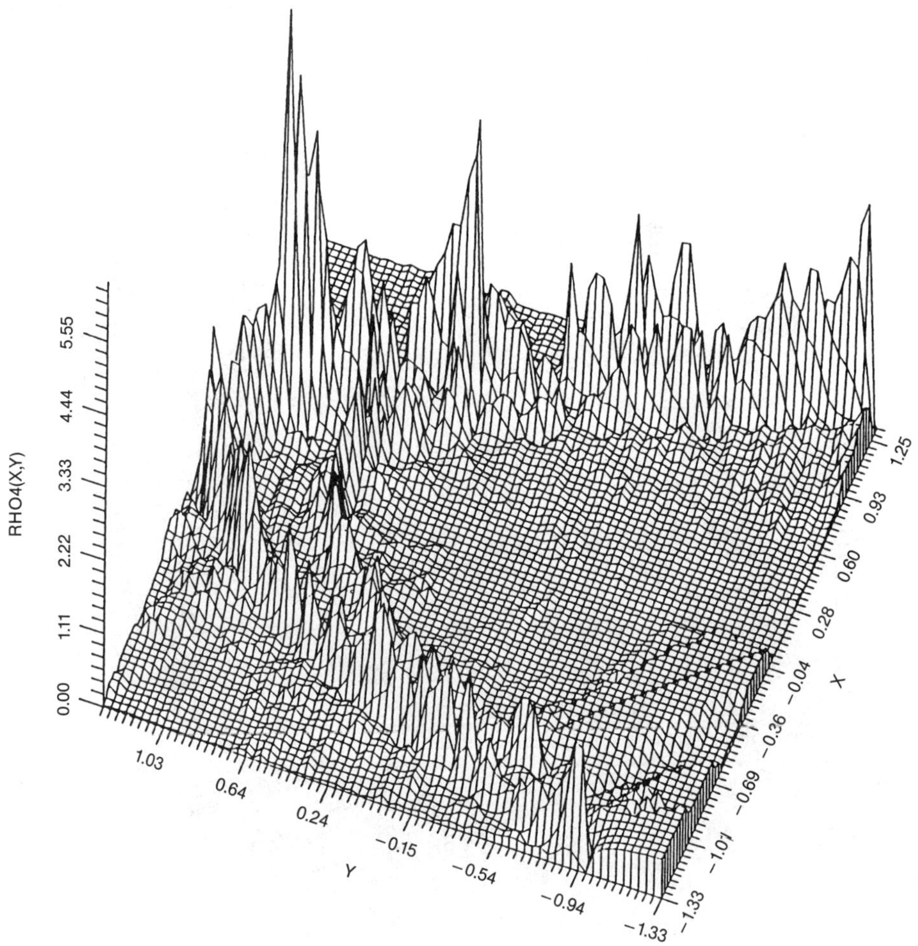

Fig. 19. Invariant density function for Hénon attractor that comes from the fourth 750 points of the 3,750-point data series. The grid size is 75×75 for the plot.

about which set of moments is the very best for classifying the attractor is hardly complete. The two sets that are strong candidates are the scaling moments $\phi(q)$, if they exist in a sensible way, and the B_μ coefficients we have just discussed. In the next section, we will use these coefficients for constraining predictive models based on properties of the attractor. What we are about to do could be done with the $\phi(q)$, but I confess to having no clear idea whether the $\phi(q)$ would work as well.

GETTING ON: PREDICTION ON STRANGE ATTRACTORS

In this section we turn from analyzing data from experiments or computer simulations for their information about the dynamical system to using the information to predict the system's future behavior. Prediction would be a matter of solving the differential equation or mapping function that generated the data, if these dynamics were known. However, this may not be a useful way to proceed. The following example illustrates the framework for this section.

The heating of the lower atmosphere may be idealized as absorption of solar radiation from below by a convection cell whose height is much less than its horizontal dimension. This Rayleigh-Bénard convection cell has been studied

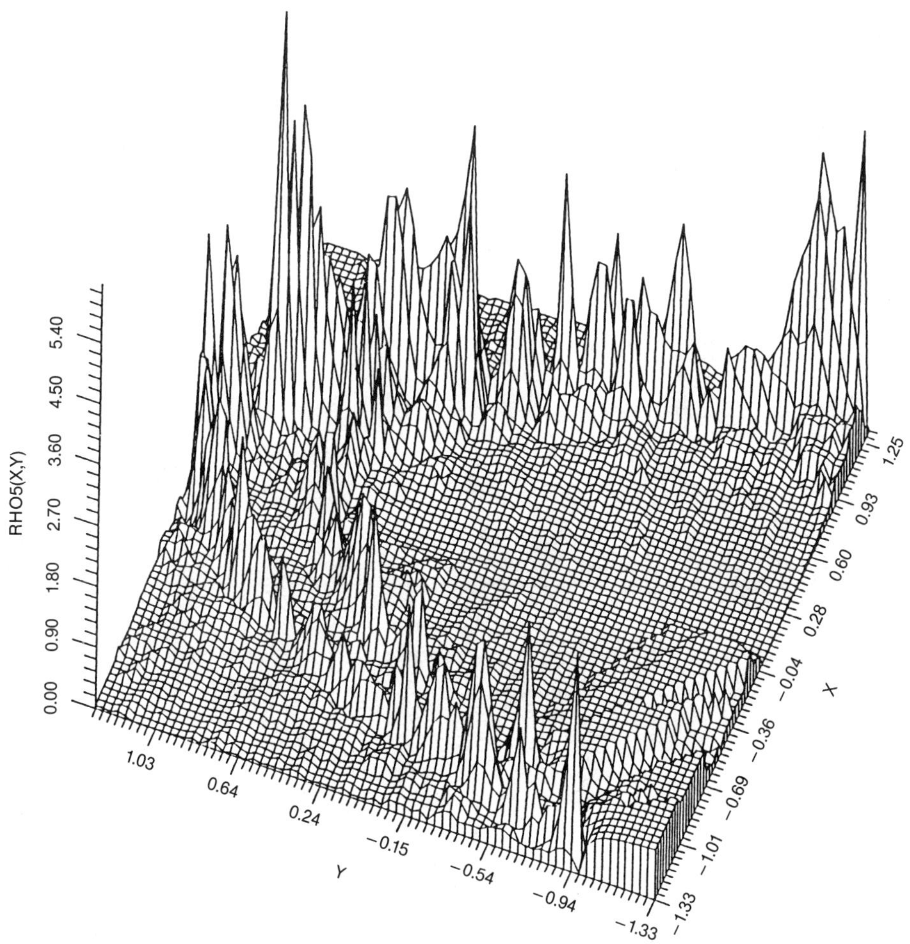

Fig. 20. Invariant density function for Hénon attractor that comes from the fifth 750 points of the 3,750-point data series. The grid size is 75×75 for the plot.

intensively in the laboratory. If the Rayleigh number, which is a dimensionless measure of the ratio of heat input to heat loss by viscosity and thermal diffusion, is not too much larger than the onset of the first convective instability away from pure conduction, then the motion of the fluid has an attractor of dimension d_a in the range of three or four. The equations of motion for the fluid are: the momentum equation

$$\rho[\frac{\partial \mathbf{v}(\mathbf{x},t)}{\partial t} + \mathbf{v}(\mathbf{x},t) \cdot \nabla \mathbf{v}(\mathbf{x},t)] = -\nabla p(\mathbf{x},t) - \rho g \hat{z} + \nu \nabla^2 \mathbf{v}(\mathbf{x},t);$$

the condition of incompressibility

$$\nabla \cdot \mathbf{v}(\mathbf{x},t) = 0;$$

and the energy equation for the temperature $T(\mathbf{x},t)$,

$$[\frac{\partial}{\partial t} + \mathbf{v}(x,t) \cdot \nabla]T(\mathbf{x},t) = \kappa \nabla^2 T(\mathbf{x},t),$$

supplemented with the equation of state $\rho = \rho_0[1 - \alpha T]$. These are partial differential equations for the fields $\mathbf{v}(\mathbf{x},t)$ and $T(\mathbf{x},t)$ and are therefore infinite-dimensional. Yet, the attractor is finite-dimensional and has a small dimension. A reasonable and common interpretation is that the infinite-dimensional system specified by the initial conditions $\mathbf{v}(\mathbf{x},0)$ and $T(\mathbf{x},0)$

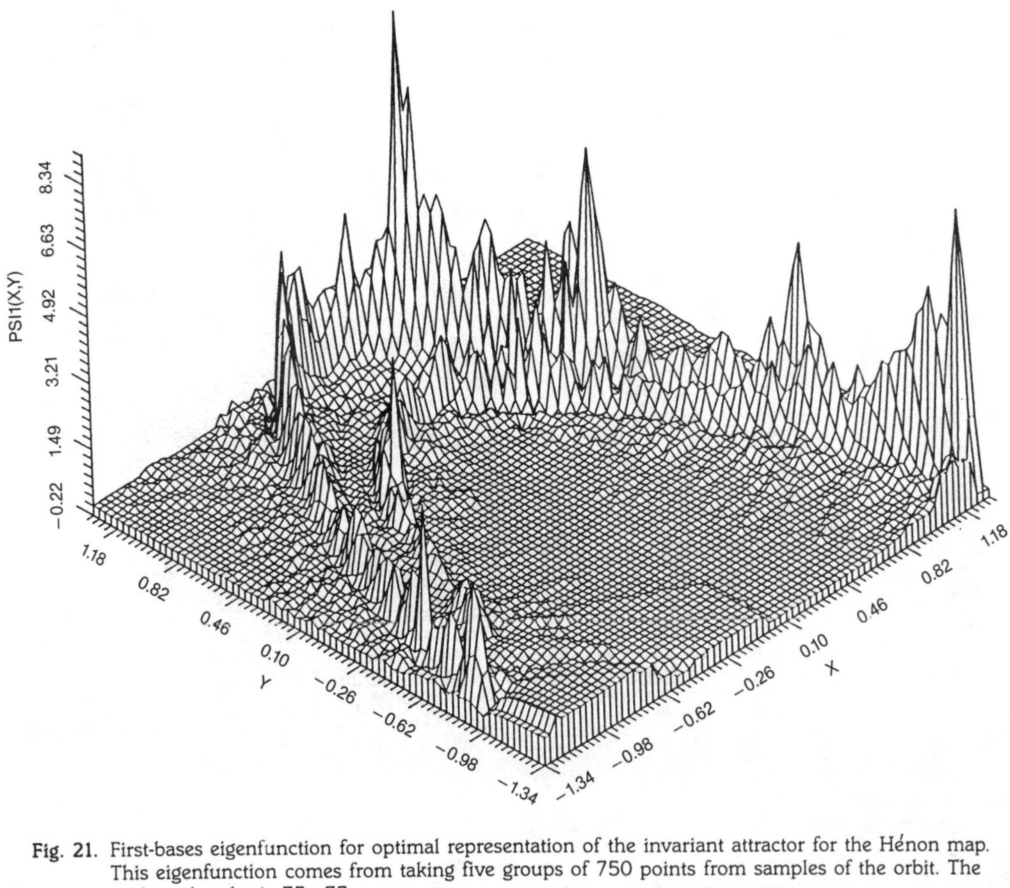

Fig. 21. First-bases eigenfunction for optimal representation of the invariant attractor for the Hénon map. This eigenfunction comes from taking five groups of 750 points from samples of the orbit. The grid on the plot is 75×75.

has contracted to a finite-dimensional strange attractor, and only a few degrees of freedom are "active" in this asymptotic state. The number of degrees of freedom among the infinity that are active is unknown, and *that is the issue!* If we could reliably project out of the partial differential equation the few variables describing the embedding space that remains after the fluid dynamics does its work, we could then take those equations, with perhaps the few parameters they contain, and continue solving until we found the behavior observed. Unfortunately, although we have observations that pass through the tests outlined above to demonstrate the presence of a low-dimension attractor, we will have to do something less fundamental than solving the correct equations if we want to predict future behavior on the basis of these observations.

Within this framework, we can proceed by means of standard engineering practice, in which observations of a time series are modeled in some way, and the parameters of the model are determined by a fitting criterion, which is typically some version of least squares. Our method for doing this emphasizes the geometrical properties we discussed earlier and puts significant stress on the fact that there is some underlying dynamical system generating the data, even if we are not yet clever enough to determine what it is.

We will proceed now as if we had acceptably answered all the questions raised in the previous sections. We presume to know the minimum embedding dimension, any Lyapunov exponents we wish, and the optimum moments of the invariant distribution $\rho(\mathbf{y})$. With the $\mathbf{y}(n)$ and the embedding space in hand, we ask the ambitious question of how we can use the $\mathbf{y}(n)$ series to predict $\mathbf{y}(N+1), \mathbf{y}(N+2)$, etc. Equivalently, we can ask, what is the evolution, under the same dynamical system that produced the $\mathbf{y}(n)$'s, of a point $\mathbf{y}$ that is on the attractor but not in the original data set? We will have answered this question when, given a data set $\mathbf{y}(1), \mathbf{y}(2), \ldots \mathbf{y}(N)$, we have identified a "reliable" map $\mathbf{F}$ from R^d to itself that is parametrized by $\mathbf{a} = (a_1, a_2, \ldots a_P)$, which takes us from $\mathbf{y}(n)$ to $\mathbf{y}(n+1)$:

$$\mathbf{y}(n+1) = \mathbf{F}\left(\mathbf{y}(n), \mathbf{a}\right).$$

If we can establish a reliable $\mathbf{F}(\mathbf{y}, \mathbf{a})$, then the evolution of a point $\mathbf{y}$ in R^d that is not a member of the measured data set would be $\mathbf{y} \to \mathbf{y}_1 = \mathbf{F}(\mathbf{y}, \mathbf{a})$; $\mathbf{y}_1 \to \mathbf{y}_2 = \mathbf{F}(\mathbf{y}_1, \mathbf{a}) = \mathbf{F}^2(\mathbf{y}, \mathbf{a}) = \mathbf{F}(\mathbf{F}(\mathbf{y}, \mathbf{a}), \mathbf{a})$; etc.

Our first view of the data $\mathbf{y}(1), \mathbf{y}(2) \ldots \mathbf{y}(N)$ is that it can be thought of as a pair of columns of vectors in R^d:

$$\begin{array}{cc}
\mathbf{y}(1) & \mathbf{y}(2) \\
\mathbf{y}(2) & \mathbf{y}(3) \\
\vdots & \\
\mathbf{y}(n) & \mathbf{y}(n+1) \\
\vdots & \\
\mathbf{y}(N-1) & \mathbf{y}(N),
\end{array}$$

and $\mathbf{F}(\mathbf{y}, \mathbf{a})$ comes from parametrically "fitting" the right-hand column of $\mathbf{y}(n+1)$ resulting from the left-hand column of $\mathbf{y}(n)$. Fitting the data then suggests making a least squares estimation of the $\mathbf{a}$, so that the *cost function*

$$\hat{C}(\mathbf{a}) = \sum_{n=1}^{N-1} \left\{ \sum_{m=1}^{d} [y_m(n+1) - F_m(\mathbf{y}(n), \mathbf{a})]^2 \right\}$$

is minimized. Our approach differs from previous work in detailed tactics and in our imposition of important geometrical structure as a constraint on the minimization of the cost function. The articles we have greatly relied on for guidance and initial impetus in our research are those by Farmer and Sidorovitch (1987) (we refer to this paper as FS in the following), Lapedes and Farber (1987, 1988), and Crutchfield and McNamara (1987).

Our main point, simply stated, is that we are not just making a fit to data with a set of functions $\mathbf{F}(\mathbf{y}, \mathbf{a})$. Rather, these functions, evaluated along the orbit, are to be related to each other in the manner of a *dynamical system*, which leads to a different view of the fitting functions than the one usually taken in trying to match data to observations. It means that the function $\mathbf{F}(\mathbf{y}, \mathbf{a})$, evaluated on the data vector $\mathbf{y}(n)$, is required to do more than reproduce $\mathbf{y}(n+1)$ as accurately as possible. $\mathbf{F}(\mathbf{y}, \mathbf{a})$ must also be a function that when *iterated* will reproduce $\mathbf{y}(n+2)$ after two applications to $\mathbf{y}(n)$, $\mathbf{y}(n+3)$ after three, etc. The treatment of $\mathbf{F}(\mathbf{y}, \mathbf{a})$ as a dynamical system also leads to modifications of the cost function, which should reflect the fact that iterations of $\mathbf{F}(\mathbf{y}, \mathbf{a})$ also yield points on the orbit. Furthermore, under our approach, geometrical properties of the dynamical system given by $\mathbf{F}(\mathbf{y}, \mathbf{a})$ are used to determine the success of the fit. The function's ability to reproduce the observed data in a least squares sense is not its only important feature. The data contain invariant information that is essential for a full description of the geometrical structure of the attractor that the data evolve on. Our key observation is that least squares fitting alone generally does not produce a map that captures the invariant characteristics of the attractor described by the data $\{\mathbf{y}(n)\}$. One must calculate from the data as many of these invariant quantities as possible and then impose them as constraints on the fit. By imposing invariants as constraints, we emphasize the fact that we are creating a dynamics, not just a fit to data. The product of our minimization of the constrained cost function is a mapping $\mathbf{F}(\mathbf{y}, \mathbf{a})$ of R^d to itself, which is not only *reliable* in that it reproduces the given data set by having a small cost function, but is also *representational* in that it has the same geometric invariants of the underlying dynamical system. Because the methods for identifying those invariants and utilizing them as classifiers for the dynamical system is a matter of some importance in itself, our emphasis on them in the earlier sections is well merited.

In this section, we discuss how to find the parameters $\mathbf{a}$ in $\mathbf{F}(\mathbf{y}, \mathbf{a})$ by minimizing a cost function subject to certain constraints. The constraints are chosen to ensure that iterations of the mapping function $\mathbf{F}(\mathbf{y}, \mathbf{a})$ give rise to values of dynamical invariants that are the same as those indicated by the experimentally measured data set $\mathbf{y}(n)$. In this way, essential geometric information about the particular attractor on which the data live will be built into the parametric mapping. Straightforward least squares does not accurately reproduce these invariants. Thus, one must perform a least squares minimization subject to the constraints that $\mathbf{F}(\mathbf{y}, \mathbf{a})$ accurately produce the Lyapunov spectra $\lambda_1, \lambda_2, \ldots, \lambda_d$ and the invariant density $\rho(\mathbf{y})$.

Choosing Maps and Predictors

Assuming for the moment that we have successfully embedded the data $x(n)$ in R^d by creating N d-dimensional vectors $\mathbf{y}(n)$, we need to choose a class of parametrized mappings, a cost function to minimize, and a means to impose the constraints on our minimization. The maps must have some way of "fitting" the data by closely reproducing one data point from the previous one by $\mathbf{y}(n+1) \approx \mathbf{F}(\mathbf{y}(n), \mathbf{a})$. Our maps are required to "look around" at the behavior of the phase-space neighbors of the point $\mathbf{y}(n)$ and make predictions according to how a cluster of these neighbors, regardless of their temporal sequence, are moved forward in time. The idea here is that one may use knowledge of the behavior of local regions of phase space as well as past points on an orbit to determine where a point will be mapped in the temporal future. The maps we choose must be sensitive to their neighborhood in phase space, and must inquire about the fate of any spatial neighbor under the map, regardless of its temporal arrival in the neighborhood. The map will then try to take any new point $\mathbf{y}$ and map it forward to some weighted average of its neighbors' forward evolution.

We take our mappings to be of the form

$$\mathbf{F}(\mathbf{y},\mathbf{a}) = \sum_{n=1}^{N-1} \mathbf{y}(n+1) g(\mathbf{y},\mathbf{y}(n);\mathbf{a}), \tag{58}$$

where $g(\mathbf{y},\mathbf{y}(n);\mathbf{a})$ is near one for $\mathbf{y} = \mathbf{y}(n)$, and vanishes rapidly for nonzero $|\mathbf{y} - \mathbf{y}(n)|$. In our case, $|\bullet|$ is a Euclidean norm, in R^d. $\mathbf{F}[\mathbf{y}(k),\mathbf{a}]$ will then be quite close to $\mathbf{y}(k+1)$.

Our choice here for $g(\mathbf{y},\mathbf{y}(n);\mathbf{a})$—one among many, of course—is

$$g(\mathbf{y},\mathbf{y}(n);\mathbf{a}) = \frac{\exp[-|\mathbf{y}-\mathbf{y}(n)|^2/\sigma][a_1 + a_2\mathbf{y}(n)\cdot(\mathbf{y}-\mathbf{y}(n)) + \sum_{k=3}^{P} a_k(|\mathbf{y}-\mathbf{y}(n)|^2/\sigma)^{m_k}]}{\sum_{n=1}^{N-1}\exp[-|\mathbf{y}-\mathbf{y}(n)|^2/\sigma][a_1 + \sum_{k=3}^{P} a_k(|\mathbf{y}-\mathbf{y}(n)|^2/\sigma)^{m_k}]}. \tag{59}$$

The parameter space $\mathbf{a}$, is P-dimensional; $\mathbf{a} = (a_1, a_2, \ldots, a_P)$. A fixed parameter σ provides a scale we can use to determine which points in the data set are "close" to $\mathbf{y}$. The m_k's are also fixed at various values. We could treat both σ and the m_k's as parameters to be optimized in the same sense as the $\mathbf{a}$'s. However, we choose not to do this, because we wish to explore other issues and to keep down the size of the parameter space over which our minimization searches were performed.

The weight function $g(\mathbf{y},\mathbf{y}(n);\mathbf{a})$ we use was arrived at after some experimentation. This choice, and many others, satisfies our general requirements, which include the following:

- The function is sensitive to the presence of near "neighbors" in phase space. Only points $\mathbf{y}(n)$ within a distance from $\mathbf{y}$ of order $\sqrt{\sigma}$ make any sizable contribution to $g(\mathbf{y},\mathbf{y}(n);\mathbf{a})$.
- It is easy to differentiate both in $\mathbf{y}$ and in $\mathbf{a}$. These derivatives are important in the minimization of the cost function using our methods. Having explicit expressions for the required derivatives in either of these independent variables makes the optimization routines run much faster.
- When $\sigma \to 0$, $g(\mathbf{y},\mathbf{y}(n);\mathbf{a})$ becomes essentially a Kronecker delta and the point $\mathbf{y}(n)$ is mapped precisely to $\mathbf{y}(n+1)$.
- In the function we have chosen, it is easy to retain many parameters of the same general form; thus, as the number of constraints on the optimization of the cost function is increased, the pattern of our searches remains the same.
- The essential function that senses neighbors, namely the exponential, can easily be replaced by other choices, such as those in Table 3.1 of the Silverman (1986) monograph. The general form of our arguments goes through without modification.

The term involving a_2 in the numerator allows this form of $g(\mathbf{y},\mathbf{y}(n);\mathbf{a})$ to satisfy, with numerical stability and accuracy, the constraints set by the Lyapunov exponents. The presence of the denominator assures us of numerical ease in making the parameters in the map $\mathbf{F}(\mathbf{y},\mathbf{a})$ meet our requirement that they produce an average over neighborhood points when projecting any phase-space point forward in time. This assurance makes the numerical algorithms we use much more efficient and accurate. The denominator serves as an approximate counter for the number of point $\mathbf{y}$ neighbors, so the numerator works less to produce the required average for the forward prediction of the point $\mathbf{y}$.

The choice of cost function is also essentially up to us. Because we are to think of $\mathbf{F}(\mathbf{y},\mathbf{a})$ as a dynamical system evolving points $\mathbf{y}(n)$ into new points $\mathbf{y}(n+1)$, we should consider asking the map to reproduce accurately from $\mathbf{y}(n)$ not only the "next" point $\mathbf{y}(n+1)$ but, via iteration, a sequence of points $\mathbf{y}(n+1), \mathbf{y}(n+2), \mathbf{y}(n+3), \ldots \mathbf{y}(n+L)$ up to some L beyond which we simply don't trust the accuracy of our algorithm $\mathbf{F}$ or of the machines we use to compute the future points $\mathbf{y}$.

The above suggests the *predictor* for future points to be a linear combination of iterated powers of the map $\mathbf{F}(\mathbf{y},\mathbf{a})$:

$$\mathbf{y}(m+1) = \sum_{k=1}^{L} X_k \mathbf{F}^k(\mathbf{y}(m-k+1),\mathbf{a}), \tag{60}$$

where $\mathbf{F}^k$ is the kth iterate of $\mathbf{F}$ as described above. If $\mathbf{F}(\mathbf{y},\mathbf{a})$ were the exact mapping, then each term in the sum over k would be $X_k\mathbf{y}(m+1)$. Thus we require

$$\sum_{k=1}^{L} X_k = 1.$$

The $\mathbf{X}$'s weight the various iterates of $\mathbf{F}$ and are used to determine which iterates are the most accurate. Typically, one would require $X_j \geq X_{j+1}$ to indicate that the lower iterates of $\mathbf{F}$ are believed to be more accurate than the higher iterates. This predictor is a natural generalization to the nonlinear problem of the common linear predictor

$$\mathbf{y}(m+1) = \sum_{k=1}^{L} X_k \mathbf{y}(m-k+1), \tag{61}$$

with the clear differences associated with the iterative nature of the map $\mathbf{F}(\mathbf{y},\mathbf{a})$.

The above predictor combines both past temporal information from times $m-k+1; k=1,2,\ldots,L$ and information from all the phase-space neighbors of the orbit points $\mathbf{y}(m-k+1)$ because of the structure of $\mathbf{F}(\mathbf{y},\mathbf{a})$. The combination of spatial and temporal information provides a significant "lever arm," which permits Eq. (60) to forecast quite accurately the forward evolution of points $\mathbf{y}$ in R^d. By utilizing the phase-space information in $\mathbf{F}(\mathbf{y},\mathbf{a})$ at each temporal step, we efficiently tap properties of the full data set.

The cost function associated with this predictor is

$$\mathbf{C}(\mathbf{X},\mathbf{a}) = \frac{\sum_{n=1}^{N-1}\left\{|\mathbf{y}(n+1) - \sum_{k=1}^{L} X_k\mathbf{F}^k(\mathbf{y}(n-k+1),\mathbf{a})|^2\right\}}{\sum_{n=1}^{N}|\mathbf{y}(n)\cdot\mathbf{y}(n)|^2}. \tag{62}$$

This type of cost function automatically contains information on the Lyapunov exponents, which themselves are expressions of the dynamics as iterations of the map. Some information on the invariant density function on the attractor is also contained in this improved cost function.

Another major consideration is the great difference in the coordinate scale of various attractors. The numerator of the cost function (Eq.(62)) is the residual of the mapped function summed over the entire trajectory, and hence gives a measure of the sum of the absolute errors over all the mapped points. Because the absolute error depends on the macro-scale of the attractor, it is more informative to rescale the final cost function value in some manner that reflects a relative error. In our examples, the scale of the Hénon attractor is on the order of unity, while that of the Lorenz attractor is of order 100. Hence, some form of rescaling of the cost function is desirable in order to have a relative measure of comparison between two systems with different macro-scales. We chose a normalization in the following straightforward manner: we simply summed the magnitudes of the position vectors of all the points on the attractor, and retained this value as a constant. Absolute values for the cost function obtained after normalization by the denominator in Eq. (62) give a more sensible relative measure of the error of our prediction function $\mathbf{F}(\mathbf{y},\mathbf{a})$.

We note that FS suggest forecasting the evolution of a point $\mathbf{y}$ by looking around at the neighbors of $\mathbf{y}$ among the data set $\mathbf{y}(n)$, and observing where these neighbors go after one iteration of the underlying map takes the $\mathbf{y}(n)$ to $\mathbf{y}(n+1)$. FS determine the future of the new point $\mathbf{y}$ by an interpolation involving the future of its neighbors. Our mapping function Eqs. (58) and (59) do precisely this, as indicated earlier. All points in the data set are given some weight in the future of $\mathbf{y}$, but if $g(\mathbf{y},\mathbf{y}(n);\mathbf{a})$ falls rapidly for large $|\mathbf{y}-\mathbf{y}(n)|$, as we shall always choose, only members of the data set $\mathbf{y}(n)$ near $\mathbf{y}$ (i.e., the neighbors) play much of a role in its future. In that sense, our $\mathbf{F}(\mathbf{y},\mathbf{a})$ is an analytic formulation of the FS idea. More or less weight can be given to the near neighbors by different choices for the function $g(\mathbf{y},\mathbf{y}(n);\mathbf{a})$. The Gaussian we work with could be replaced by a Lorentzian, or other choices that weight neighbors more.

Optimization of the Constrained Cost Function: Parameter Determination

Analysis for the Hénon Map. Our first application of the methods described above is to data generated by the Hénon map of the plane to itself. Data were created by iterating the map from some initial condition and discarding the first 50 points of that data set. Two data sets of $x_1(n)$ were created this way. The first had $N' = 3,750$ points, which we divided into five groups of 750 points each. These groups were used to create the densities $\rho_\alpha(\mathbf{y})$ shown in Figs. 16-20, and the phase-space correlation function among groups used to generate the eigenfunction shown in Fig. 21. The second data set was used to select samples of length $N = 750$, $1,200$, and $1,752$ for our analysis.

We first studied the distribution of Euclidean distances among the two vectors $\mathbf{y}(n) = (x_1(n), x_1(n+1)); n = 1,2,\ldots,N-1$ formed from the data set. On the natural scale of the attractor, which is order unity, the minimum distance was always order 10^{-5}–10^{-4}. This distance led us to choose $\sigma = 5\text{x}10^{-6}$ for the parameter σ in our maps, so that each data point, at least for $N \geq 500$, would have neighbors. We varied σ by a factor of ten or so, with no qualitative differences in our results. A thorough parameter search would vary σ in the constrained minimization of the cost function.

Next we chose to use four parameters $\mathbf{a}$ in our set, and took the powers m_3 and m_4 in $\mathbf{F}(\mathbf{y},\mathbf{a})$ to be $m_3 = 4, m_4 = 5$. We did not further vary these parameters. Our choice of four $\mathbf{a}$ rested on our knowledge that we would be constraining our cost functions by only the largest Lyapunov exponent λ_1 and the projection B_1 of $\rho(\mathbf{y})$on the eigenfunction $\psi_1(\mathbf{y})$ with the largest eigenvalue. Four seemed a reasonable minimum number of parameters, particularly since the amount of work required to search large parameter sets can become significant. In effect, we had two free parameters in $\mathbf{F}(\mathbf{y},\mathbf{a})$ when the values of λ_1 and B_1 were specified.

Our final *a priori* choice was on the values of X_j in the predictor Eq. (60). Here we took three terms, a conservative choice based on the number of iterations of the map $\mathbf{F}(\mathbf{y},\mathbf{a})$ we felt we could trust. Further reflecting our sense that iterations of $\mathbf{F}(\mathbf{y},\mathbf{a})$ could become unreliable was our choice of $X_1 = 0.8$, $X_2 = 0.1$, and $X_3 = 0.1$. Once again, the $\mathbf{X}$ could be parameters that vary in our constrained minimization. We found that varying the $\mathbf{X}$ by 20% or so did not qualitatively change our results. In the case of the Lorenz attractor study, which will be discussed in the next section, we report results for $X_1 = 0.5, X_2 = 0.3, X_3 = 0.2$, and note that the cost function changes by $\approx 20\%$.

For the purposes of this paper, we chose simply to fix the $\mathbf{X}$. However, the $\mathbf{X}$ clearly can be varied along with the $\mathbf{a}$, σ, and $m_3,\ldots,m_P$, if one wishes. Ours is a first try in "fitting" the data and meeting the invariant constraints with the $\mathbf{F}(\mathbf{y}, \mathbf{a})$ we have chosen. The feasibility of accomplishing this seemed daunting enough when we set out, but we expect to include many more parameters in future work in this area.

One additional important matter deserves note before we proceed to the discussion of our numerical results. As we carry out our search over the parameters $\mathbf{a}$, the maps $\mathbf{F}(\mathbf{y}, \mathbf{a})$ have very little ability to fit reliably the given data for most $\mathbf{a}$. Only when we arrive near a (constrained) minimum of the cost function can we be very confident that our map is reasonable. Until the map is near the optimum, points in the data set are quite often mapped far off the attractor. For numerical stability in our search algorithms, we need a method to identify orbits that are leaving the attractor for non-optimal values of $\mathbf{a}$, and return them to the neighborhood of the attractor.

With maps of the form we have chosen, points far off the attractor, as defined by the data set itself, are mapped to $\mathbf{y} = 0$. There is no reason to expect the origin of coordinates to lie on an attractor that has $d_A < d$ and is quite sparse in R^d, but we choose always to translate our data set so that one of its points is the origin. This approach changes nothing about the signal processing issues we address in this paper, and makes our parameter searches numerically sensible. When the origin is translated, those orbits generated by $\mathbf{F}(\mathbf{y}, \mathbf{a})$ when $\mathbf{a}$ is not optimal that try to depart significantly from the attractor are sent back to $\mathbf{y} = 0$, which is now on the attractor. When $\mathbf{a}$ is near its optimal values, this feature is operationally unimportant, because the map is tracking the data very accurately.

Our experience here indicates that when one is trying to create global maps $\mathbf{F}(\mathbf{y}, \mathbf{a})$, some form of "orbit reinjection" will be required to give numerical sense to the whole process of searching parameter space, in order to minimize the cost function. The problem becomes more important as d grows, because the attractor of dimension $d_a < d$ occupies "less and less" of the full volume of the phase space. However, if one is making "fits" to the data by numerous local or nearly local polynomial maps, as in the work of FS, the issue raised here is absent. Global maps have an economy of parameters and a potential ease of interpretation; local maps appear to have a calculational speed advantage. We have no overall judgment on how to choose between these alternatives.

Our results for data from the Hénon map are shown in Table 1. The parameter searches were carried out using the FORTRAN package NPSOL (Gill et al., 1986). One of its authors, Professor Philip Gill, was kind enough to consult with us extensively on its use and on the interpretation of its output. For each value of the number of points in the data set—namely, $N = 750, 1,200$, and $1,752$—we report seven quantities for each of three cases: (1) unconstrained minimization of the cost function; (2) minimization constrained by λ_1 ($\lambda_1^{map} = 0.408$); and (3) minimization constrained by both λ_1 and B_1. In each case we report two values: the value of the cost function normalized by the sum of the squares of the Euclidean lengths over all data vectors; and the values of the $\mathbf{a}$ at the minimum cost function, as well as the deviations $\Delta\lambda_1$ and ΔB_1 from the values of λ_1^{data} and B_1^{data} determined by the data. The allowed tolerances on these deviations are set in NPSOL by the user. We typically required the relative magnitudes of $\Delta\lambda_1$ to λ_1, and the same for b_1, to be in the range 0.5 to 5%. This range is not a limitation of NPSOL, but it seemed accurate enough for our purposes.

A look at Table 1 reveals a consistent pattern. Unconstrained optimization resulted in a cost function with an RMS deviation of our predictor from the data of 0.1% or smaller! Not surprisingly, when we track the data so accurately, the value of B_1 comes out precise. The value of λ_1^{map} for this best least squares fit is remarkably bad. Indeed, in our examples this quantity was actually *negative*, which indicates the absence of chaos for the parametric map!

Table 1. Henon map data. Optimization results. C(X,a) is shown with and without invariant constraints.

$\mathbf{F}(\mathbf{y},\mathbf{a}) = \sum_{j=1}^{N-1} \mathbf{y}(j+1) g(\mathbf{y},\mathbf{y}(j);\mathbf{a}))$							
$C(\mathbf{X},\mathbf{a}) = \frac{\sum_{k=L}^{N-1} \lvert\mathbf{y}(k+1)-\sum_{j=1}^{L} X_j \mathbf{F}^{j}(\mathbf{y}(k-j+1),\mathbf{a})\rvert^2}{\sum_{n=1}^{N} \mathbf{y}(n)\cdot\mathbf{y}(n)}$							
$X_1 = .8$		$X_2 = .1$			$X_3 = .1$		
Number of points=750			$\lambda_1^{data} = 0.408$				$B_1^{data} = 3.4739$
	$C(\mathbf{X},\mathbf{a})$	a_1	a_2	a_3	a_4	$\Delta\lambda_1^{map}$	ΔB_1^{map}
Unconst.	$4.016\text{x}10^{-7}$	7.5347	1.3289	-0.7041	0.1485	-2.0098	$-5.70\text{x}10^{-3}$
λ_1	$4.77\text{x}10^{-6}$	6.6855	20.6948	-0.1714	0.0956	$1.6\text{x}10^{-4}$	0.223
λ_1, B_1	$2.06\text{x}10^{-5}$	0.3422	1.1169	0.3766	-0.05586	$4.23\text{x}10^{-2}$	0.140
Number of points = 1,200			$\lambda_1^{data} = 0.408$				$B_1^{data} = 3.388$
	$C(\mathbf{X},\mathbf{a})$	a_1	a_2	a_3	a_4	$\Delta\lambda_1^{map}$	ΔB_1^{map}
Unconst.	$3.41\text{x}10^{-6}$	7.5217	2.9658	-0.3145	0.07502	-1.676	$-9.85\text{x}10^{-3}$
λ_1	$1.1297\text{x}10^{-5}$	8.4520	26.7454	0.2686	0.01177	$-6.32\text{x}10^{-4}$	0.214
λ_1, B_1	$2.38\text{x}10^{-5}$	6.6093	19.6341	0.08362	-0.01087	$-1.1\text{x}10^{-4}$	0.198
Number of points = 1,752			$\lambda_1^{data} = 0.408$				$B_1^{data} = 3.369$
	$C(\mathbf{X},\mathbf{a})$	a_1	a_2	a_3	a_4	$\Delta\lambda_1^{map}$	ΔB_1^{map}
Unconst.	$3.5359\text{x}10^{-7}$	8.4093	6.0546	-0.1497	0.02315	-1.211	$-5.80\text{x}10^{-3}$
λ_1	$1.7284\text{x}10^{-5}$	3.1671	9.8576	-0.1120	0.02743	$-6.512\text{x}10^{-3}$	0.2605
λ_1, B_1	$2.54\text{x}10^{-5}$	5.8314	18.44612	0.1832	-0.02818	$1.005\text{x}10^{-4}$	0.2466

When the λ_1 constraint is imposed, the parameters **a** change, but we regard their specific values as of incidental interest here. More important is the observation that the RMS value of the cost function—the bare measure of the quality of the fit—remains about 0.5%, while the Lyapunov exponent is now accurate to about 1% or better. Of course, having moved away from the very best point-to-point least squares tracking of the data, the accuracy of B_1 degrades to $\approx 10\%$. Finally, imposing both constraints, we achieve about 0.5% in the RMS error for the cost function, highly accurate λ_1, and somewhat better B_1 values.

The message of these calculations is that the procedure we outlined in this paper is both feasible and highly accurate. The few scalar numbers—the cost function, λ_1, and B_1—do not tell the whole story. One can take the map with the optimum **a** and calculate a new orbit starting from some new phase-space point, $\mathbf{y}^{new}(1)$: $\mathbf{y}^{new}(1), \mathbf{y}^{new}(2), \ldots$, and compare the new orbit to that generated by the Hénon map, starting with the same initial point. The data so generated looks the same when plotted as a sequence of two vectors, but because this temporal representation contains very little useful information, we do not show it.

What is more important is that our predictor

$$\mathbf{y}(m+1) = \sum_{k=1}^{L} X_k \mathbf{F}^k(\mathbf{y}(m-k+1), \mathbf{a}) \tag{63}$$

predicts accurately. We have taken numerous points from our data set and evolved them forward by use of the predictor. We find we are able to track the actual data to the 1% level seven to ten steps along the orbit all around the attractor. This means that iterates of our optimum map $\mathbf{F}^k(\mathbf{y}, \mathbf{a})$ are accurate to $k \approx 7 \sim 10$, far beyond our original safe choice of $k = 3$. The implications of this remarkable accuracy for prediction and control of nonlinear chaotic systems are transparent.

Prediction for the Lorenz System. We now turn to the application of our methods to the Lorenz system, defined by Eq. (3). These equations were originally motivated by an attempt to model atmospheric phenomena using a dynamical system with only a few degrees of freedom. It was one of the first systems known to exhibit an attractor of fractal dimension or a strange attractor, and, consequently, to connect this attractor with the apparent chaotic motion of the resultant dynamics. In modifying our previous techniques for use on the Lorenz system, we have two primary concerns:

1. The jump to a three-dimensional embedding space, which will require much longer time series to fill out the attractor properly.
2. The large difference in the macro-scale of the two attractors, which will require the rescaling of some of the variables we have previously defined.

Before addressing these concerns, however, we will give a short review of some of the characteristics of the Lorenz system.

For the parameter values $\sigma = 16.0$, r $= 45.92$, and b$= 4.0$, the Lorenz system possesses a strange attractor that has become one of the classic examples of nonlinear science. The structure consists of two nearly flat lobes that are connected, roughly at a point, and are angled somewhat with respect to one another. Hence, the local dimension of the attractor is essentially two; however, the minimum embedding space required is three. Note that the motion of the phase-space orbits for the Lorenz system is continuous (i.e., a flow) as opposed to that of the Hénon system, which is a mapping. The discreteness of the Lorenz orbits after phase-space reconstruction, and the density of points along an orbit, are therefore due to the choice of a sampling rate in the measurement of the data time series. This sampling rate therefore can be thought of as a time scale set in the reconstructed picture of the attractor. The time scale, in turn, determines the time-delay values for the method of phase-space reconstruction used above, the evolution times for Lyapunov exponent calculations, etc. A discussion of optimal ranges of sampling rates, and the problems that occur when sampling rates are too large or small, is given at some length by Mayer-Kress (1987). In practical applications, of course, one often has no control over the data set, although too-frequent sampling can often be remedied by simply throwing away data.

In order to investigate the behavior of our prediction technique on a system with a somewhat larger embedding space, we chose the Lorenz system as a test case with known parameters, as was done with the Hénon system. Using the parameter values listed above, an "experimental" time series was generated for the Lorenz equations (Eq. (3)) by a Runge-Kutta numerical integration scheme with a fixed time-step of 0.03. A data set of the x_1 variable consisting of approximately 20,000 points (after transients) was generated. We used the same time series for all of our numerical runs.

As stated above, we chose an embedding dimension of three. Note that in an actual experimental situation, a more cautious choice of four would also be reasonable, although our computational requirements would increase by a significant amount. For the choice of the delay-time constant τ, a number of different choices could be made. Because the delay-time reconstruction is rather weakly dependent on this constant—provided one is within certain limits—there is no unique choice for this variable. Our final choice, which was motivated by the desire to have the reconstructed attractor look most like the original Lorenz attractor, resulted in a delay of two time steps. A feel for the required density of points can also be obtained by calculating the minimum nearest-neighbor distance (and perhaps the frequencies of a range of somewhat larger neighbor distances), and comparing this with the "macro-scale" of the attractor (i.e., the maximum ranges of the coordinates of an attractor).

The embedding dimension and delay time comprise the two parameters necessary to reconstruct the dynamics of the systems attractor correctly; determining these is the first step in setting up the prediction method. We now consider the changes to the numerical algorithm that are necessary when we consider the Lorenz system.

The most significant difference between the prediction models for the Hénon and Lorenz systems is that of the size of the time series required for the phase-space reconstruction. Because the dimensionality of the embedding space increases from two to three, the number of phase-space points required to perform our procedure increases dramatically. The reason for this is clear. Our prediction function $\mathbf{F}(\mathbf{y}, \mathbf{a})$ requires that most points have a significant number of nearby neighbors (i.e., points within distances of a few $\sqrt{\sigma}$ values), so that a good "mapping" of the local phase space around a particular region is obtained. Additionally, nearby neighbors give good numerical approximations to the gradients of the objective and constraint functions. Because the number of points required to yield a given-mean, nearest-neighbor distance is considerably larger for a volume than for an area, the number of points required to fill out the attractor properly is much greater for a three-dimensional embedding space. In the section on phase-space reconstruction, we presented general methods for determining the number of data vectors needed for a given embedding dimension d. For our particular analysis of the Lorenz attractor, we found the minimum number of points that gave reasonable results to be about 6,000. For the numerical experiments reported in Table 2, we used data sets with 6,000 and 8,000 points.

One final change in the numerical parameters for the prediction code is in the number of matrices that are to be multiplied together to obtain the Lyapunov exponent from the mapping function. Because each iteration of the Hénon map represents a significant evolution of the system, the multiplication of 500 Jacobian matrices for the Lyapunov calculation represents a good average over the phase space, and results in a fairly accurate final value. However, each step of the time series for the Lorenz system represents much less evolution time for the dynamics. By experimenting with the number of matrices, we found that about 1,000 matrix products gave a reasonably good convergence to the final value, without being excessively computationally intensive.

To complete the formulation of the prediction model for the Lorenz data, it is necessary to pick the exact form of the mapping and cost functions that are to be minimized. We first discuss the choice of the polynomial terms that multiply the exponential in the mapping function. These terms are defined, as for the Hénon analysis, with the intention of giving the exponential form in the mapping function a longer "tail" by adding to it multiplicative polynomial terms. As for the Hénon analysis, we chose to use four polynomial terms in the mapping function, and hence have four variables in the minimization fit. The first coefficient is, of course, the constant term, and the second multiplies the linear term that expresses some dependence of the mapping function on the Lyapunov exponent. Therefore, the powers of the last two polynomial terms remain to be determined.

Table 2. Lorenz Attractor data. Optimization results. C(X,a) is shown with and without invariant constraints.

$$\mathbf{F}(\mathbf{y},\mathbf{a}) = \sum_{j=1}^{N-1} \mathbf{y}(j+1) g(\mathbf{y},\mathbf{y}(j);\mathbf{a})$$

$$C(\mathbf{X},\mathbf{a}) = \frac{\sum_{k=L}^{N-1} |\mathbf{y}(k+1) - \sum_{j=1}^{L} X_j \mathbf{F}^j(\mathbf{y}(k-j+1),\mathbf{a})|^2}{\sum_{n=1}^{N} \mathbf{y}(n)\cdot\mathbf{y}(n)}$$

	$X_1 = .8$		$X_2 = .1$		$X_3 = .1$	
Number of points=6,000			$\lambda_1^{data} = 1.51$			
	C(X,a)	a_1	a_2	a_3	a_4	$\Delta\lambda_1^{map}$
Unconst.	2.51672x10^{-5}	37.8328	-0.03587	0.08338	0.02050	-1.9758
λ_1	2.5170x10^{-5}	37.8177	0.2563	0.08334	0.3438	-9.996x10^{-3}
Number of points=8,000			$\lambda_1^{data} = 1.51$			
	C(X,a)	a_1	a_2	a_3	a_4	$\Delta\lambda_1^{map}$
Unconst.	1.87056x10^{-5}	57.7936	0.2683	10.2508	0.06132	-0.3875
λ_1	1.87087x10^{-5}	57.7944	-0.3883	10.2504	0.4681	-1.856x10^{-2}
	$X_1 = .5$		$X_2 = .3$		$X_3 = .2$	
Number of points = 6,000			$\lambda_1^{data} = 1.51$			
	C(X,a)	a_1	a_2	a_3	a_4	$\Delta\lambda_1^{map}$
Unconst.	3.2237x10^{-5}	37.8223	-0.001847	0.08326	-0.003749	-0.4942
λ_1	3.2244x10^{-5}	37.8156	0.2359	0.08338	0.5205	-9.323x10^{-2}
Number of points = 8,000			$\lambda_1^{data} = 1.51$			
	C(X,a)	a_1	a_2	a_3	a_4	$\Delta\lambda_1^{map}$
Unconst.	3.22745x10^{-5}	77.8247	0.4492	10.0837	11.9902	-0.1763
λ_1	3.22751x10^{-5}	77.8223	0.5301	10.0834	11.9986	-1.024x10^{-2}

In choosing the values of the exponents of the remaining two polynomial terms, we recall that we wish to elongate the "tail" of the exponential term in the mapping function to make it "feel" more of the surrounding neighbors. However, we do not wish to make these exponents so large that the scale increases well beyond that which we set by σ. After some experimentation, we chose $m = 3$ and $m = 6$ as the two powers for the polynomial terms, although these are by no means the only possible choices.

The second set of parameters of the minimization procedure that need to be chosen are the $\mathbf{X}$'s, which appear in the definition of the cost function in Eq. (62). These coefficients weight the different iterates of the map $\mathbf{F}(\mathbf{y},\mathbf{a})$ and essentially determine how many iterates forward we wish the map to reproduce the data accurately. For the Hénon analysis, we chose three $\mathbf{X}$'s with values (0.8, 0.1, 0.1). Our choice indicates a desire to weight the first forward iterate very heavily, while giving the second and third iterates only minimal importance. This set of values was chosen primarily because the Hénon system is a mapping, and each iterate represents a large step in evolution for the original system. On the other hand, the

Lorenz system produces a flow in phase space, and the time step we chose for each iterate of the time series represents a rather small amount of forward evolution of the system. Thus, we weighted some of the multiple iterates of the map more heavily than we did for the Hénon system. We have therefore presented data for the Lorenz system with two different sets of values for these parameters. In one case, we used the original weights of the Hénon system, (0.8, 0.1, 0.1). For the other case, we weighted the multiple iterates more heavily, namely, (0.5, 0.3, 0.2). Note that we could have easily chosen to take more than two multiple iterates of the system. However, for the sake of simplicity and comparison, we chose to use two, as for the Hénon system. We also note that the $\mathbf{X}$'s, like the $\mathbf{a}$'s, could be made variables in the minimization search—a tactic we will use in our further work in this matter.

Finally, to determine a value of the parameter σ, which sets a characteristic scale of distance over which the mapping function "feels" its neighbors, it is necessary to experiment with different values by actually doing a number of minimization runs. One can, however, make an *a priori* guess by considering two factors. The largest value that σ can possibly have will be the scale of the linear regime for the system, or roughly 1% of the attractor's macro-scale, as mentioned previously. Hence, σ should be considerably smaller than this value. Additionally, the smallest value that σ can possibly attain is given by the smallest neighbor-distance of the data set, and should be at least one to two orders of magnitude larger than this value. Within this range, σ must be chosen with some experimentation. We have found that, typically, the value of the $C(\mathbf{X}, \mathbf{a})$ at its minimum will be relatively large for larger values of σ, and will decrease until a threshold in σ is crossed. For values of σ smaller than the threshold value, the minimum of $C(\mathbf{X}, \mathbf{a})$ becomes a great deal less, sometimes by an order of magnitude or more. We recommend that σ be somewhat smaller than the threshold value, but not too much smaller, as it is still desirable to have each mapped orbital point "feel" as much of the surrounding phase space as possible. For our experiments on the Lorenz system we used $\sigma = 1.0 \times 10^{-4}$.

Using the parameter values given above, a search for the minima of Eq. (62) in the parameter space $\mathbf{a}$ was conducted using the NPSOL (Gill et al., 1986) package. Because no general numerical method is known for determining the absolute minimum of a function, one generally proceeds by finding the minimum after iteration for each of a large number of initial conditions, while attempting to cover a large representation of the phase space. In practice, a number of local minima are usually found, all of which have "basins of attraction" of varying sizes. After a number of runs, one often gains some intuition as to which regions of the parameter space evolve to which local minima. When sufficiently confident that a large region of the parameter space has been investigated, we label the minimum with the lowest cost function value the "absolute" minimum. However, generally speaking, one can never be sure that one has found the actual global minimum.

Using the time series for the Lorenz data and the given parameter values, the NPSOL routine was able to find a number of minima of the cost function $C(\mathbf{X}, \mathbf{a})$. The cost function values ranged over as much as two orders of magnitude, with the lowest value in the neighborhood of 1.87×10^{-5}, as indicated in Table 2. Three minima had almost this same value in the preliminary analysis. However, a more detailed analysis found that after many iterations of the search routine two of these minima actually evolved into the third. Using better error tolerances in NPSOL, this point was found actually to have a slightly lower minimum. It should be noted that even though the three minima had cost functions that agreed very closely, their resulting values for the $\mathbf{a}$'s were much different. The close agreement of the minima is in keeping with our observation that, for a large range of parameter values around these minima, the cost function was very "flat" with respect to the parameters; i.e., $C(\mathbf{X}, \mathbf{a})$ varied very little over a large range of $\mathbf{a}$'s. This flatness had the unfortunate effect of considerably slowing the iteration procedure, because the minima were very shallow, and a large number of iterations were required to achieve the optimal solution. One possible conclusion from this is that if a purely least squares fit of the map to the data were of interest, any of the parameter sets in this range were nearly as good as the optimal solution.

After the analysis just described, we performed another, with the **X**'s changed to (0.5, 0.3, 0.2). These parameter values weight the later iterates of the map more heavily and were chosen in an attempt to make the map predict farther into the future. We did not impose the B_μ constraints on the Lorenz system, but used this system to explore both the variations of the cost function and our ability to reproduce the largest Lyapunov exponent as we changed the weights X_j in the predictor. The results of these minimization searches are presented in Table 2; 6,000 and 8,000 points on the attractor are used in our example. As expected, the cost function for these minima are about 1/3 higher than for the previous system, because the later iterates, which must be inherently less accurate, now give a much larger contribution to the cost function. In terms of relative fitting error, however, these minima are still surprisingly low. Although significantly different from the previous system, the final parameter values are still similar enough to give the same general character to the fitting function.

One noticeable difference between the two different values of **X** was in the fitting of the map using the Lyapunov constraint. The iteration procedure for the (0.5, 0.3, 0.2) system went far more quickly than for the (0.8, 0.1, 0.1) system. This difference can probably be interpreted in terms of the fact that if later iterates of the map are weighted more heavily, the parameters result in more sensitivity of the map to the Lyapunov constraint, which usually requires longer evolution times to manifest itself for flows.

SUMMARY AND CLOSING REMARKS

This chapter has been directed toward an exposition of currently available methods—some well tested, some rather speculative—for the analysis of time series arising from nonlinear systems that behave chaotically. We discussed the idea of using an embedding dimension and phase-space reconstruction as a way of "unfolding" a scalar data set to reveal its geometric structure. We also discussed ways of classifying this geometric information in terms of invariants of the motion. Invariants are quantities independent of initial conditions and thus not subject to the same wild fluctuations as individual orbits. These two subjects allow us to identify and get a quantitative hold on properties of interesting dynamical systems that could be of interest to meteorological studies. Indeed, starting with the 1963 Lorenz model (Lorenz, 1963), and continuing through the analysis of geophysical data sets as represented by the work of Fraedrich (1986), there is a continuing tradition of using dynamical system techniques to understand properties of the atmosphere-ocean system.

In the last section, we moved beyond describing the dynamical system whose signals we had measured, and took on the more ambitious task of processing those signals for purposes of predicting from given data sets what the future behavior of the system would be. Our nonlinear time series analysis consists of a set of procedures that may be used to process signals $x(n), n = 1, 2, \ldots$ having a broadband power spectrum. Using numerically generated data from the Hénon map and from the Lorenz equations, we explicitly demonstrated the feasibility of our procedures. Processing a signal means that from the time series $x(n)$ we do the following:

- find an integer-dimensional embedding space of time-lagged d-vectors

$$\mathbf{y}(n) = [x(n), x(n+\tau_1), \ldots, x(n+\tau_{d-1})]$$

that fully exposes the geometric structure of the attractor on which the data evolves. The attractor has dimension d_A, which may be fractional. Choosing the integer $d > 2d_A + 1$ is guaranteed to be sufficient for this purpose, but smaller d may often work.

- find invariants of the evolution $\mathbf{y}(1), \mathbf{y}(2), \ldots \mathbf{y}(N)$ in R^d—specifically the Lyapunov exponent spectrum $\lambda_1, \lambda_2, \ldots, \lambda_d$ and selected optimum moments $B_1, B_2, \ldots, B_G$ of the invariant density $\rho(\mathbf{y})$ on the attractor.

- use these vectors $\mathbf{y}(n)$ and invariants to construct a parametrized map of R^d to itself $\mathbf{y} \rightarrow \mathbf{F}(\mathbf{y}, \mathbf{a})$, which minimizes a certain constrained least squares cost function based on the residual errors of a nonlinear predictor $\mathbf{y}(m+1) = \sum_{k=1}^{L} X_k \mathbf{F}^k(\mathbf{y}(m-k+1), \mathbf{a})$ involving iterates $\mathbf{F}^k$ of the map.

The output of the signal processing is a parametrized map $\mathbf{F}(\mathbf{y}, \mathbf{a})$—both its form and the parameters $\mathbf{a}$—and the coefficients X_j in the predictor.

A map $\mathbf{F}(\mathbf{y}, \mathbf{a})$ and a predictor that give very small least squares residuals when evaluated on the data we call ***reliable***. We have explicitly demonstrated in this paper that even a reliable $\mathbf{F}(\mathbf{y}, \mathbf{a})$ does not necessarily reproduce invariants such as the λ_a and the B_μ discussed by us. The reason is that a least squares tracking of a data set $\mathbf{y}(n)$ by a map $\mathbf{y}(n+1) \approx \mathbf{F}(\mathbf{y}(n), \mathbf{a})$ does not necessarily provide a good evaluation of the local tangent space mapping $M_{ij} = \frac{\partial F_i(\mathbf{y})}{\partial y_j}$. A map that is reliable and also gives the correct invariants we call ***representational***. Our maps are representational because we constrain least squares minimization by the invariants. A map that closely tracks data but does not yield the dynamical invariants misses the essential ingredients that classify or identify the dynamical system underlying the data.

In our constrained optimization procedure, the cost function used in determining the map should not be composed only of the squares of the residuals in the predictor. It should also contain terms that measure the residuals in matching the invariants determined by the data, and the same quantity determined by the maps. NPSOL and other contemporary optimization routines essentially do this by adding a combination of Lagrange multiplier and quadratic penalty terms to the least squares cost function. This point of view suggests that we should not focus on the size of $C(\mathbf{X}, \mathbf{a})$ as our goodness-of-fit criterion, but on $C(\mathbf{X}, \mathbf{a}) + \sum_a (\Delta\lambda_a)^2 + \sum_\mu (\Delta B_\mu)^2$. In Tables 1 and 2 we have reported the values of each of these quantities separately, but the sum should measure the merit of our maps.

If we remain with our general form, which we do not insist upon, then we must address properties of $g(\mathbf{y}, \mathbf{y}(n); \mathbf{a})$, as should be clear from our discussion of the structure of the parametrized map $\mathbf{F}(\mathbf{y}, \mathbf{a})$. Our choice in this paper has been to use scalar products of $\mathbf{y}$ and $\mathbf{y}(n)$ in forming g. These products are insensitive to directional information on the attractor. The structure of neighborhoods of phase-space points near the orbit $\mathbf{y}(n)$ is not isotropic, so much of the information in our data may not be used in our present choice of g. Because we want g to provide direction-sensitive weights, we might wish to build in some of the local phase-space structure on the attractor. Some of this information is contained in the correlation function among points in the neighborhood of the orbit. If an orbit point $\mathbf{y}(n)$ has N_b neighbors $\mathbf{y}^\beta(n)$ within $\sqrt{\sigma}$, the correlation function is

$$W_{ij}(n) = \frac{1}{N_b} \sum_{\beta=1}^{N_b} [y^\beta(n) - y(n)]_i [y^\beta(n) - y(n)]_j. \tag{64}$$

Following a suggestion of Fukunaga (1972), we use the local correlation matrix in our $g(\mathbf{y}, \mathbf{y}(n); \mathbf{a})$ by making the replacements

$$|\mathbf{y} - \mathbf{y}(n)|^2 \rightarrow \sum_{i,j=1}^{d} [y - y(n)]_i W_{ij}^{-1}(n) [y - y(n)]_j \tag{65}$$

and

$$\mathbf{y}(n) \cdot (\mathbf{y} - \mathbf{y}(n)) \rightarrow \sum_{i,j=1}^{d} y(n)_i W_{ij}^{-1}(n) [y - y(n)]_j . \tag{66}$$

This now emphasizes directions in phase space along the attractor where the correlation is larger.

I hope to have set a tone in this chapter that builds on ideas in the familiar linear signal processing literature, but moves the emphasis from linear fits. Linear fits are easy to work with because they give algebraic problems in the frequency domain to nonlinear analysis in the time domain, which takes into account the fundamental geometry of the attractor on which the signals evolve. I also hope to kindle the realization that treating the model-building as the creation of a dynamical system is critical.

One of the comments that flows throughout this chapter in an implicit fashion concerns the inadequacy of traditional linear methods of signal analysis to treat interesting behaviors of nonlinear systems, such as coupled ocean-atmosphere dynamics. A very salient comment on this is found in Lorenz's statement regarding his original model (Lorenz, 1984). In an informal discussion, Lorenz mentions the issues that led him to make a model in the first place; that is, what lessons he wanted to learn about the atmosphere by studying a mathematical representation of it, which surely missed most of the details. The matter of most concern to Lorenz was the irregularity of the atmosphere-ocean system, and that is what he was investigating:

> One might wonder why irregularity should be one of the atmospheric properties worth reproducing with a model. My own interest arose some 25 years ago in connection with statistical weather forecasting, i.e., forecasting by means of empirically established formulas. The idea had circulated among the statistical forecasting community, apparently as a misinterpretation of a paper by Wiener (1956), that suitable linear prediction formulas, where predicted future values of observable quantities are expressed as linear combinations of present and past values, would perform as well as any other attainable formulas. I found the idea implausible, and proposed to test it by choosing a system of nonlinear equations, solving it numerically, and treating the numerical output as observational data, to which standard methods of determining optimum linear prediction formulas could be applied.
>
> ....
>
> It soon became apparent that the model would also have to possess an aperiodic general solution, since the linear prediction of a periodic variable by simple extrapolation is a trivial matter. I eventually discovered a 13-variable model whose solutions were unmistakably aperiodic. ...... The model clearly revealed that linear forecasting had its limitations.

This would seem to be a good note on which to end this description of where our efforts have led us 25 years after Lorenz's pioneering inquiries.

ACKNOWLEDGMENTS

The work I have reported on in this chapter has been done in conjunction with R. Brown, J. B. Kadtke, and E. N. Lorenz over a period of many months. Discussions with P. Bryant, J. "Sid" Sidorowich, H. Levine, T. Kailath, and other colleagues at UCSD and elsewhere have been very important in formulating the ideas here. My work in this area has been supported by the U. S. Department of Energy, Division of Engineering and Geosciences, Office of Basic Energy Sciences, under Grant No. DE-AS03-84ER13165.

REFERENCES

Abarbanel, H. D. I., 1990, Prediction in chaotic nonlinear systems: Methods for time series with broadband Fourier spectra, *Phys. Rev. A.*, **41**:1,782.

Abarbanel, H. D. I., Brown, R., and Kadtke, J. B., 1989a, Prediction and system identification in chaotic nonlinear systems: Time series with broadband spectra, *Phys. Lett. A*, **138**:401-408.

Abarbanel, H. D. I., Brown, R., and Kadtke, J. B., 1989b, Prediction in chaotic nonlinear systems: Methods for time series with broadband Fourier spectra, UCSD/INLS Preprint INLS-1014, to appear in *Phys. Rev. A*.

Abarbanel, H. D. I., and Kadtke, J. B., 1989, Information theoretic methods for determining minimum embedding dimensions for strange attractors, UCSD/INLS Preprint, in preparation, September, 1989.

Abarbanel, H. D. I., and Latham, P. E., 1982, Finite resolution approximation to the asymptotic distribution for dynamical systems, *Phys. Lett. A*, **89**:55-58.

Akaike, H., 1974, A new look at the statistical model identification, *IEEE Trans. Automat. Contr.*, **AC-19**:716-723.

Benettin, G., Froeschle, C., and Scheidecker, J. P., 1979, Kolmogorov entropy of a dynamical system with an increasing number of degrees of freedom, *Phys. Rev. A*, **19**:2,454-2,460.

Broomhead, D. S., and Jones, R., 1989, Time-series analysis, *Proc. Roy. Soc. Lond. A*, **423**:103-121.

Broomhead, D. S., and King, G. P., 1986, Extracting qualitative dynamics from experimental data, *Physica D*, **20**:217.

Brown, R., 1989, Invited talk at the Annual Meeting of SIAM, UCSD/INLS Preprint, August.

Bryant, P., 1989, Private communication and UCSD/INLS Preprint, October.

Cramér, H., 1946, "Mathematical Methods of Statistics," Princeton University Press, Princeton, NJ.

Crutchfield, J. P., and McNamara, B. S., 1987, Equations of motion from a data series, *Complex Systems*, **1**:417-452.

Eckmann, J.-P., Kamphorst, S. O., Ruelle, D., and Ciliberto, S., 1986, Liapunov exponents from time series, *Phys. Rev. A*, **34**:4,971–4,979.

Eckmann, J.-P., and Ruelle, D., 1985, Ergodic theory of chaos and strange attractors, *Rev. Mod. Phys.*, **57**:617-656.

Farmer, J. D., and Sidorovitch, J. J., 1987, Predicting chaotic time series, *Phys. Rev. Lett.*, **59**:845.

Fraedrich, K., 1986, Estimating the dimensions of weather and climate attractors, *J. Atmos. Sci.*, **43**:419-432.

Fraser, A. M., 1988, Information and entropy in strange attractors, Ph.D. Dissertation, University of Texas at Austin, May, 1988.

Fraser, A. M., 1989, Reconstructing attractors from scalar time series: A comparison of singular system and redundancy criteria, *Physica D*, **34**:391-404.

Fraser, A. M., and Swinney, H. L., 1986, Independent coordinates for strange attractors, *Phys. Rev. A*, **33**:1,134-1,140.

Fukunaga, K., 1972, "Introduction to Statistical Pattern Recognition," Academic Press, New York.

Gill, P. E., Murray, W., Saunders, M. A., and Wright, M. H., 1986, "User's guide for NPSOL (Version 4.0): A FORTRAN package for nonlinear programming," Technical Report SOL 86-2, Systems Optimization Laboratory, Stanford University.

Grassberger, P., and Procaccia, I., 1983a, Characterization of strange attractors, *Phys. Rev. Lett.*, **50**:346-349.

Grassberger, P., and Procaccia, I., 1983b, Measuring the strangeness of strange attractors, *Physica D*, **9**:189-208.

Greene, J. M., and Kim, J. S., 1987, The calculation of Lyapunov spectra, *Physica D*, **24**:213-225.

Grossmann, S., and Thomae, S., 1977, Invariant distributions and stationary correlation functions of one-dimensional discrete processes, *Z. Naturforsch.*, **32A**:1,353-1,363.

Hénon, M., 1976, A two-dimensional mapping with a strange attractor, *Commun. Math. Phys.*, **50**:69.

Lapedes, A. S., and Farber, 1987, Nonlinear signal processing using neural networks: Prediction and system modeling, Preprint LA-UR-87-2662, Los Alamos National Laboratory.

Lapedes, A. S., and Farber, 1988, How neural nets work, Preprint LA-UR-88-418, Los Alamos National Laboratory.

Lawley, D. N., 1956, Tests of significance of the latent roots of the covariance and correlation matrices, *Biometrica*, **43**:128-136.

Lorenz, E. N., 1963, Deterministic aperiodic flow, *J. Atmos. Sci.*, **20**:130.

Lorenz, E. N., 1984, Irregularity: a fundamental property of the atmosphere, *Tellus*, **36A**:98-110.

Lorenz, E. N., 1989a, Can chaos and intransitivity lead to interannual variability?, MIT Preprint, Winter, accepted for publication in *Tellus*.

Lorenz, E. N., 1989b, Private communication, UCSD, Winter.

Lorenz, E. N., 1989c, Private communication, UCSD/INLS, Winter.

Mañé, R., 1980, On the dimension of the compact invariant sets of certain nonlinear maps, *in*: "Dynamical Systems and Turbulence, Warwick, 1980," D. Rand and L. S. Young, eds., Lecture Notes in Mathematics **898**, Springer-Verlag, Berlin.

Mayer-Kress, G., 1987, Application of dimension algorithms to experimental chaos, Preprint LA-UR-87-1030, Los Alamos National Laboratory.

Orfanidis, S. J., 1988, "Optimum Signal Processing: An Introduction," Second Edition, Macmillan, New York.

Oseledec, V. I., 1968, A multiplicative ergodic theorem, Lyapunov characteristic numbers for dynamical systems, *Trudy Mosk. Mat. Obsc. Moscow Math. Soc.*, **19**:17.

Paladin, G., and Vulpiani, A., 1987, Anomalous scaling laws in multifractal objects, *Phys. Repts.*, **156**:147-225.

Poincaré, H., 1899, "Les méthodes nouvelles de la méchanique céleste," Gauthier-Villars, Paris, Three Volumes.

Press, W. H., Flannery, B. P., Teukolsky, S. A., and Vetterling, W.T., 1986, "Numerical Recipes," Cambridge University Press, Cambridge.

Renyi, A., 1970a, "Foundations of Probability," Holden-Day, San Francisco.

Renyi, A., 1970b, "Probability Theory," North Holland, Amsterdam.

Rissanen, R., 1978, Modeling by shortest data description, *Automatica*, **14**:465-471.

Root, W. L., 1987, Remarks, mostly historical, on signal detection and signal parameter estimation, *Proceedings of the IEEE*, **75**:1,446-1,457.

Rosenfeld, A., and Kak, A. C., 1981, "Digital Picture Processing," Second Edition, Academic Press, New York.

Sano, M., and Sawada, Y., 1985, Measurement of the Lyapunov spectrum from a chaotic time series, *Phys. Rev. Lett.*, **55**:1,082.

Schmidt, R. O., 1979, Multiple emitter location and signal parameter estimation, *in*: "Proc. RADC Spectral Estimation Workshop," Rome, New York.

Schwartz, G., 1978, Estimating the dimension of a model, *The Annals of Statistics*, **6**:461-464.

Shimada, I., and Nagashima, T., 1979, A numerical approach to ergodic problems of dissipative dynamical systems, *Progress of Theoretical Physics*, **61**:1,605-1,616.

Silverman, B. W., 1986, "Density Estimation for Statistics and Data Analysis," Chapman and Hall, London.

Smith, L. A., 1988, Intrinsic limits on dimension calculations, *Phys. Lett. A*, **133**:283.

Stone, M., 1979, Comments on model selection criteria of Akaike and Schwartz, *J. R. Statist. Soc. B*, **41**:276-278.

Takens, F., 1981, Detecting strange attractors in turbulence, *in* "Dynamical Systems and Turbulence, Warwick 1980," D. Rand and L. S. Young, eds., Lecture Notes in Mathematics **898**, Springer-Verlag, Berlin.

Theiler, J., 1989, Estimating fractal dimension, MIT Lincoln Laboratory Preprint, April.

Wax, M., and Kailath, T., 1985, Detection of signals by information theoretic criteria, *IEEE Trans. Acoust., Speech, Signal Processing*, **ASSP-30**:387-392.

Wiener, N., 1956, Nonlinear prediction and dynamics, *Proc. 3rd Berkeley Sympos. Math. Stat. and Prob.*, **3**:247-252.

Wolf, A., Swift, J. B., Swinney, H. L., and Vastano, J. A., 1985, Determining Lyapunov exponents from a time series, *Physica D*, **16**:285-317.

THE USE OF SIMULATED ANNEALING TO SOLVE EXTREMELY LARGE AND COMPLEX PROBLEMS

Bjarne Andresen

Physics Laboratory
University of Copenhagen
Copenhagen, Denmark

INTRODUCTION

Simulated annealing is an optimization technique (Kirkpatrick et al., 1983) that exploits an analogy between combinatorial optimization problems and the statistical mechanics of physical systems. The analogy gives rise to an algorithm for finding near-optimal solutions to any given problem by simulating the cooling of the corresponding physical system. Just as nature, under most conditions, can cool a macroscopic system to, or very close to, its ground state in a short period of time (even though its degrees of freedom are of the order of Avogadro's number), so simulated annealing rapidly finds a good approximation of the posed problem's solution. Even though the original class of problems under consideration (Kirkpatrick et al., 1983) was combinatorial optimization, excellent results have also been obtained by applying simulated annealing to seismic inversion (Jakobsen et al., 1987), pattern recognition (Hansen, 1990), and neural networks (Hansen and Salamon, 1989), to name just a few. Simulated annealing is an efficient general-purpose method for handling extremely complicated problems, and as such holds great promise for the areas of climatology and ecology.

THE ALGORITHM

Simulated annealing is based on the Monte Carlo simulation of physical systems. It requires the definition of a state space $\Omega = \{\omega\}$ with an associated cost function (physical analog: energy) $E : \Omega \to R$, which is to be minimized in the optimization. At each point of the Monte Carlo random walk in the state space, the system may jump to a neighboring state; this set of neighbors, known as the move class $N(\omega)$, must also be specified. The only control parameter of the algorithm is the temperature T of the heat bath in which the corresponding physical system is immersed.

The random walk inherent in the Monte Carlo simulation is accomplished by the Metropolis algorithm (Metropolis et al., 1953), which states that:

1. At each step t of the algorithm, a neighbor ω' of the current state ω_t is selected at random from the move class $N(\omega_t)$ to become the candidate for the next state.

2. A neighbor actually becomes the next state, with probability $P_{acceptance} = 1$,

only if $\Delta E \leq 0$, and $\exp(-\Delta E/T_t)$ if $\Delta E > 0$, where $\Delta E = E(\omega') - E(\omega_t)$. If this candidate is accepted, then $\omega_{t+1} = \omega'$; otherwise, the next state is the same as the old state, $\omega_{t+1} = \omega_t$.

The only thing left to specify is the sequence of temperatures T_t appearing in the Boltzmann factor in $P_{acceptance}$—the so-called annealing schedule. As in metallurgy, the cooling rate has a major influence on the final result. Too rapid a quench often leaves the system stranded in metastable states high above the ground state/optimal solution. Slow annealing produces the best result but is computationally expensive.

This completes the description of the simulated annealing algorithm, which, in principle, is simply repeated numerous times until a satisfactory result is obtained.

CONNECTIONS TO FINITE-TIME THERMODYNAMICS

In the following, we carry the analogy to statistical mechanical systems one step further than did Kirkpatrick et al. (1983) in their original work. Utilizing our knowledge of the minimum dissipation path for a given total process that is calculated by finite-time thermodynamics (Salamon and Berry, 1983; Feldmann et al., 1985; Salamon et al., 1988), we arrive at the optimal annealing schedule. This optimal path is one of constant thermodynamic speed (Salamon and Berry, 1983; Feldmann et al., 1985; Salamon et al., 1988), which is easily translated into a particular annealing schedule $T(t)$:

$$dT/dt = -vT(t)/\varepsilon(T)\sqrt{C(T)},$$

where v is the (constant) thermodynamic speed, and $\varepsilon(T)$ and $C(T)$ are the current relaxation time and heat capacity of the system, respectively. An equivalent prescription for calculating the annealing schedule is to select $T(t)$ such that

$$(< E > -E_{eq}(T))/\sigma = v$$

at all times. An ensemble of random walkers is most useful (Andresen et al., 1988) for calculating the current average energy $< E >$ and the standard deviation σ, as well as ε and C. E_{eq} is the energy of the system when it is in equilibrium with the environment at temperature T.

RESULTS

Using this approach, we have considered such problems as partitioning of necklaces (Ruppeiner et al., 1989) and random graphs, and inversion of seismic reflection data (Jakobsen et al., 1987). The figure below results from careful annealing of the equipartitioning of a [20,2] necklace and shows the inherent relaxation time of the system at a range of inverse temperatures. Simple Arrhenius dynamics would have followed the upper curve, Boltzmann-to-Boltzmann dynamics would have followed the lower. Clearly, the necklace does not behave like either, and a simple, predetermined annealing schedule across the cusp would have been disastrous. An annealing schedule with constant thermodynamic speed automatically slows down at such places.

Even simple combinatorial problems have state spaces with 10^{100} states, whereas the number of state spaces in industrially important problems (e.g., chip layout) runs as high as 10^{1000}. In all cases, simulated annealing has performed well without the need for special design or fine tuning. It should therefore also be a powerful method for the extremely complex problems encountered in climatology and ecology.

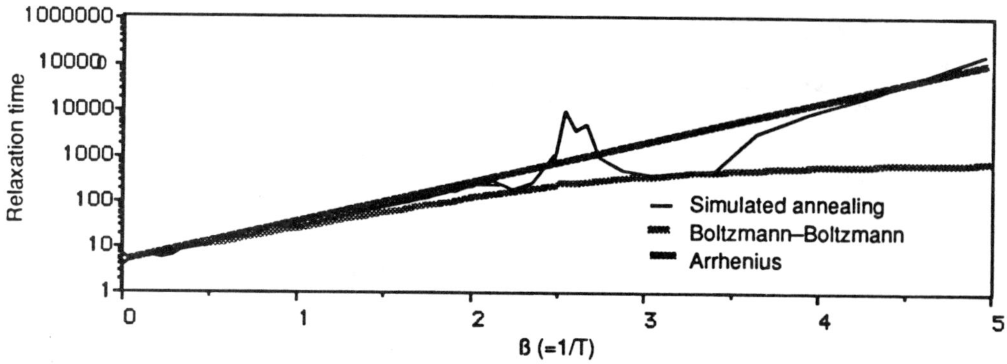

Figure 1

ACKNOWLEDGMENTS

Much of the work and ideas underlying this brief presentation have been contributed by Drs. Jacob Mørch Pedersen, Peter Salamon, and Jim Nulton. Their participation is greatly appreciated.

REFERENCES

Andresen, B., Hoffmann, K. H., Mosegaard, K., Nulton, J., Pedersen, J. M., and Salamon, P., 1988, On lumped models for thermodynamic properties of simulated annealing problems, *J. Phys.* (France), **49**:1,485.

Feldmann, T., Andresen, B., Qi, A., and Salamon, P., 1985, Thermodynamic lengths and intrinsic time scales in molecular relaxation, *J. Chem. Phys.*, **83**:5,849.

Hansen, L. K., and Salamon, P., 1989, Neural network ensembles (accepted by *IEEE Trans. on Pattern Analysis and Machine Intelligence*).

Hansen, L. K., 1990, On collective decisions of neural network ensembles, presented at Symposium on Applied Statistics, UNI.C, Copenhagen (in preparation).

Jakobsen, M. O., Mosegaard, K., and Pedersen, J. M., 1987, Global model optimization in reflection seismology by simulated annealing, the Mathematical Geophysics Fifth International Seminar on Model Optimization in Exploration Geophysics, Berlin.

Kirkpatrick, S., Gelatt, C. D., and Vecchi, M. P., 1983, Optimization by simulated annealing, *Science*, **220**:671-680.

Metropolis, N., Rosenbluth, A. W., Rosenbluth, M. N., Teller, A. H., and Teller E., 1953, Equation of state calculations by fast computing machines, *J. Chem. Phys.*, **21**:1,087-1,092.

Ruppeiner, G., Pedersen, J. M., and Salamon, P., 1989, Ensemble approach to simulated annealing (submitted to *J. Phys.* (France)).

Salamon, P., and Berry, R. S., 1983, Thermodynamic length and dissipated availability, *Phys. Rev. Lett.*, **51**:1,127.

Salamon P., Nulton, J., Robinson, J., Pedersen, J. M., Ruppeiner, G., and Liao, L., 1988, Simulated annealing with constant thermodynamic speed, *Comp. Phys. Comm.*, **49**:423.

PREDICTABILITY AND DIMENSIONALITY OF A SIMPLIFIED ATMOSPHERIC MODEL

Anna Trevisan

FISBAT CNR
Bologna, Italy

INTRODUCTION

There is general agreement that the behavior of the atmosphere is chaotic: its sensitivity to initial conditions is such that predictability beyond a certain range is lost. Chaotic behavior can be found in systems with only a few total or relevant dynamical variables. It has recently been suggested, however, that a few variables may be sufficient to model the dynamics of the atmosphere. This hypothesis is supported only by studies aimed at estimating the dimensionality of the climate system through analysis of observational data (Fraedrich, 1986; Keppenne and Nicolis, 1989).

Phase space can be reconstructed from one variable record by means of the embedding theorem (Takens, 1981) and the method devised by Grassberger and Procaccia (1983), and used to estimate the correlation dimension. Although these authors find low fractal dimensions, their results are controversial (Procaccia, 1988); it is not clear that the amount of data is sufficient or the sampling time adequate to draw firm conclusions. Keppenne and Nicolis (1989) also estimate the first few Lyapunov exponents using recently developed algorithms (Sano and Sawada, 1985; Eckmann et al., 1986) and find the estimated doubling time for small errors to be on the order of 20 days.

This result contrasts with the results of Lorenz (1982), which are based on the statistical evaluation of atmospheric analogs deduced from analysis and forecasts by the European Central Meteorological Weather Forecasting Center. Lorenz finds the upper bound on the doubling time of small errors to be approximately two days. Because the method implemented by Keppenne and Nicolis (1989) requires a large amount of data, the discrepancy may be due to the shortness of the record used in their computations.

As an alternative to data analysis, one may resort to the use of simple models to gain insights into the behavior of the atmosphere and the applicability of the techniques described above. One example of such a model is the recent work by Malguzzi et al. (1990), which parallels many previous studies, starting with Lorenz (1963). Malguzzi et al.'s model is a quasi-geostrophic model with thermal forcing and dissipation and a high resolution in the meridional direction, which allows a good description of nonlinear equilibration of baroclinic waves.

The statistical and predictability properties of this model resulted from thorough study. The model obtains non-periodic solutions that present zonal flow and wave amplitude variability in the range observed in the atmosphere. The algebra of linearized equations allows the Lyapunov exponents to be computed

Global Climate and Ecosystems Change, Edited by G. MacDonald and L. Sertorio, Plenum Press, New York, 1990

with confidence. The doubling time for small error is deduced from the first Lyapunov exponent, which appears to be nearly independent of the level of truncation and is on the same order of magnitude as Lorenz's estimate for the atmosphere.

Given the model's performance characteristics with regard to these basic properties, we address the problem of dimensionality. Two experiments were performed with 96 and 192 degrees of freedom, respectively. These experiments showed the Lyapunov dimension, which strongly depends upon the level of truncation, to be about half as large as the phase-space dimension. The correlation dimension, computed in the total phase space for the experiment with 96 degrees of freedom, was at least 20. Increasing the number of points used in the computation of Malguzzi et al. to 20,000 showed that convergence was not well established. Much more critical is the scarcity of points in the estimate of the correlation dimension performed in reconstructed phase space: with 20,000 points used, no sign of convergence was observed.

Before drawing firm conclusions regarding the real atmosphere, it would be desirable to use more realistic models. However, the above results raise some doubts as to the reliability of the low dimensionality estimates deduced from real data.

REFERENCES

Eckmann, J.P., Ollifson Kamphorst, S., Ruelle, D., and Ciliberto, S., 1986, Lyapunov exponents from time series, *Phys. Rev.*, **34A**:4,971-4,979.

Fraedrich, K., 1986, Estimating the dimensions of weather and climate attractors, *J. Atmos. Sci.*, **43**:419-432.

Grassberger, P., and Procaccia, I., 1983, Measuring the strangeness of strange attractors, *Physica*, **9D**:189-208.

Keppenne, C.L., and Nicolis, C., 1989, Global properties and local structure of the weather attractor over Western Europe, *J. Atmos. Sci.*, **46**:2,356-2,370.

Lorenz, E.N., 1963, The mechanics of vacillation, *J. Atmos. Sci.*, **20**:448-464.

Lorenz, E.N., 1982, Atmospheric predictability experiments with a large numerical model, *Tellus*, **34**:505-513.

Malguzzi, P., Trevisan, A., and Speranza, A., 1990, Statistics and predictability for an intermediate dimensionality model of the baroclinic jet, *Ann. Geoph.*, **8**:128-143.

Procaccia, I., 1988, Weather systems: Complex or just complicated?, *Nature*, **333**:498-499.

Sano, M., and Sawada, Y., 1985, Measurement of the Lyapunov spectrum from a chaotic time series, *Phys. Rev. Lett.*, **53**:1,082-1,085.

Takens, F., 1981, Detecting strange attractors in turbulence, in "Dynamical Systems and Turbulence," Springer-Verlag, Berlin.

INDEX